The History and Philosophy of Technology

The History and Philosophy of

TECHNOLOGY

Edited by George Bugliarello
and Dean B. Doner

With an introduction by
Melvin Kranzberg

UNIVERSITY OF ILLINOIS PRESS
Urbana Chicago London

Publication of this volume has been made possible in part by a grant from the Joseph and Helen Regenstein Foundation.

Manufactured in the United States of America

Library of Congress Cataloging in Publication Data

Symposium on the History and Philosophy of Technology, Chicago, 1973.
The history and philosophy of technology.

Sponsored by the College of Engineering and the College of Liberal Arts and Sciences of the University of Illinois at Chicago Circle.
Includes bibliographical references.
1. Technology—History—Congresses. 2. Technology—Philosophy—Congresses. I. Bugliarello, George. II. Doner, Dean B. III. University of Illinois at Chicago Circle. College of Liberal Arts and Sciences. IV. University of Illinois at Chicago Circle. College of Engineering. V. Title.
T15.S96 1973 609 78-26846
ISBN 0-252-00462-0

Contents

Editors' Preface, *George Bugliarello and Dean B. Doner* vii

Introduction: Trends in the History and Philosophy of Technology, *Melvin Kranzberg* xiii

Part I: The History of Technology

1. Problems of the Data Base, *Donald S. L. Cardwell* 3
2. Toward a Social History of Technological Ideas: Joseph Black, James Watt, and the Separate Condenser, *Arthur L. Donovan* 19
3. Remarks on the Discovery of Techniques and on Sources for the Study of Their History, *Cyril Stanley Smith* 31
4. Chinese History of Technology: Some Points for Comparison with the West, *E-tu Zen Sun* 38
5. The Engineer and the Historian, *George Bugliarello* 50
6. What Can the History of Technology Contribute to Our Understanding? *Harold L. Burstyn* 57
7. Technology, Economy, and Values, *Nathan Rosenberg* 81
8. What Should We Ask of the History of Technology? *Steve M. Slaby* 112
9. What Do We Ask of the History of Technology? *David Joravsky* 128
10. The History of Technology: An Anthropological Point of View, *Heather Lechtman and Arthur Steinberg* 135

Part II: The Philosophy of Technology

11. Philosophy and the History of Technology, *Carl Mitcham* 163

12. Technology from an Encyclopedic Point of View, *Jean-Claude Beaune* 202

13. *Praxis* and *Techne*, *Peter Caws* 227

14. The Structure of Technological Revolutions, *David Wojick* 238

15. Philosophical Inputs and Outputs of Technology, *Mario Bunge* 262

16. On the Relationship between Philosophy and Technology in the German-speaking Countries, *Werner Koenne* 282

17. The Technological Challenge to Political Theory, *Frances Svensson* 294

18. Technological Metaphor and Social Control, *David Edge* 309

19. Philosophy of Technology as a Philosophy of Man, *Henryk Skolimowski* 325

Part III: The Future of Technology

20. Besieging the Fortress, *Ion Curievici* 339

21. Evolution of Building Technology, *Rafael López Palanco* 344

22. Where Do We Go from Here? *Heinz Von Foerster* 358

23. Is There Anyone Else? *Iraj Zandi* 371

Notes on Contributors 381

George Bugliarello and Dean B. Doner

Editors' Preface

Every day, explicitly or implicitly, consciously or unconsciously, fateful decisions are being made about technology and thus about our lives. For the human race to survive and thrive, it is absolutely essential that we understand the nature of the process called technology. We must be able to place the problems posed by technology into the mainstream of philosophical inquiry. A society of which technology has become an essential part but which lacks a deep philosophical understanding of technology's nature certainly invites disaster. Part of this understanding must also be historical. Our society will be a disturbed and schizophrenic one unless it can relate technology, which is concerned with the future, to history, which is concerned with the analysis and interpretation of the past. Our future—not only its shape, but indeed its very existence—depends critically on our ability to relate our technology to our humanity.

Today these problems seem to be viewed as marginal in the universities. Engineering schools continue to train generation after generation of possibly the most powerful agents of change that our planet has ever produced, yet engineering curricula rarely consider and never require the study of historical and philosophical aspects of technology. With similar insensitivity, schools of liberal arts and sciences exclude what should, on the contrary, be mandatory: curricula that consider what benefits technology offers humanity and what problems it can pose.

All our questions of value are cross-disciplinary. If a machine is indeed an extension of the human organism, what are the criteria of validity for machines? When can we accept and when must we reject certain machine forms? To what extent is our morality to emerge from and be sanctified by the past, and to what extent must it be continuously reshaped by the vistas of the future that are opened by technology?

Do we not see one of these vistas of the future in the BIOSOMA [1] —that now feasible combination of biological organisms, society, and machines? What conditions favor the development of technology? What can we do with technology to enhance our society? And what can we *not* do without deeply affecting certain basic views and tenets of society? How does the problem pose itself in different times and in different societies?

We clearly need a methodology and viewpoint for answering questions such as these. We need them as much as we need the answers themselves. By and large, philosophers and engineers have only peripherally considered the philosophical problems of the machine and of technology. In this volume they confront shared problems together. Similarly, engineers and historians have traditionally viewed history from nonoverlapping perspectives. Here they search together for the point where history and technology meet—where the infinity of the past and the infinity of the future come together to shape man's destiny.

The chief task of this book has been to pose certain key questions. What, in the first place, is the nature of technology? Is it a logical and inevitable feature of the taxonomy of man himself? Is it something man could not help developing? Many see technology as reflecting the image of man himself, as both tool maker and tool user: man as the animal of innate curiosity, creator, artist, searcher, visionary. In this view there is no choice; where there is man there is technology—and a direct line links the first chip made in the first handworked stone to the development and use of the computer. Technology is an unstoppable force; no system—political, economic, or social—can take any other road. We may thus paraphrase Louis XIV: *La technique, c'est moi.*

No one really disputes this. No one suggests that technological development can be stopped, much less reversed. But can it be directed? Are we able, with respect to the development of technology, to order our priorities? Or is technology not so much a force as an animal that will satisfy its own appetite—one that may well declare its favorite food to be people? One author in these pages has rephrased the technological imperative: we must do what technology can do.

Born out of man's innate curiosity, technology is developed through discovery. Discovery has no methodology. The critical moment when technology is advanced through discovery is frequently serendipitous and not subject to an ordering of priorities, but no one would claim that technological discovery is entirely without sequence or logic. The vacuum tube preceded the transistor; the internal combustion engine preceded the jet engine. Discovery itself, however, is without method and occurs outside priorities. As Professor Wojick points out, the social implications we need to understand in order to develop priorities come only after the fact.

Only when the effects of pollution are measured do we have the evidence needed to take action and develop technology for improved sanitation. The question is real: what is the hard evidence on which we based decisions about priorities with respect to the Alaskan pipeline? The SST is perhaps a more relevant case in point; for the first time, an effort has been made to evaluate, control, and divert the advances of technology prior to obtaining hard evidence. One mark of the present stage of technology is our effort to understand its nature and that of our own priorities —and to see how they work together.

Technology, then, is an unstoppable force which we adapt or do not adapt. Insofar as we do adapt it, we make up our own values as we go along. How do we evaluate the values of technology? One answer argues that technology is a value system coherent with Western thought, immersed within the world view of Western man, and therefore incapable of being evaluated; one cannot change a conceptual system from within. One writer in these pages believes that technology has been a natural growth of Western man rather than of man himself and that alternatives to the conceptual system that produced technology are "soft." Mysticism and renunciation are not viable alternatives. He says that Western man has no ground upon which to erect a critique of technology and is unable to profit by critiques offered from outside Western thought.

This book is not without dissent from such a view. We do, after all, know the values of Western man and therefore know those of technology. The value placed on technology, according to one essay, is its impact on social and political development. In Germany—and elsewhere—technology was not valued until it had a value in war. Since war appears to be valued by men, from the days of the spear thrower to the present time technology has been valued as it has advanced the cause of war. Or consider affluence. The development of technology, another author holds, is based on the values in a social structure that have dictated its allocation of resources. These are not values we do not know or have not evaluated. We do not have to stand outside our system to develop critiques of war or poverty, and we are therefore able to evaluate technology in terms of our known and recognized values.

Some who write here hold technology to be neutral with respect to values. Technology, they say, is not itself a value. Either they see curiosity and discovery as natural and neutral "givens," or they see technology as "things." Arrowheads, catapults, steam engines, atomic bombs are simply objects of no intrinsic significance that acquire meaning and value by the place accorded them within man's system of values. Let me know what a man believes, and I will tell you what his technology means. An examination of the values of a culture's technology is simply an examination of its values.

It thus becomes a key question whether technology itself carries values or creates them. If technology is a methodology, the question is whether methodologies create values; if technology is simply the proliferation of things, do things create values?

Ours is sometimes called the Age of Anxiety. Technology affects our values through the speed of change to which contemporary man is subjected. If values are manifested and perhaps even developed through choice, then it is in choosing that man declares his values, strengthens them, and codifies them into social systems that become his heritage and history. The development of rapid technological innovation, imposing adjustments on people faster than they can absorb them, breaks choice down. When technological innovations break too rapidly upon man, he can order his technological priorities only in theory; in fact he simply does not have enough time. Our sense that technology is in control rather than man grows out of insufficient time for making choices, deciding values, and ordering priorities. Not only do man's own power and control seem to be slipping away from him, but also man, unable to make those meaningful choices he senses he should be making, feels even his history slipping away. He feels he can no longer declare, strengthen, and codify his values into a system to pass on to his heirs.

It has been asked what factors influence the speed at which new technologies are adopted and diffused throughout society. Does it turn out, finally, that one of the values of man is the adoption of the new at a rate faster than he can absorb without anxiety. One writer declares that we all suffer from a lack of imagination about the conditions of historical times. Are we so bemused by our own age of anxiety that we have become insensitive to the impact caused by technological innovations in earlier times? How has man absorbed the technology of his own time in all the ages past?

These and a host of other questions led, in a spirit of deep and urgent concern, to the calling of an international Symposium on the History and Philosophy of Technology, May 14-16, 1973, at the Chicago Circle campus of the University of Illinois. Supported by a generous grant from the Joseph and Helen Regenstein Foundation of Chicago, the symposium was organized as a joint endeavor of the College of Engineering and the College of Liberal Arts at Chicago Circle.

The symposium was an end unto itself, an opportunity for historians, engineers, and philosophers to exchange views and become sensitized to the contributions that their own disciplines could make to the other disciplines. It was felt, however, that the questions raised and the issues discussed in the symposium were deserving of broader exposure. Thus, an effort was undertaken to convert a number of the papers presented into a publication—the present volume—which highlights the themes of the symposium.

The process of preparing this publication was unfortunately slow for a variety of reasons, including the fact that shortly after the symposium both of the editors moved to new and absorbing positions. We regret that the discipline imposed by the conversion of the lively proceedings of the symposium into a written document led to the omission of the discussions as well as to the omission of a number of essays, more general in theme or more anecdotal in content than the rest, which had proven stimulating in the context of the symposium. All participants, whether or not their presentations are included in this volume, contributed most significantly to the success of the symposium. To them go the acknowledgments and the warm thanks of the editors.

Notes

1. This extended acronym was first coined by G. Bugliarello in 1971.

Melvin Kranzberg

Introduction: Trends in the History and Philosophy of Technology

Until some fifteen years ago, there was no such thing as the history of technology as an academic discipline. Today it is in a healthy juvenile stage of growth; its sister, the philosophy of technology, is still in swaddling clothes. This symposium is a landmark in the development of both.

Two decades ago, courses in the history of engineering or the history of technology were occasionally offered at a few institutions. These courses were usually entrusted to an "obsolete" engineering professor on the grounds that, since he had lived through many great technological developments in the twentieth century, he should be able to tell students something about them.

This approach to the history of technology is exemplified by a story Simon Ramo tells about returning to his alma mater for an alumni reunion. Visiting the classroom of his mentor, a professor only a few years from retirement, Ramo sat in on a course entitled "Applications of Electricity." Much to his surprise, he heard exactly the same things being taught as when he took the course many years earlier: the first polar generator, the first electric light bulb, power-line transmission, and the like. When the lecture was over, he went up to the dais and warmly greeted his old teacher, but he could not resist saying: "I was a little bit surprised to hear your lecture because it was so much like the material taught me when I sat in this very classroom over thirty years ago. But now we have radar and very high voltage transmission lines and diesel electrics and color television; yet you're still teaching these very old applications of electricity. Do the students really want to learn about these things; don't they want to hear about the up-to-date applications?" The professor replied, "You know, I think you've got something there." So the next year

Delivered as keynote address at the symposium on May 14, 1973.

he changed the title of the course to "The History of Applications of Electricity."

Two decades ago about the only published works touching on the history of technology were popularizations of the "gee whiz!" variety: "Look at the wonders of technology, and see how far we have come." In terms of serious scholarly research, there were only a few pioneer works, those by Abbott Payson Usher, Lewis Mumford, and Roger Burlingame.[1]

Today, however, the history of technology is one of the fastest-growing fields of scholarship. Along with its younger sister discipline, the philosophy of technology, it is running counter to the appeal for zero growth, which seems to have hit academia instead of its real targets.

One interesting thing to note about the growth of the history of technology is that it has within a very short time developed an imposing institutional matrix—although, as we will see, it has yet to develop a conceptual and theoretical framework to match. For one thing, several major universities now offer higher degrees in the history of technology, specifically labeled as such, with other institutions offering occasional degrees through their history or history of science departments. In a short time there will undoubtedly be analogous degree opportunities available in the philosophy of technology. Let us hope institutions planning these programs will learn from our experience with the history of technology, in which some promising programs developed so quickly that they outpaced professional capabilities and became early casualties. Now we have settled down to surer and steadier growth.

As evidence of its scholarly stature, there are now several learned journals in the history of technology. The largest is *Technology and Culture,* published by the Society for the History of Technology, which has achieved a circulation of over 2,300. In Britain there is the older *Transactions of the Newcomen Society;* in Italy, *Il Machine;* in Germany, *Technikgeschichte,* published by the Verein Deutscher Ingenieure, and more recently *Kultur und Technik,* published by the Deutsches Museum; and there are publications devoted to the history of technology in Poland, Russia, and other countries. Even sub-branches of the history of technology have begun their own publications, for instance *In-*

dustrial Archaeology in Britain and the newsletter and annual of the Society for Industrial Archaeology in this country.

Organizational growth has been equally impressive. The Society for the History of Technology is constantly growing in size and activity, as are similar organizations throughout the world. There is also an international organization, International Cooperation in History of Technology Committee (ICOHTEC), which holds scholarly symposiums and which forms a scientific section of the International Union of the History and Philosophy of Science.

The growth of scholarly interest in the history of technology is reflected in the programs of the triennial International Congresses of the History of Science. Whereas in 1959 at the International Congress in Barcelona and Madrid only five papers were offered in the history of technology, in Moscow in 1971 there were over seventy papers, not counting an additional thirty-five in a specialized branch, the history of space technology, in which the Russians are particularly interested. The same trend continued in the 1974 International Congress in Tokyo.

Another manifestation of growth is the phenomenal increase in research and corresponding publications. In 1964 when *Technology and Culture* began the publication of an annual current bibliography in the history of technology, the first contained only eight and a half double-columned pages. Even with the application of rigorous screening procedures to cut out trivial and transient publications, by 1974—just ten years after the first publication—the bibliography had grown to some seventy double-columned pages.

A similar growth is now occurring in the philosophy of technology. A few years ago Carl Mitcham and Robert Mackey, who are at this conference, sent me a fifty-page typescript labeled "Bibliography of the Philosophy of Technology" for consideration for publication in *Technology and Culture*. When encouraged to make their bibliography more comprehensive, they responded so enthusiastically that it required an entire supplement of *Technology and Culture* and was then published separately as a hard-bound book of some 205 double-columned pages.

In both the history and philosophy of technology, therefore, it would seem that there is now sufficient scholarly activity to justify the convocation of this conference. It is time to take stock. Our symposium therefore asks several questions about the history of technology: Where are we? What do we ask of the history of technology? Where do we go from here?

It is indeed time to ask these questions, and it is timely that this be done in the context of similar questions directed at the philosophy of technology. Above all, we must ask questions about the relationships between the history and the philosophy: what are these relationships, and what can and should they be?

However, I must confess that I am somewhat distressed to find that our symposium begins with the data base and methodology of the history of technology. Although this arrangement reflects, I suppose, the logic of the engineering mind, it does not necessarily correspond to any internal logic in the history of technology.

Discussing the data base and methodology is useless, I think, unless the discussion is related to questions of substance. Indeed, we run the danger of pursuing the data base and methodology as ends in themselves rather than as means and instruments for answering questions. In other words, the data base and methodology should grow from the questions asked, not vice versa. Furthermore, we must not think of these in the singular; instead, we should refer to data bases and methodologies, for these may differ according to the problem, the period under discussion, the technology, and so forth. Once we have established what questions the history of technology should ask, then we may see what data base is available and what methodology might be most appropriate.

Still another group of questions presents itself when we talk about data bases and methodologies. How do those in the history of technology differ from those employed in other kinds of history—literary, social, intellectual, scientific, economic, legal, military? How are they similar? In either case, why? Once we have asked these and related questions, we open up a whole new series of relationships of technology to other human endeavors, which may shed light on the questions of where we are and where we should be going.

In looking at the major questions we might find it instructive to compare certain trends in the history of technology with those in another sister discipline, the history of science. There the work of Alexander Koyré and his followers, with their intense concentration upon the internal workings of science, helped develop serious, in-depth scholarship in the history of science. Lately some younger historians of science have rebelled against the Koyréan internalist approach and have sought to emphasize the external factors, paying a great deal of attention to the intellectual and social milieu in which scientific developments have occurred.

The history of technology has been spared this polite scholarly struggle between "internalists" and "externalists." From its very beginnings, the history of technology has been interdisciplinary in scope and largely externalist in approach. Perhaps this is because technology usually arises from responsiveness to social needs and is sometimes responsible for social wants. Although new technology may proceed largely from old technology, it has always been related to human work and play. Therefore, technology only makes sense if it is regarded in terms of its relationships with the external world of thoughts, values, and institutions.

Nevertheless, there has always been—and there will continue to be—an important role for the internal, or "hardware," history of technology. While internalists recognize the need for placing the hardware in the context of economic, political, and other factors, externalists are conscious of the fact that the technological element itself might determine, sometimes subtly, the external forces. In other words, no one can be completely an internalist or externalist in the history of technology; one must be something of both.

While professional historians of technology seem to be getting more and more involved in the complex interrelationships of technological developments with social, economic, political, intellectual, legal, and military ones—as is evidenced by the papers in this symposium—there is nevertheless a flourishing movement in the internal history of technology which focuses on a new field of study called "industrial archaeology." Beginning in England slightly over a dozen years ago under the leadership of an energetic journalist, Kenneth Hudson, industrial archaeology has focused on the artifacts of the Industrial Revolution in Britain. Interest in this subject in Britain, as in the United States, is partly historical and partly antiquarian; its practitioners include professional historians and amateur hobbyists.

Industrial archaeology, according to Kenneth Hudson, is the same as "technological monuments." What he means by it is the "material relating to yesterday's manufacturing and transport which has survived, more or less intact, on its original site." [2] In Britain, industrial archaeology has achieved a great vogue, with both the publication of a journal and an academic home at the University of Bath.

In the United States the development of industrial archaeology has followed the British model. A combination of professional historians of technology and enthusiastic amateurs—and I mean this in the best sense—banded together in 1971 to form the Society for Industrial Archaeology. It holds annual meetings, usually in places where there are significant industrial artifacts of the past, and the papers given at its meetings demonstrate the industrial archaeologist's special concern for artifacts. Here are a few typical titles: "The Adjustable Wrench, 1831-41," "A Study of the Barrackville Covered Bridge," "Nineteenth-Century Stove Foundries in Troy and Their Preservation," and "Wood-Burning Lime Kilns in the Clove Valley, Montague, New Jersey."

The activities of the Society for Industrial Archaeology have been stimulated by the bicentennial anniversary of the United States. In collaboration with the Historic American Engineering Record of the National Park Service, industrial archaeologists have been engaged in surveying significant industrial monuments of the American past.

The fact that many of those involved in the pursuit of industrial archaeology are enthusiastic amateurs rather than academic scholars does not mean that we should be contemptuous of their efforts. Quite the contrary. There are simply not enough professional historians of technology to do all the jobs which must be done to obtain meaningful knowledge and understanding of technological history. By providing basic studies of the technological elements themselves, industrial archaeologists perform a great service for historians of technology. Their preservationist tendencies also help to maintain the role of historians as custodians of the past as well as expounders of historical developments. For their part, industrial archaeologists are eager to enlist the support of professional historians who can tell them what kind of technical information is of importance in historical explanation and who can help upgrade the scholarly level of industrial archaeology. The importance of their artifactual approach can be seen in some of the historical papers in this conference; although not specifically industrial archaeology, the kind of detailed and specific technical information which is employed is exactly the kind of thing which the industrial archaeologist "unearths" and preserves for broader historical treatment.

After this digression into internalist, "hardware" history, we can look at some other scholarly concerns of historians of technology. We will find that these involve both internalist and externalist approaches. First among these are the question of innovation and the problem of science-technology relationships.

The search for the wellsprings of creativity is one of the chief intellectual commitments of the twentieth century. In the history of technology, this interest manifests itself in studies of the processes of invention and innovation and, as a by-product, which is itself a chief intellectual concern, a study of the relationships between science and technology.

Although much has already been written on these subjects, our researches have barely scratched the surface. We have succeeded in dispelling, at least for scholars if not for the general public, the notion of invention as a simple "flash of intuition." We are in the process of showing, by numerous case studies of individual inventions and inventors, the problems arising from the act of invention itself and the even more complex questions deriving from the innovative process, whereby the invention becomes a part of our social matrix. This subject is so fascinating, both in its human and technological dimensions and in its economic and social relationships, that interest in it will certainly continue undiminished. Several of the papers of this conference deal very directly with this question of invention—what was invented, why, and, most particularly, how.

We have also done away with the old maxim that technology is simply applied science and that the process of innovation is a simple linear progression from basic research to application. Models of the innovative process have been provided, for instance those of Abbott Payson Usher and Joseph Schumpeter, but all of them have proved unsatisfactory in showing the relationships between science and technology in innovation. I, too, have tried my hand at models and have arrived at a series of push-pull diagrams describing a few different kind of relationships. But nobody has come up with one overall model encompassing all the parameters of science-technology relationships or the even more complex elements involved in innovation. It now appears that innovation most nearly resembles an ecological process and requires a dynamic systems model.[3]

Another major concern of historians is the transfer of technology. Interest in this field has been growing along two lines: the movement of technology from one sector of the economy to another, and from one country to another. The former centers in the "spillover," or "spinoff," of military and space technology to the civilian economy. Because of our tremendous military and aerospace expenditures, there is a tendency to seek justification of these costs in the light of their historical contribution to the everyday, down-to-earth requirements of society. Historical interest in the transfer of technology from one country to another derives from the demand for industrialization by the less-developed countries to provide at least minimal subsistence for their burgeoning populations.

Although the matter of "spinoff" has been scarcely touched by historians, much research is being conducted on the transfer of technology from industrialized nations to the less-developed countries. International interest in this subject has been manifested in a series of colloquiums sponsored by ICOHTEC. The first, held in France in 1970, discussed the "acquisition of technology by noninitiating countries in the 18th and early 19th centuries"; [4] its purpose was to trace the spread of the Industrial Revolution from mid-eighteenth-century Britain, where it had originated, to other Western European nations and to America up to the mid-nineteenth century. A second ICOHTEC colloquium dealt with the transfer of technology in the latter part of the nineteenth century; it was held in Moscow in 1971 as part of the thirteenth International Congress of the History of Science. Another conference, one directed at the transfer of industrial technology to the agricultural countries of Central Europe in the nineteenth century, took place in Poland in 1973;

in 1974, at the fourteenth International Congress in Tokyo, problems of the transfer of technology in the twentieth century were considered.

As historians begin tracing the diffusion of modern technology to nonindustrialized nations, they are discovering that they must become familiar with cultural anthropology, geography, demography, and the like. Such studies require a broader outlook and wider training on the part of historians of technology. This is because, as we have come increasingly to learn, the transfer of technology is not simply a technical matter—it involves political science, sociology, anthropology, and many other "ologies." This synthesis of the history of technology with other branches of knowledge makes it part of the latest movement in the field of history, namely, that associated with the French journal *Annales:Economies, Societes, Civilisations.* Under the leadership of Lucien Febvre and Mark Bloch, and now under their disciple Fernand Braudel, the *Annales* "school" has turned away from the old preoccupation with political and narrative history; instead it investigates the conjuncture with history of information and methodologies drawn from many different fields—economics, linguistics, sociology, geography, anthropology, and what have you. In addition, it attempts to inject a historical orientation into the social and "human" sciences.[5]

Closely related to studies of the transfer of technology are those of the industrialization process itself. All such studies perforce take as their starting and reference point the Industrial Revolution in Britain, the classic example. Not surprisingly, the Industrial Revolution with its landmark inventions preoccupies many scholars of the history of technology, very much as the Scientific Revolution remains a major preoccupation—if not an obsession—of historians of science, and very much as the Civil War continues to occupy the attention of American historians.

While there will be no diminution of studies of the Industrial Revolution—indeed, we have some papers dealing with major developments of that era in this conference—historians are beginning to look farther afield and to study the process of industrialization in places other than Britain and the United States. Such comparative studies will force historians of technology to know a great deal about the history of areas where the process of industrialization is taking place. Here again I see the education of historians of technology becoming broader so that they can deal meaningfully with the introduction of industrial technology into foreign areas and alien cultures.

Comparative studies of the industrialization process require scholars to ask why and how the same technologies develop differently in different countries. These studies point up the fact that technological development is much more than a technical phenomenon: it is a social phenomenon.

Such investigations have a great deal of current relevance. Management experts, for example, wish to know why automobile assembly plants in Germany, Britain, and the United States—employing the same machine tools and equipment, presumably utilizing the same management goals and techniques, and engaged in producing exactly the same or interchangeable products—vary in productivity per worker. Are there "national" styles in technology, as there are in literature, music, art, and perhaps science?

Such comparative investigations will require a broad knowledge of the history of areas whose technological "style" the historian will compare with other regions. Beyond the internal hardware and mechanism of technical development, the historian of technology must be acquainted with economic, political, social, and cultural forces which provide the matrix within which technology operates and changes.

In trying to define the role of technology in society, we begin to ask such philosophical questions as, "Can technology be guided or controlled, and if so, how?" This brings us to the very practical issues which George Bugliarello pointed out in introducing this symposium: the relevance of technology to current concerns; the new art (or science?) of technological assessment; [6] how to stimulate innovation; how to transfer technology to underdeveloped areas; and how to utilize it for economic growth.

Consideration of the innovation process and of industrialization in different societies emphasizes the externalist approach to the history of technology. The entire question of the relationships of technology and society thus becomes focused on one of the central concerns of this conference, where the interests of historians and philosophers of technology converge. This is the question of technological determinism. Already much has been written about technological determinism; various hypotheses have been posed, cults have been formed, and some hypotheses have already been discarded, at least in their most extreme form.

Since the question of technological determinism is one which concerns both historians and philosophers of technology, a brief historical outline of the problem may be in order. As a scholarly question, the problem did not raise its head until the nineteenth century, when industrialism began to affect the lives of large numbers of the population. Saint-Simon, in early nineteenth-century France, was perhaps the first social thinker to recognize the important role which industrialism was to play in human affairs. However, it was Karl Marx who seized upon the relations of production as the essential element in social development and made them the key factor in class divisions and his dialectic of social evolution. His was a "hard" determinism, for he saw the entire superstructure of society—laws, religion, institutions, values, and so forth—deriving from the productive modes and their ownership. As he stated it, "The hand-mill gives you society with the feudal lord; the steam-mill, society with the industrial capitalist." Ever since then, the Marxists have been the chief proponents of a rigorous technological determinism. Lenin extended the analysis to the twentieth century by equating socialism with electrification; to him, the dynamo meant the inevitable Communist society.

Interestingly enough, social thinkers, for instance Herbert Spencer, on the opposite side of the fence—defenders of the competitive capitalist system—also recognized technology as a central element in the evolution of society. Both Spencer and Marx were sociologists as well as philosophers; indeed, consideration of the problem of technological determinism has been as much in the realm of sociology as of philosophy. This is perhaps because the philosophy of technology, if it had to be categorized, would probably come under the rubric of "social philosophy."

Opposed to the "hard" determinism of the Marxists is "soft" determinism, perhaps best represented in the views of Lynn White, Jr., a distinguished medieval historian and a former president of both the American Historical Association and the Society for the History of Technology. In his book *Medieval Technology and Social Change,* White states, "Technology . . . opens doors, but it does not compel one to enter." This view of technology as a neutral, enabling instrument raises more questions than it answers. Who determines which doors to open? Besides, an open door is an invitation to enter; and once one passes through the door, does technology determine the contours of the room and the chamber which one enters? And can one back out the door after entering?

Moreover, if technology is viewed simply as a means, there is the question of who determines the ends—and there is also the danger of the means becoming the ends themselves. Indeed, some of the popular social thinkers of our times believe that this has already happened. Jacques Ellul, the French legal philosopher, believes that *"la technique"* controls our lives: we are more concerned with technical mechanisms than goals, and the mechanisms have become goals themselves.[7] Lewis Mumford has postulated a similar view: we have lost sight of human goals in our relentless pursuit of technological advance, without asking what human purposes are served thereby.[8]

My own view on this matter is embodied in what I call "Kranzberg's Law." (Although I am certain that others have thought of it before I did, I am so enamored of it that I hope it will bear my name.) Kranzberg's Law is simply stated: "Technology is neither good nor bad—nor is it neutral." By that I mean that technology interacts with society in ways which do not seem necessarily inherent in the technology itself. Thus, for example, by the very scale of its use the automobile has "locked" American society into certain spatial distribution patterns, life-styles, and economic activities. Values become attached to particular technologies and hence serve to determine the lines of future political, social, and, yes, technological action itself. The task of both history and philosophy is to make clear the interactions which transform technological instrumentalities to social phenomena and value-laden decisions and activities.

Only recently have professional philosophers entered into this subject—and we are indeed fortunate that most of those who are professionally concerned with the problem are participants in this conference. The thought which immediately comes to mind is whether these philosophers will truly concern themselves with the "big questions" regarding the interrelations of society and technology or whether they may bog down in a morass of definitional problems without ever reaching the major battleground where the clash of ideas must take place. Can they disentangle themselves from the logical and semantic underbrush and emerge from the swamp to converse with the historians of technology, who must also avoid the tangle of the internalist-externalist conflict?

This kind of squabbling has already occurred in the philosophy of science, and it has hindered fruitful dialogue between historians and philosophers of science. Instead, the philosophers of science have often immersed themselves in questions of logic and methodology, and only a very few individuals have been able to bridge the gap between the philosophers and the historians.

That gap has not yet appeared between historians and philosophers of technology, and we must not allow it to appear. This conference can be the major starting point for a continuing effort to bring historians and philosophers of technology together so that they can contribute their knowledge and wisdom to one another.

There remain two other major trends in the history of technology which deserve brief mention in these introductory remarks. One, I am afraid, will not receive sufficient attention at this meeting; the other provides one of the focal points of this conference so that it will not suffer from neglect, but perhaps only by mistreatment. The first is a changing historical methodology, which is already being felt in the history of technology. The second is a closer relationship of the history of technology to other disciplines, resulting in what I call "participatory history." Let me explain both of these.

During the past few years there have appeared new historical methods and approaches. One of these, revolving around the application of quantitative methodology, is sometimes called "Cliometrics"—that is, the use of measurement in historical studies. Although still viewed with suspicion by some historians, quantification is being widely accepted as a necessary tool of historical investigation.

It should be noted that some nonhistorians have also attempted to apply quantification to historical data. In the recent study *The Limits to Growth*, a group of systems specialists used—or rather, misused—historical data to arrive at certain hypotheses regarding what they describe as the "predicament of mankind." [9]

We can leave aside such aberrations and look at the employment of quantification by historians themselves. Like most trends, this has some roots in past practices. Many historians of earlier generations, including the great Frederick Jackson Turner, made significant use of quantitative methods and materials, but contemporary use of quantification by historians differs in important ways from such earlier manifestations. One obvious source of difference derives from technology itself, namely, the availability of computers. These permit application of more diverse and more powerful statistical procedures and the utilization of larger and more varied bodies of quantitative and quantifiable data than would have been conceivable even a few short years ago. Computers can facilitate both intensive case studies and research across cultures, nations, and long periods of time without the loss of empirical detail and without the heavy reliance upon impressionistic methods and evidence that formerly characterized such grand-scale efforts. Thus, contemporary quantitative studies in history differ sharply from the efforts of Turner's generation because of greater scope, methodological diversity, and sophistication, as well as the use of empirical data on a much larger scale.

For many, perhaps most, historians, quantification will be at best no more than a supplementary tool for the pursuit of traditional interests and problems. They will not deviate from the humanist's concern for the unique properties of individuals and events. However, quantitative methods and materials can permit historians to achieve new levels of precision and comprehensiveness in the reconstruction of the past.

Cliometrics is of special value to historians of technology. Because this is one of the areas where economic history and the history of technology converge, historians of technology have perhaps more experience with applying quantitative data to their research.

There have been two landmark books applying econometric methods to historical phenomena. The first was Robert W. Fogel's *Railroads and American Economic Growth*.[10] Its conclusions not only ran counter to conventional wisdom and past historical interpretations—a fact which I would be willing to accept—but also ran counter to evidence taken from a wider body of materials than the limited economic data utilized by Fogel. Nevertheless, Fogel's groundbreaking efforts assumed even greater importance when he and Stanley L. Engerman applied the methodology to the economics of Negro slavery in the South before the Civil War. The result was a monumental book, *Time on the Cross*,[11] which upset many traditional historical interpretations regarding the institution of slavery in the American South. It is a book of economic history, technological history, and social history, with repercussions which have been felt in all those disciplines. More detailed investigations of both the sources and methodologies employed by Fogel and Engerman have led to the virtual destruction of their thesis, as well as all the supporting apparatus.[12] *Time on the Cross* is wrong, but it is a monumental failure. Although a few historians might be deterred by this failure from further pursuing quantitative methodology, most will recognize the underlying value of such an approach. The failure of *Time on the Cross* will enable them to sharpen their methodological tools and their interpretive analysis in order not to repeat those same errors. After all, engineers do not stop building bridges and dams just because some of those structures break under unanticipated stresses and strains; instead, they learn from their failures and build better and stronger bridges and dams the next time. So, too, historians can improve their selection of source materials and advance their methodologies as a result of the disintegration of the structure fabricated by Fogel and Engerman.

One of the major omissions in this conference, I am afraid, is that there is no representative of this new quantitative approach to history. Although our first section deals with data base and methodology, our authors have approached this task from other than the quantitative point of view, and they have employed a methodology which varies greatly from those used by the Cliometricians. This does not mean that the data base and methodology employed by Dr. Cardwell, Dr. Donovan, and others is without validity. Far from it. They have used the methods and data which are appropriate to the kinds of problems to which they have addressed themselves. Cliometrics has by no means made all past historians and historical approaches obsolete. Nevertheless, Cliometrics is a tool which can serve to enlarge our historical understanding, and it is of particular importance to historians of technology, although it is not adequately represented at this symposium.

Another new field of history—or rather, another new historical approach which has just begun to emerge—is known as "psycho-history." Its chief exemplar is Bruce Mazlish, of M.I.T., who has done some writing in the history of technology but has never applied the new tool of psycho-history to it. One can easily see how psycho-history might be applied in the study of the history of invention and innovation. We have had many biographies of inventors, but little has been done to apply the tools of modern psychiatric analysis to these studies. Instead, we are left with the Victorian homilies of hard work and enterprise, as embodied in Samuel Smiles's *Lives of the Engineers.* In today's collective enterprise of research and development, we might find it difficult to uncover the contribution of the creative individual to the research effort; the application of psycho-history to investigations of this nature might shed much light upon this new institutionalized approach to innovation.

On the question of the collaboration of historians with other disciplines, in order to contribute actively to policy formation for current problems, this conference is more promising. Up to now, I have spoken only of the collaboration of historians and philosophers of technology in this conference. However, there is another —and extremely significant—field of study represented here: engineering. Indeed, one of the remarkable things about this conference is that it is sponsored not by a historical organization nor a philosophical organization nor a hybrid of the two, but by an engineering college. Furthermore, the engineers are not here as a passive audience. Instead, they are here as inquiring and active participants. It is the engineers—the practical technologists —who evince their concern about the part which the history and philosophy of technology might play in their own professional development and roles in society.

The presence of the engineers highlights the opportunity for historians and philosophers of technology to participate in the great events of our times. For example, I believe that historians and philosophers of technology can be drawn more and more into technology assessment. This movement attempts to look beyond the first-order effects of technological advance—those which have formerly been considered have usually been limited to immediate economic profits or military feasibility—to second- and third-order consequences. Such studies require a historical input, and historians of technology can help in this new approach to social policy. The same holds for the philosophers of technology; technology cannot be assessed or evaluated without some reference to human goals and social values. Philosophers, too, will have an active role to play in this new concern of society.

We will be historical activists, or participatory historians. This does not imply revisionism of historical interpretations, nor "politicizing" of historical associations, as the rebels at the American Historical Association convention attempted to do several years ago. Instead, historians can use their scholarly expertise to inform decision makers in matters of grave social import and political policy. I am basically a historical activist in this sense, in my belief that the study of history has meaning and importance in showing how the present came to be what it is and in providing some guidance for the future.

Thus historians can participate actively in a major trend of today: the current movement toward "accountability." Corporations, government, and even teachers—all are increasingly being held accountable for their actions by the public. The trend toward participatory democracy requires accountability and responsibility of every element in society. Science and technology are also being called to account—and the history of technology and the philosophy of technology can do much to provide public understanding of the actual roles and workings of technology in the past.

We cannot do that job by ourselves. We need help from many areas of the scholarly world. That is why it is important to have interdisciplinary conferences which bring together groups from varying fields to discuss problems which concern all of them and, indeed, all of society.

It would be logically symmetrical if we could compartmentalize this conference by saying that it deals with what technology has done in the past, what it is doing now, and what it should be doing—and then we might say that we have the historians to tell us what technology has done in the past, the engineers to inform us of what it is doing now, and the philosophers to advise us what it should do. However, neither life nor technology is so simple; nor are the fields of study represented here so tightly compartmentalized. Instead, what we have done, are doing, and should do represent interlocking questions, just as technology consists of social, political, economic, and cultural elements which go far beyond the view of engineering as a narrow technical subject.

All of us need the enlightenment which the other fields represented at this meeting can provide. All of us need the understanding which only a broad interdisciplinary approach involving thinkers and doers, theory and fact, idealism and reality can bring to discussions dealing with the technology-society interface, one of the most significant issues of our own and future times.

Alfred North Whitehead once said, "It is the business of the future to be dangerous." We can face that risk with composure and confidence if we can arrive at some understanding of technological forces and their interrelations with society, for these will form a most significant component of that dangerous future. Our deliberations can help remove some of the dangers and perhaps point the way to a better future.

Notes

1. Abbott Payson Usher, *A History of Mechanical Inventions* (Cambridge, Mass.: Harvard University Press, 1929); Lewis Mumford, *Technics and Civilization* (New York: Harcourt, Brace, 1934); Roger Burlingame, *March of the Iron Men* (New York: Scribner, 1938), *Engines of Democracy* (New York: Scribner, 1940), and *Backgrounds of Power* (New York: Scribner, 1949).
2. Kenneth Hudson, *Industrial Archaeology* (Chester Springs, Pa.: DuFour, 1964). See also Kenneth Hudson, "The Growing Pains of Industrial Archaeology," *Technology and Culture*, 6 (1965):621-26.
3. See Patrick Kelly and Melvin Kranzberg, eds., *Technological Innovation: A Critical Review of Current Knowledge* (San Francisco: San Francisco Press, 1978).
4. *L'Acquisition des techniques par les pays non-initiateurs* (Paris: Centre national de la recherche scientifique, 1973).
5. A full discussion of the *Annales* school is to be found in Traian Stoianovich, *French Historical Method: The* Annales *Paradigm* (Ithaca, N.Y.: Cornell University Press, 1976).

6. A pioneering conference showing how the historical approach can be utilized in technology assessment was sponsored by Carnegie-Mellon University in 1976. See Joel A. Tarr, ed., *Retrospective Technology Assessment* (San Francisco: San Francisco Press, 1977).

7. Jacques Ellul, *The Technological Society* (New York: Knopf, 1964).

8. Lewis Mumford, "Authoritarian and Democratic Technics," *Technology and Culture*, 5 (1964):1-8.

9. Donella H. Meadows, Dennis L. Meadows, Jørgen Randers, and William H. Behrens III, *The Limits to Growth* (New York: Universe Books, 1972).

10. Robert William Fogel, *Railroads and American Economic Growth: Essays in Econometric History* (Baltimore, Md.: Johns Hopkins University Press, 1964).

11. Robert William Fogel and Stanley L. Engerman, *Time on the Cross*, 2 vols. (Boston: Little, Brown, 1974).

12. The most devastating analysis is to be found in Herbert G. Gutman, "The World Two Cliometricians Made," *Journal of Negro History*, 60 (1975):53-227. (Also published as *Slavery and the Numbers Game: A Critique of* Time on the Cross [Urbana, Ill.: University of Illinois Press, 1975].) See also Charles Crowe, "Slavery, Ideology, and 'Cliometrics,'" *Technology and Culture*, 17 (1976):271-85.

Part I The History of Technology

Donald S. L. Cardwell

1

Problems of the Data Base

When I began to take an interest in the history and philosophy of science and in the history of technology it struck me forcibly that there was an almost complete gap between these two branches of scholarship. The first was concerned with the achievements of Copernicus, Galileo, Newton, and Lavoisier and their contemporaries as well as with the ideas of Kant, Mach, Duhem, and their successors. The historians of technology devoted themselves to measuring deserted warehouses, peering down abandoned mineshafts and reevaluating Matthew Boulton and James Watt's Soho Foundry. I speak of course of English historians of technology and I hope I am not being unfair to them: I am concerned with the overall picture, not with detailed assessment.

This culture gap seemed to me to be as surprising as it was unjustified. It was inconsistent with my education as a physicist, which naturally contained a strong technological component, and also I believe with the experience of the majority of scientists of my generation. I am not concerned with how this culture gap came about, although that in itself might be an interesting study. I am concerned with the relationship between science on the one hand and technology on the other.

All the dictionaries that I have consulted indicate that technology is the scientific study either of the industrial arts or of the application of science to industry. Such definitions do not take us very far, so we have to turn to the contexts and the dates at which the various terms entered common use. *Technology* was first used in the English language in the seventeenth century, but it did not enter into common usage until the nineteenth century. This is confirmed by such publications as Crabb's *Technological*

Dictionary (1817), which appeared almost contemporaneously with the first volumes of the *Bulletin des sciences technologiques.* On the other hand the word *technologist* was not coined until 1854, which suggests to me that there was no clearly recognizable or self-conscious group of technologists until the nineteenth century.[1] The *engineer* was recognized much earlier. The medieval engineer was a military man and his engines were weapons of war. Consider the words Christopher Marlowe puts in the mouth of the hero of *Tamburlaine:*

> I will, with engines never exercised,
> Conquer, sack and utterly consume,
> Your cities and your golden palaces,
> And with the flames that beat against the clouds,
> Incense the heavens and make the stars to melt. . . .

A frightening prevision, perhaps, of nuclear war; but for my purposes enough to demonstrate the meaning of the word *engine* in the sixteenth century.

The civil engineer seems to have become conspicuous in the seventeenth and eighteenth centuries, while mechanical, electrical, and chemical engineers emerged in the nineteenth century. On the other hand, although the words astronomer, chemist, and mathematician are of respectable antiquity, the word *physicist* was coined by Whewell as recently as 1840, the year in which he invented the word *scientist.* A translation of the French word *physicien* would, he felt, have caused confusion with the established English word *physician.* The autonomous science of physics, with its program of interpreting the phenomena of electricity, magnetism, light, heat, and the properties of matter in terms of mechanical mathematical law, was I believe founded at the beginning of the nineteenth century in France. However, the important doctrines of energy and of field theory were evolved later, in the middle of that century. In some quarters in England an Aristotelian interpretation of the scope of physics lingered on into the second half of the century. Thus in Walter Bagehot's *Physics and Politics* the "physics" in question is the Darwinian theory of evolution by natural selection.

As far as I know, Francis Bacon was the first man to discuss at length the relationship between science, as he understood it, and technology, and to propose a clear-cut program for the advancement of technology. Bacon's writings are freely available, and since many people, including me,[2] have discussed his ideas, I will offer only the briefest of outlines.

Bacon asserted that the noblest ambition a man could have was to increase the empire of all mankind over nature. He was not tainted by racial arrogance—that was a failing of future generations—for he recognized that the Europeans had been able to achieve supremacy, military and political if not moral, over the other peoples of the world because they had made use of three sovereign inventions: the mariner's compass, the printing press, and the gun. Such is the power of inventions. But, he thought, they should be used to benefit all mankind; how, then, could we increase their number?

At this point we note a distinction made by Bacon between the two main types of invention. We can, he said, take the printing press as representative of the first type. Its principles could easily have been understood by Archimedes, say, for he would not have required any further information than the knowledge he already possessed. Inventions like the printing press could, in theory, have been made at any time. But the other two inventions, the compass and the gun, would have mystified Archimedes for he had no knowledge of the properties of black powder, a mixture of sulfur, saltpeter, and charcoal that, if compressed in a strong container, can explode with such devastating force. Nor did he know anything about the north-seeking property of a freely suspended and magnetized steel needle. These things had to be discovered before the compass and the gun could be invented. Stretching a point, we may say that such inventions are science-based and can be made only if the appropriate knowledge is available. Both types of invention, the empirical and the science-based, continue to be made at the present time. It would be easy to draw up long lists of both types, although the science-based tend to be more sensational.

Evidently if we can augment the stock of knowledge, of science, we may reasonably expect to increase the number of science-based inventions. I do not, of course, suggest that Bacon had in mind systematic Galilean and post-Galilean science. I believe he was thinking of some kind of down-to-earth know-how, general knowledge about the world and the things in it that a shrewd, economically active community might expect to acquire —a community that would not be unduly inhibited by superstitious, political, or social taboos; a community that knew well enough how to acquire or imitate the technical accomplishments of others. Imitation, after all, is an important factor in the spread of technology.

Parenthetically, I would like to point out that Manchester, one of the cradles of the industrial revolution, exemplifies Baconian ideas in its technological and scientific history. It all began with the invention of new textile machines: the water-frame and the spinning jenny. Machines of this type can be traced back to the Italian Renaissance engineers and are illustrated in works such as that of Zonca. They are essentially empirical inventions, and it is easy to explain their operation to those without scientific knowledge. But, as a consequence of the economic growth that these machines made possible, a wealth of scientific ideas began to be incorporated into the textile industry. And early in the nineteenth century, a remarkable voluntary education movement—the Mechanics' Institute movement—sprang up with the avowed aim of increasing scientific knowledge among skilled workers so that, on Baconian grounds, a great increase in science-based inventions might be expected.[3]

I shall not say anything about Bacon's science, his famous reformed inductive method. That has been discussed quite adequately by a sufficient number of commentators. It has been pointed out, quite fairly, that Bacon was unable to show how his science related to technology, by what formal means the scientific knowledge gained by his method could be applied systematically to invention. And that, perhaps, is a problem that is still with us, even though we may not accept Bacon's scientific methodology.

I have, I think, said enough to show how it came about that technology, concerned with the relationship between science and industrial innovation, came to be recognizable by the seventeenth century. My incursion into lexicography at the beginning was therefore justified. The influential Bacon recognized the importance of inventions and innovation, and he had some understanding of the role of new knowledge in enabling additional inventions to be made. The more new knowledge, the more new inventions we may expect. And finally, as we all know, the seventeenth century was the time of scientific revolution, when physical science was established and made its great initial triumphs.

I do not want to exaggerate the numbers and the importance of the innovators. Many of them would be "projectors," men who innovated by introducing inventions made in other parts of the world. Original inventors were widely scattered and formed more or less isolated groups; they were instrument makers, medical men, soldiers, surveyors, miners and metal workers, millwrights and smiths, and occasionally, very occasionally, a teacher or a university man. Technology was not an established social institution. Accordingly there were no technological societies (the early Royal Society notwithstanding) and no specialized journals for technologists. Furthermore, the men who made innovations, whether as projectors or as original inventors, were usually engaged in trade or industry; innovation was thus an incidental part of their activities. They were, in other words, only part-time technologists.

It follows from this absence of institutionalization that the sources for our knowledge of the history of technology before, let us say, 1800 are different from the sources for our knowledge of the history of science. Very briefly, and rather obviously, these are the actual relics themselves (usually few in number and often very incomplete in representation), manuscript sources such as notebooks, letters, plans and drawings as well as travelers' accounts and legal documents, printed sources such as patent specifications, pamphlets, broadsheets, official reports, articles in scientific journals (such as the *Machines et Inventions* of the Académie Royale des Sciences), books of various sorts, papers in technical journals (such as the *Journal of the Society of Arts*), and reports of conferences and committees. May I therefore suggest that some industrious student might consider doing for technological societies and journals the same sort of job that Martha Ornstein did for the scientific societies of the seventeenth century? The rise of the professional, full-time engineer is surely also worth investigation.

The wide variety of our sources is unavoidably associated with a no less wide diversity in the standards of the information available to us. There are several reasons for this, and it further complicates the task of interpreting past technology.

In the first place, verbal communication seems to have played a much larger part in the transmission of technological information than it did in science. This may well have been because it was rarely felt necessary to give exhaustive accounts of the nature or the functioning of particular inventions save to those actually involved in the art. Another factor here was the need, actual or imagined, for secrecy. This was often felt to be important in those branches of industry in which there was an element of aesthetic accomplishment. In the manufacture of dyestuffs, for example, the surviving recipe books are usually marred by significant omissions of certain important steps; these omissions could be rectified verbally in the dyehouse itself. In the second place, we must remember that engineering drawing originated only in the seventeenth century with the work of Desargues and was not firmly established until the work of Monge at the end of the eighteenth century.[4] For this and earlier periods we are often dependent on artist's impressions and these can be misleading, either because the artist was incompetent or perhaps because he was required to keep certain details secret. If he did not fully understand the invention, he may have judiciously hidden the components he did not understand behind a human figure or a convenient bit of smoke; it must be remembered that the artist may not have seen the engine in question or conversed with the engineer who built it. Finally, inventors have tended to be boasters and to have put forward claims that often will not bear examination. The temptations to cut corners are no doubt very strong, and between the claim to have achieved a long-sought breakthrough and its actual, effective accomplishment there may well be a very big gap indeed. I am thinking here of such men as the egregious Dud Dudley who claimed to have achieved the smelting of iron by means of coal long before the actual breakthrough by Abraham Darby. Other very obvious examples are provided by the many claims to have discovered the principle of the perpetual motion machine.

I am not trying to decry the written and printed record which must obviously form the main source for our knowledge of past technology, but it has its shortcomings and the problem is, what can we do about it? Perhaps I can suggest some answers by mentioning research which my colleagues and I, in my department at Manchester, are now carrying out.

I have already published a brief outline of some of the knowledge we have gained from building and operating a one-third scale model Newcomen engine for our science museum.[5] Although a model, our engine is about twelve feet high. We learned a number of things that, with hindsight, now seem fairly obvious but which we had not foreseen and which are not mentioned in any written or printed source that we know of. We also found that one or two features that we had regarded as unimportant or unnecessary details were in fact essential for the operation of the engine. However, possibly the most interesting lesson we learned was that it took our skilled and enthusiastic technician several months to get our beautifully made engine to work. I have learned how to operate this engine, but I am not very good at it, and it is a sobering thought that despite my Ph.D. in physics and subsequent practical experience I would only just be qualified to be an engineman in 1712.

The next thing we want to do is to see what differences in performance will be caused by variations and interchange of the valve mechanisms. What were the reasons, we wonder, for using different valve mechanisms to control the steam and the condensing water supplies?

After we have finished our first series of experiments with the Newcomen engine I intend to try to see if the Savery engine can be made to work. Personally I am very doubtful that it can ever have worked and I am certain that, if it did, it can only have done so in a most unsatisfactory way. I am not thinking of the modified and partial Savery engine described by Musson and Robinson,[6] which did work after a fashion, but of the engine that Savery described and patented. I do not propose to build a large-scale model Savery engine; all I want to do is to carry out a series of experiments to test the basic principles and see if the thing will actually work. May I remind you that Savery was another boaster with an inflated idea of his own achievements, as anyone who cares to read *The Miner's Friend* can confirm for himself. But he boasted to such effect that contemporaries like Belidor and Leupold took him at his face value and ascribed the invention of the Newcomen engine to him, although to be sure Belidor did know that a Mr. Newcomen must be credited with making certain improvements to the engine.

Turning now to the field of chemical technology, my colleague Wilfred V. Farrar is studying *inter alia* the history of dyestuffs. He has been fortunate enough to secure the collection of samples that formerly belonged to H. E. Schunck, an able nineteenth-century organic chemist who lived in Manchester and was a close friend of James Prescott Joule. Among these samples is one of the original mauveine, which was the first synthetic dyestuff, produced by W. H. Perkin in 1856. Dr. Farrar is comparing the composition of this original dyestuff with mauveine made by modern methods and from modern materials. He intends to determine the nature of the original manufacturing process and of the materials Perkin used. A similar project to determine the composition of murexide, an early science-based dyestuff made from uric acid, has had to be put aside as we have not yet been able to obtain samples of Peruvian guano, the original raw material used in the manufacture of murexide 130 years ago.

Another member of my department, Alan R. Williams, is interested in the composition and manufacture of medieval and Renaissance armor.[7] Suits of armor are, generally speaking, the most numerous large items of iron and steel to have survived into the modern age. Their manufacture covered a long period of technological development, a period during which an early arms race took place. As the offensive weapons improved—the longbow with bodkin-tipped arrow, the handgun, the musket—so armor was improved, culminating ultimately with armor made from steel. The problem is, simply, how was this done? How could the Renaissance iron workers produce steel if they were unable to melt the iron? Cast iron could be melted; thanks to its high carbon content it has a low melting point which was within the competence of the Renaissance iron workers to achieve; but steel is a different matter.

To throw some light on this Mr. Williams has built his own furnace and is engaged in smelting iron ore using charcoal to see how far he can approximate Renaissance techniques. The fuel, charcoal, is easy to get because it is imported from Kenya to supply the considerable garden barbecue market. The main obstacle we have met so far is that a few museums are reluctant to let Mr. Williams examine, and carry out nondestructive tests on, the suits of armor in their collections.

I have mentioned three of the research projects being carried out in my department. I myself am qualified to comment on the first one only. But I can emphasize—and here I think I speak for all of us—that wherever the written or printed record is deficient it is not necessary, lovingly and completely, to restore each invention or process before it can be put to practical test. Generally speaking it is only a few direct questions that we need to answer: did it work? how did it work? how well did it work? In most cases a simple thought experiment should be enough to answer the questions.

For instance, in his masterly *Science and Civilisation in China* Dr. Needham mentions such ingenious devices as a paddlewheel boat that could travel faster than a horse could charge.[8] Although I do not have the facts before me, I am pretty certain that it was not until the middle of the nineteenth century that it became *possible* for a paddlewheel boat to travel faster than a horse could gallop. Unless, that is, Chinese horses used to gallop extremely slowly.

It is, I think, permissible to use modern scientific and technological knowledge to amend and/or to criticize claims made by the technologists of the past. Indeed it may be informative to do so and may tell us a great deal about the standards of judgment in use in past times and the attitudes of the people concerned toward technological claims. I think too that Hume's argument against miracles is exactly applicable in many of these cases. Few of us, I imagine, would accept at their face values the claims of alchemists and seekers after perpetual motion.

I should like now to return to my discussion of the simple Baconian classification of inventions into two groups: empirical and science-based. The century following Bacon, the eighteenth, witnessed the introduction of one, and possibly two, new modes of innovation. The first was that of systematic improvement, associated in Britain with the name of the engineer John Smeaton; the second was that of scientific research in pursuit of specific innovation or invention. Modern technology is unthinkable without these two modes of innovation.

Smeaton's technique of applying experimental procedures to get the optimum performance out of an engine owed something to the experimental science established in the seventeenth century and exemplified in various works of Newton. It may seem fairly obvious that the components of an engine should be so proportioned and designed that the optimum performance is obtained, but in fact the very idea of *optimum*—and therefore *measurable*—performance is the outcome of the application of Galilean principles to power technology. I do not think that the idea of optimum performance could have been conceived of without prior knowledge of measurable work (or effect, *effet*), independent of the particular task that the engine is used for. For this we must surely thank Galileo and his successors.

Antoine Parent had opened the debate in 1704 with his paper on the greatest possible "perfection" of machines. He calculated the optimum efficiency of a friction-free waterwheel, using the calculus in his work (he was perhaps the first engineer to use the calculus). The debate arose from certain erroneous assumptions he made that vitiated his answer, an answer that practical engineers were to find must be wrong. Smeaton's paper of 1759, which won him the Copley Medal of the Royal Society, constituted an effective experimental refutation of Parent's calculation but at the same time a confirmation of the importance of the theme Parent had proposed.

It would not, I think, be an exaggeration to say that before the middle of the eighteenth century waterwheels and Newcomen engines were generally built individually, to suit particular purposes. Comparisons between different engines would therefore be difficult. But toward the end of the century economic growth may, I suspect, have made comparisons much easier and ultimately essential. Hence I assume that Smeaton's procedures became institutionalized, although I do not know the steps by which this may have happened. We shall have to await the first definitive biography of this man for further enlightenment. We do know that there was something approaching a power shortage in the textile areas around Manchester at the end of the eighteenth century,[9] and we know that a long sustained campaign to improve the performance of Cornish pumping engines began in 1811 through the efforts of Joel Lean. It was subsequently carried on by his two sons.[10] The breakdown of the old individualistic view of engines, brought about by economic necessity and the acceptance of the principles of Galilean mechanics, contributed very largely I believe to the progressive clarification of the concepts of energy in the nineteenth century. However, this field is still largely unexplored, and a history of the development of industrial power, together with a critical account of its scientific rationale, is an obvious need of modern scholarship. Civilization, Liebig remarked, is economy of power [11]—an exaggeration no doubt, but not without a substantial kernel of truth.

Another instance of the quantification of industrial processes resulting from the advance of science is provided by the chemical industry in the decades following Lavoisier's chemical revolution and the quantification of his system made possible by the scientific atomic theory from 1808 onward. In principle it became possible to calculate the exact amounts of different substances required for chemical processes and thus to eliminate waste. At the same time—and this was probably at least as important—unnecessary stages and unnecessary ingredients, hitherto sanctioned by custom, could be cut out. The optimization of industrial chemical processes seems to be analogous to the optimization of power in the processes of mechanical industry, but as this is well documented I shall say nothing more about it.[12]

As for the use of scientific research in aid of specific invention, James Watt's investigations of the properties of steam and into the specific heat capacities of different materials may be taken as an early example, but it can hardly be taken as typical; there were not many Watts around. In fact, the institutionalization of industrial research seems to have begun a hundred years after Watt's research and in connection with the synthetic dyestuffs industry in Germany. Conditions in Germany a hundred years ago seem to have been most favorable for such a development. The new azo dyes opened up prospects of a virtually unlimited range of new coloring matters, but the potential could only be actualized by those with some knowledge of organic chemistry. The German universities were by then producing an abundant supply of young research-trained chemists. Hence the first industrial research laboratories, which employed salaried research scientists, were set up a hundred years ago in Germany.[13] The success of the organic chemicals industry—dyestuffs and later pharmaceuticals—undoubtedly stimulated other industries into setting up research laboratories. These are now, I suppose, universal in industries in modern industrialized countries. It would not be unreasonable to suggest that, in some material respects at least, the Baconian dream has now come true.

Just what is meant by "industrial research" in these days is a matter of some doubt. Such is the range of industrial activities and so limited is one's understanding that it is very difficult to form coherent and compact ideas about it. For one thing, I understand that in the development of pesticides and fungicides a good deal of the research is straightforward Baconian inductivism: that is, in the fairly safe knowledge that if you try out a sufficient number you will probably find at least one that is a winner, one compound after another is tried out to see which ones work. On the other hand, my personal experience suggests that in the electrical and electronics field research consists in inventing arrangements of components to achieve a desired result. That is, knowing the relevant laws and the properties of the units available, to couple them together in such a way as to solve the problem set or envisaged. In other circumstances research may consist of investigations into the behavior and properties of substances or processes that may be or become of practical importance. In this case we get, perhaps, closer to what most of us would accept as a rough definition of science. However, I suspect that only in relatively few cases, and then perhaps only in the largest laboratories, do we find research of such a nature that its results could be published in a journal devoted to "pure" science. As the practical value of such research can only be conjectural, it would follow that only the wealthiest corporations or the best-endowed state laboratories could afford it.

I have now covered the main aspects of technology as, it seems to me, they have evolved over the last 300 years. No doubt other modes of innovation can be recognized and no doubt the descriptions I have put forward can be corrected and amplified. The institutionalization of technology has proceeded *pari passu* with its complexity and sophistication. From the individual inventor and the artist-craftsman of the Renaissance we pass on to recognize the first professional engineers of the eighteenth century, acknowledging that they often worked in collaboration with businessmen, because such a symbiotic partnership made for efficient innovation. By the end of the nineteenth century the salaried technologist and applied scientist had both appeared, celebrating, in Whitehead's words, the triumph of the professional man.

As a result of my own limited qualifications, I have not discussed such things as medicine (considered as a technology), agriculture, and what we may call social technology, even though these have histories at least as long as that of physical technology.

Before I conclude I would like to mention two points that I think are of some interest. The first concerns the problem of backlog. Thus, for example, Galileo's theory of the strength of beams, propounded in *Two New Sciences,* remained of limited academic interest and virtually no practical interest for about 160 years. It suddenly became of great importance when at the end of the eighteenth century increasingly abundant supplies of iron made it feasible to use that metal as a building material, particularly for industrial buildings. At the same time iron was increasingly used in the construction of machines, particularly steam engines and industrial waterwheels. In this work Galileo's theory was widely used. The famous error did not matter for in the comparative way in which the engineers applied the theory the error was cancelled out. So Galileo had a very direct impact on the industrial revolution even though it took place long after his death.[14]

Instances such as this—and many others could be quoted—prove how very difficult it is to foresee the time, place, and nature of the practical impact of a particular scientific advance. This I think must have its consequences for any discussion of the role of the scientist, as distinct from the technologist, in modern society.

My second, and I think more important, point concerns the effects of technology on science. Just as technology draws on science so, I believe, science draws on technology, and not just for an assured supply of high-grade instruments. Science is surely based on human experiences, which characteristically are common to all men. Judgments may validly differ over many things ranging from feminine charm to musical appreciation, but they can hardly differ over the coordinates defining the position of a planet. I think the advance of technology has contributed to experiences of this sort. The industrialization of the West, the spread of water and steam power, and the growth of transportation made for an enormous change in the world of man, and of its nature invited critical appraisal, as much perhaps as the mechanism of the heavens did in the years between Copernicus and Newton. This is not a particularly original statement. Osborne Reynolds argued that while the first steam engines were comparatively remote from everyday life, the spread of rail systems in the middle of the nineteenth century brought the "obtrusive steam locomotive" to everybody's attention,[15] epitomizing for even the least imaginative the immense power to be obtained from heat. Indeed I believe that from his industrial experiences man created much of his nineteenth-century physical science. The doctrines of energy and the laws of thermodynamics carry with them unmistakable signs of their technological origins. Not for nothing were the men who formulated these theories either engineers or scientists with strong engineering biases. As James Clerk Maxwell put it: "The cultivation and popularisation of correct dynamical ideas since the time of Galileo and Newton has effected an immense change in the language and ideas of common life, but it is only within recent times and in consequence of the increasing importance of machinery that the ideas of force, energy and power have become accurately distinguished from each other" [16]

Force, energy, power: I do not think anyone would deny the importance today of the concepts in their philosophical, scientific, or economic contexts. In short, then, technology has itself effected vast changes in science and beyond that in the world of ideas. This is a field for study that has not, as far as I know, been more than very tentatively examined.

Notes

1. The definitions are from *A New English Dictionary on Historical Principles*, ed. J. A. H. Murray et al., 20 vols. and supplements (Oxford: Clarendon Press, 1884-1928). Webster's *International Dictionary* and all the other authorities I have been able to consult confirm these definitions.
2. See, for example, D. S. L. Cardwell, "The Development of Scientific Research in Modern Universities: A Comparative Study of Motives and

Opportunities," in *Scientific Change*, ed. A. C. Crombie (London: Heinemann, 1963), p. 661.

3. Mabel Tylecote, *The Mechanics' Institutes of Lancashire and Yorkshire* (Manchester: University Press, 1957), pp. 34, 42.

4. P. J. Booker, *A History of Engineering Drawing* (London: Chatto & Windus, 1963).

5. D. S. L. Cardwell, *Turning Points in Western Technology* (London: Heinemann, 1972), pp. 71-72.

The importance of the Newcomen engine cannot be exaggerated. It was the first successful heat engine to be invented (1712) and therefore marks the origin of modern power technology (apart, of course, from water power).

Some idea of the success of the early Newcomen engines can be gained from the fact that 112 of them had been erected on the Tyne coal field alone by 1785. This is a clear indication also of the wide diffusion of engineering skills by that time. (I am indebted to J. D. Swallow, a student in the department of mechanical engineering, University of Manchester Institute of Science and Technology, for information about the Tyne coal field.)

6. A. E. Musson and E. Robinson, *Science and Technology in the Industrial Revolution* (Manchester: University Press, 1969), pp. 396 ff. The authors quote John Farey as stating that Joshua Wrigley, or Rigley, had made an engine that "usually raised the water about 16 to 20 feet high; and the water descending again gave motion to an overshot waterwheel. . . . Mr. Rigley contrived his engine to work without an attendant; the motion of the waterwheel being made to open and shut the regulator, and injection cock, at the proper intervals." The performance of this engine, excluding as it does the pressure phase, is far more modest than that claimed—by implication!—by Savery for his engine in *The Miner's Friend* of 1702.

7. A. R. Williams, "Metallographic Examination of Sixteenth Century Armour," *Bulletin of the Historical Metallurgy Group*, vol. VI, no. 2 (1972), p. 15. The author points out that King Henry VIII of England ordered no fewer than 3,900 suits of armor on one occasion.

8. J. Needham, *Science and Civilisation in China: Mechanical Engineering*, vol. 4 (Cambridge: University Press, 1965), p. 159. Dr. Needham says, "There is no need to take these stories *au pied de la lettre*, but we need not doubt that there was some substance behind them" (why not?). For the account of the paddlewheel boat, see p. 418.

9. R. L. Hills, *Power in the Industrial Revolution* (Manchester: University Press, 1970), pp. 89 ff., pp. 134 ff.

10. Thomas Lean & Brother, *Historical Statement . . . of the Steam Engines in Cornwall* (London, 1839). "The Monthly Engine Reporter" was published from 1812 to 1880. There is a complete set in the Public Library, Redruth, Cornwall, England.

11. Quoted by W. S. Jevons, *The Coal Question* (London: 1865), p. 125. See also Liebig's *Familiar Lectures on Chemistry.*

12. A. Clow and N. Clow, *The Chemical Revolution* (London: Batcheworth Press, 1952), p. 110. "[Muspratt] acquired an understanding of the theory on which Leblanc's manufacture was based and from calculations

of the quantities of materials used and products obtained he concluded that it would be a lucrative operation. Thus we see that the quantitative atomic theory . . . soon enabled the manufacturer to cost his processes with increased accuracy."

13. D. S. L. Cardwell, *The Organization of Science in England* (London: Heinemann, 1957, 2nd ed., 1972). See also J. J. Beer, "Coal Tar, Dye Manufacture and the Origin of the Modern Industrial Research Laboratory," *Isis*, vol. 49 (1958), pp. 121-31.

14. I am grateful to my friend and former colleague A. J. Pacey for the information in this paragraph.

15. Osborne Reynolds, "Memoirs of James Prescott Joule," *Memoirs and Proceedings of the Manchester Literary and Philosophical Society*, vol. 6., 4th series (1892), p. 21.

16. James Clerk Maxwell, *A Treatise on Electricity and Magnetism*, vol. 1 (Oxford: 1873), p. 142.

Arthur L. Donovan

Toward a Social History of Technological Ideas: Joseph Black, James Watt, and the Separate Condenser

I

History is essentially an articulation of the meaning of the past. It uses concepts, symbols, and myths to establish significant relationships between past events and the present and thereby specifies, clarifies, and interprets the collective experience of its audience. If received ideas, the common sense of the day, provide an adequate account of the particular social activity, then it is unlikely that the history of that activity will be marked by intense debate. But when the received ideas with which we interpret experience and construct our expectations appear radically inadequate, a thorough reexamination of historical accounts shaped by those ideas becomes a matter of some urgency. Such, I believe, is the situation in the history of technology today.

A deep and threatening fissure has appeared in our received ideas about the nature of technology. For nearly a century analysis of social change in the modern world has taken as its point of departure technical change in the means of production. The commonplaces of our language and the common sense of our culture point to technology as the engine of industrialization and modernization. East and West, in the industrialized nations and in the so-called underdeveloped nations, in the socialist countries as well as in those ruled by corporate capital, technological innovation has been seen as both the key to our understanding of the modern world and the Aladdin's lamp whose genie has the power to fashion singlehandedly a better life for all. This intellectual inheritance has been criticized in the past but has survived, like an estate defended by a well-drawn will, down to the present.

But now, like laissez-faire, the doctrine of *laissez-innover* is being challenged.[1] Must we continue to encourage the growth of technology while passively accepting the cultural devastations and social dislocations which attend the unregulated introduction of new techniques of production? More and more the answer is in the negative. Widespread concern with environmental impact, the revolt of the "third world" against economic dependence, and the flourishing of communitarian social movements in industrialized nations indicate that the relationship between technology and social change is being radically revalued. In the future I think the development and utilization of new techniques of production will be increasingly governed by the consensus of all those directly affected by the changes involved. If in fact the rate and direction of technological change are brought under democratic control, then our theories about the relationship between technological innovation and social change will have to be reconstructed. We will no longer be able to assume that in seeking to explain social and cultural change one should begin with an account of changes in the mode of production.

Historians of technology can play a constructive part in this ongoing reexamination and revaluation of the relationship between technical and social change, but only if they address certain questions. Now that we are beginning to insist that those most affected be allowed to control the introduction of new technologies, we would like to know if similar demands have been raised and satisfied in the past. We need to identify the nontechnological aspects of society which have influenced the development and application of technical change in the past. In the realm of ideas, we want to know if the main normative and descriptive assumptions utilized by innovators were in fact among the dominant ideas in the societies in which they lived. Can we write a social history of the ideas which encouraged the development of techniques, which made the discovery of specific innovations possible, and which guided the uses to which these new techniques were put?

The relationship between Joseph Black the chemist and James

Watt, the inventor of the separate condenser for the steam engine, provides a case in which some of these questions can be examined in a particular historical setting. But we must not expect to find that certain key ideas were clearly and purposefully transmitted from the scientist to the inventor or vice versa. Rather, we should take note of the general character of the society which brought these men together and supported their activities. We will then be better able to identify those areas of shared understanding which were their common property and which each man utilized in his own way. By proceeding in this manner we will be able to emphasize the uniqueness of each individual's achievement without artificially separating his experience as a man of his time and place from his brilliance as a discoverer or inventor.

II

What role did contemporary science play in the invention of the separate condenser, that modification which has been called the most important technical development in the Industrial Revolution? According to John Robison, one of Watt's closest friends, Watt invented the separate condenser by applying Joseph Black's theory of latent heat to the study of the steam engine. Watt himself objected to this account of his invention, for while he was grateful to Black for his help and encouragement, he did not view his achievement as a rather straightforward application of a general principle. But Watt's opposition did little to slow the spread of Robison's story. Science discovers the laws of nature and technology applies them to the service of man, to paraphrase the motto inscribed in the dome of Chicago's Museum of Science and Industry, and this notion was so pervasive in the nineteenth century that Robison's story was accepted precisely because it conformed to the general understanding of the relationship between science and technology. Robison's account sprang from a desire to honor Black, his teacher and predecessor. It was the nineteenth-century view of technology that transformed Robison's homage into the legend which Donald Cardwell has so aptly described and criticized.[2]

The growth of the history of science in the last few decades has necessarily resulted in a good deal of demythologizing. Armed with Watt's own statements and aware of the physics of the steam engine, several historians have pointed out that Robison's story should be transferred from the history of technology to the history of enthusiasm.[3] And yet there is a danger involved in this kind of revision. Robison's story contains an important kernel of truth, namely, that there was a close relationship between Black's science and Watt's invention. While we now know Robison misrepresented the essence of this relationship, we must be wary of accepting a revision which postulates too great a distance between Black's and Watt's achievements. In our own day it is all too easy to think of technology as an autonomous activity, one which generates whatever kinds of understanding it needs. But this supposed autonomy is neither beyond question in our society nor was it characteristic of the setting in which Watt worked. Rather than abandoning the Robison legend, we should modify it in a way which is both true to the historical and physical facts and which illuminates our own emerging understanding of technology.

James Watt arrived in Glasgow in 1754.[4] Eighteen years old and eager to make his own way in the world, he spent his first year living with his mother's relatives, the Muirheads, while seeking a position in which he could acquire the skills needed to become a mathematical instrument maker. His choice of trade reflected both his own mathematical proclivities, which seem to have been a characteristic of the Watt family, and his intimate knowledge of the nautical apparatus and instruments sold by his father.

In Glasgow, Watt had the good fortune to be introduced to several members of the university faculty through his relative George Muirhead, who was then professor of Latin. Although James never attended a course of lectures given in the university, he did become friendly with Dr. Dick, the young professor of natural philosophy who had previously been Black's closest companion when he was a student in Glasgow. When Professor Dick heard of Watt's interest in mathematical instruments, he strongly advised him to learn his trade in London and further encouraged him by supplying several letters of introduction. In June, 1755, Watt set off on horseback for London. At first he encountered considerable difficulty in finding a master, for the few skilled instrument makers in London were reluctant to dispense with the customary seven-year apprenticeship. With Dick's letters, however, Watt gained the attention and then the friendship of James Short, a Fellow of the Royal Society and a celebrated maker of reflecting telescopes, and through Short he obtained a position with John Morgan, a maker of Hadley quadrants. After a year of arduous work and frugal living, Watt returned to Scotland to practice his newly acquired skills. Shortly after Watt's return to Glasgow, Professor Dick asked him to repair a set of astronomical instruments which had been given to the university, and a year later Watt was given a shop and an apartment in the university buildings and was designated "mathematical instrument maker to the university."

Watt met John Robison soon after he moved into his new accommodations. Robison was a bright young man who had recently taken his degree in Glasgow and was evidently at loose ends. He soon discovered that Watt was a man of the highest ability and an enthusiastic investigator of mechanical questions, and the two young men quickly formed a friendship which lasted until Robison's death.

Late in 1756 Joseph Black returned to Glasgow to become professor of anatomy and botany and lecturer in chemistry. He was, no doubt, soon introduced to Watt by their mutual friend Professor Dick. Black was greatly impressed by Watt's quick scientific mind and unusual mechanical ability, and he soon asked him to construct several pieces of experimental apparatus for him. Black enjoyed visiting Watt's workshop and it was there that Robison first met his future mentor. Throughout 1758 these three young philosophical bachelors thrived upon each other's company. Black was then trying to think of ways to extend and test his theory of latent heat, Watt was struggling to make a go of it as an instrument maker, and Robison was casting about for a situation which would oblige him to fasten his agile mind to a concrete set of problems. Robison later recalled an example of the way in which Watt threw himself into challenges he found appealing: "A Mason lodge in Glasgow wanted an Organ. The office bearers were acquaintances of Mr. Watt. We imagined that Mr. Watt could do any thing, and although aware that he did not know one musical note from another, he was asked if he could build this organ. . . . He said yes. But he began by building a very small one for his intimate friend Dr. Black." [5]

Late in 1758 Robison left Glasgow. When he returned after an absence of four years, he was immediately welcomed back into the philosophic circle. As Robison later said of his friendship with Watt: "I had had the advantage of a more regular education: this frequently enabled me to direct or confirm Mr. Watt's speculations, and put into a systematic form the random suggestions of his inquisitive and inventive mind. This kind of friendly commerce knit us more together, and each of us knew the whole extent of the other's reading and knowledge. I was not singular in this attachment. All the young lads of our little place that were any way remarkable for scientific predilection were acquaintances of Mr. Watt; and his parlour was a rendezvous for all of this description." [6] Although Black was not one of the "young lads," Robison said that upon his return he found Watt "most intimately connected with Dr. Black, and continually speaking of his lectures and doctrines." [7] Watt himself told Robison that "every thing I learnt from him [Black] was in conversation and by doing small mechanical jobs for him. These conversations and those I had with you served to give me true notions in science and to develope the powers of my mind." [8]

What significance should we attach to the friendship between Black, Robison, and Watt? First, it should be pointed out that all the statements given above refer to their relations before 1765, the year in which Watt invented the separate condenser. Second, it is clear that these men were drawn together by their common interest in a particular set of problems and that discussion of those problems dominated their conversations. These three men constituted a self-selected and self-supporting research team of extraordinary historical importance. Surely it would be artificial on our part to try to erect barriers between the conceptual foundations of Black's natural philosophy and Watt's mechanical investigations. We can, of course, distinguish between Black's evident desire to construct a general theory of chemistry and Watt's clearly stated ambition to be a successful craftsman and engineer, but insofar as their activities depended upon an understanding of nature, we must assume that they were well informed about each other's thoughts. The question is not, therefore, whether Black's scientific knowledge played a part in Watt's invention, but whether we can correctly identify and specify the part which it played.

Sometime in 1758 Robison turned Watt's attention to the steam engine, a device, Watt admitted, "of which I was then very ignorant." [9] Watt built a small model which proved disappointing and the investigation was soon abandoned. Five years later John Anderson, who had succeeded Dr. Dick as professor of natural philosophy in Glasgow, asked Watt to repair the university's demonstration model of the Newcomen engine. At first, Watt later recalled, "I set about repairing it as a mere mechanician, and when that was done and it was set to work, I was surprized to find that its boiler could not supply it with [enough] steam, though apparently quite large enough." [10] Surprise quickly became a consuming curiosity as Watt pressed on in his analysis of the model steam engine. He puzzled over "the imense Quantity of fuel it consumed in proportion to its Cylinder" [11] and wondered where all the heat given off by that fuel was going. Through his study of the operation of the model Newcomen engine Watt recognized this need for a heat analysis of the steam engine, and it was the approach to the problem which enabled him to invent the separate condenser. (See Figure 1.)

But how else could he have analyzed the steam engine? For us, the steam engine is a heat engine by definition and its operation is explained by the laws of thermodynamics, but Watt's predecessors thought of it as a mechanical engine and they explained its operation in terms of the pressure exerted by the elastic atmosphere.[12] Different sides of the same coin, you may say, but the improvements suggested by the two views are fundamentally different. Watt's decision to undertake a heat analysis of the steam engine thus stands as the fundamental intellectual shift in the discovery of the separate condenser. However, I suspect that Watt was not even aware of the novelty of his approach. For years he had been listening to Joseph Black talk about the central role of heat in chemistry and he knew that Black and his teacher William Cullen both believed that chemistry was the most practical branch of natural philosophy. Thus when Watt began studying the steam engine, he naturally saw it as a heat

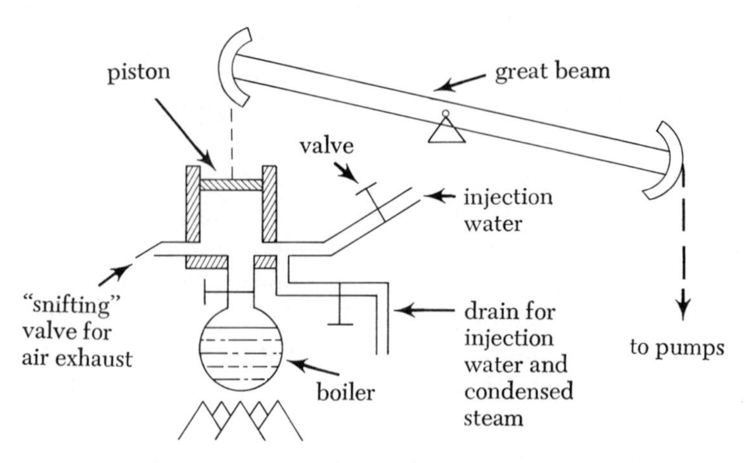

Figure 1. Simplified diagram of the Newcomen Engine. The piston of the engine was connected to one end of a great beam, the other end of which was attached to the chains that operated the pumps. The beam was so balanced that the weight of the pump chains drew the piston upward in the cylinder. As it traveled upward, the piston drew steam from the boiler into the cylinder. When the piston reached the top of its stroke, cold water was sprayed into the cylinder to condense the steam. This condensation created a partial vacuum in the cylinder, thereby causing the piston to be driven downward by the pressure of the atmosphere.

engine. He found this approach a natural one not because of the nature of the engine itself, but because it was the dominant approach to the study of physical nature in the immediate context of his life. Here, in the largely unconscious social determination of Watt's conceptual framework, we find the crucial link between science and technology.

There is also evidence that Watt had Black's specific contributions to the study of heat in mind during his initial analysis of the steam engine. To reduce the amount of heat being lost Watt sheathed his boiler with wood and constructed a wooden cylinder for his engine. He also used the results of Black's study of specific heats to determine the quantity of heat absorbed by different metals when heated over a given range of temperature.[13] Watt discovered that a great deal of heat is lost to the water used to condense the steam in the cylinder of the Newcomen engine. The condensing water would not drain out of the cylinder until it had become nearly boiling hot, and so long as it remained in the cylinder, the steam admitted to start the next cycle was weak.[14] He suspected that a great deal of steam was being condensed before it had done any work, and he tested his suspicion by measuring the volume ratio between steam and water. Desaguliers, one of the leading English natural philosophers of the first half of the eighteenth century, had concluded that "water in boiling is expanded 14,000 times," [15] and William Cullen, Black's teacher and the leading chemist in Scotland during the middle decades of the century, had used Desaguliers's figure in his lectures without questioning its correctness. On the basis of his own experiments, however, Watt determined "that there was a great error in Dr. Desaguliers' calculations . . . on the bulk of steam" and that the correct figure was between 1,600 and 1,800, not 14,000.[16] Watt then calculated the volume of steam produced by his boiler and concluded that "the quantity of steam used in every stroke of the [Newcomen] engine [is] several times the full of the cylinder." [17] He had determined, in other words, that a great deal of heat was being lost through the condensation of steam in the first part of the cycle.

Watt knew that most of this lost heat was being used to heat up the heavy metal cylinder which had been cooled by the condensing water at the end of the previous cycle. Although he had a rough idea of the amount of steam needed to run the engine, he was still puzzled because "the Quantity of water used for Injection in fire engines was much greater than I thought was necessary to Cool the Quantity of water contained in the Steam to below the boiling point." [18] He therefore performed two more experiments. In the first he mixed measured quantities of boiling hot water and cool water and noted the temperature of the mixture. In the second he condensed steam by passing it into a measured amount of cold water and then noted the resulting rise in temperature and the increase in the quantity of water. When he compared the results of these experiments, he discovered that water in the form of steam contains a great deal more heat than water heated to the boiling point. Upon asking Black what he made of this fact, Watt was told that he had stumbled upon one of the instances of latent heat, a phenomenon which Black had been studying and describing in his lectures for several years. Unlike Black, however, Watt was looking for a way to improve the heat efficiency of the steam engine. His independent discovery of the latent heat of steam simply meant that there was even more heat being lost in the condensation of steam than he had suspected.

Our record of Watt's thoughts during this period is sketchy but we do know that the invention of the separate condenser did not follow directly from his discovery of the latent heat of steam. Watt realized that the amount of heat trapped in the latent state is determined by fixed laws of nature rather than by a correctable flaw in the steam engine. He therefore again turned his attention to the heat lost in cooling the cylinder during the condensation of the steam. Why, he apparently wondered, must the cylinder of a fully loaded engine be cooled down to around 100° Fahrenheit? This is far below the boiling point of water, yet if the flow of condensing water is shut off before this temperature is reached, the engine can only be operated under reduced load. We of course know that so long as the temperature of the cylinder remains above 100°, the water in the cylinder continues to boil in the vacuum created by the condensation of the steam, thereby creating a back pressure which opposes the atmospheric pressure acting on the external face of the cylinder. But how was Watt to know about boiling under reduced pressure? His own experiments were hardly of the sort that would have brought this phenomenon to his attention and he never claimed to have discovered it himself. This phenomenon had, however, been the subject of intense study by Cullen, Black, and Robison in the decade preceding Watt's invention, and Watt himself said that for his knowledge of boiling under reduced pressure he "was indebted to the Experiments of Dr. Cullen on the boiling of Ether *in Vacuo* and some conversations I had on the subject of that experiment with Dr. B[lack] and Mr. Robison." [19]

Once he had grasped the relationship between the temperature of the cylinder, boiling under reduced pressure, and back pressure on the piston, Watt was able to formulate his fundamental problem clearly. "To make a perfect steam engine," he said, "it was necessary that the cylinder should be always as hot as the steam which entered it, and that the steam should be cooled down below 100° in order to exert its full powers. The gain by such construction would be double: first, no steam would be condensed on entering the cylinder; and secondly the power exerted would be greater as the steam was more cooled."[20] Such an engine would be perfect because it would extract as much mechanical work from a given quantity of heat as the laws of nature would allow. But how could one have a cylinder which stayed hot and yet was cooled down below 100°? The answer, as Watt discovered during his famous Spring walk on the Glasgow Green, was to add a second cylinder to the engine: a separate condenser.

I have tried to illustrate how, in this particular historical instance, the invention of an important technical innovation depended upon a close relationship between a man interested primarily in natural philosophy and another interested primarily in practical improvement. At the risk of seeming to generalize from a single example, I will go further and admit that I attach a great deal of significance to other aspects of this case. Black and Watt lived in a society that supported a rather subtle understanding of the relationship between the pursuit of understanding and the pursuit of improvement. They had the good fortune to be born into a culture which did not elevate the philosopher and demean the mechanician, or vice versa.[21] Although no utopia, Scotland during the latter half of the eighteenth century nourished an integrated and balanced cultural life. Education at the primary level was designed to insure that all members of the nation were equipped with the elementary skills needed to defend their liberties and to make their way in a society open to talent. At a more advanced level, the universities were committed to the development of those systems of political and social thought and those areas of specialized knowledge upon which the future growth and guidance of the nation depended. I believe that the world of Black and Watt, although swept away by the dual revolutions—French and Industrial—that erupted during the last years of their century, remains an instructive historical example of the happy integration of technology, society, and culture. If historians today wish to play a part in articulating the new understanding of technology and society now emerging, they should pay particular attention to the comparable integrative aspects of the subjects they choose to study.

Notes

1. On the concept of *laissez-innover*, see John McDermott, "Technology: The Opiate of the Intellectuals," *New York Review of Books*, vol. 13, no. 2 (July 31, 1969), pp. 25-35.
2. D. S. L. Cardwell, *From Watt to Clausius; the Rise of Thermodynamics in the Early Industrial Age* (Ithaca, N.Y., 1971), chap. 2.
3. See Donald Fleming, "Latent Heat and the Invention of the Watt Engine," *Isis*, vol. 43 (1952), pp. 3-5, and W. A. Smeaton, "Some Comments on James Watt's Published Account of His Work on Steam and Steam Engines," *Notes and Records of the Royal Society of London*, vol. 26 (1971), pp. 35-42.
4. For biographical information on Watt, see James Patrick Muirhead, *The Life of James Watt*, 2nd. ed. (London, 1859).
5. Eric Robinson and Douglas McKie, eds., *Partners in Science: Letters of James Watt and Joseph Black* (London, 1970), pp. 259-60.
6. Ibid., pp. 257-58.
7. Ibid., p. 379.
8. Ibid., p. 416
9. Ibid., p. 410.
10. From Watt's notes, published in John Robison, *A System of Mechanical Philosophy*, ed. David Brewster, 4 vols. (Edinburgh, 1822), II:113.
11. Robinson and McKie, p. 434.
12. See Cardwell, pp. 49-50, 54.
13. See Robinson and McKie, pp. 435-36.
14. Ibid., p. 436.
15. J. T. Desaguliers, "An Attempt to Solve the Phenomenon of the Rise of Vapours, Formation of Clouds and Descent of Rain," *Philosophical Transactions of the Royal Society of London*, vol. 36, no. 407:6-22 (1729-30), pp. 16-17.
16. Robison, II:115.
17. Robinson and McKie, p. 437; Robison, II:116.
18. Robinson and McKie, pp. 438-39.
19. Ibid., p. 418; cf. D. S. L. Cardwell, *Steam Power in the Eighteenth Century* (London, 1963), p. 42. See also William Cullen, "Of the Cold Produced by Evaporating Fluids, and of Some Other Means of Producing Cold," in *Essays and Observations, Physical and Literary*, vol. 2 (1756), pp. 145-56, and Joseph Black, *Lectures on the Elements of Chemistry*, ed. John Robison, 2 vols. (Edinburgh, 1803), I:161-62.
20. From Watt's "Plain Story," in Muirhead, p. 87.
21. For an extended discussion of these themes, see A. L. Donovan, *Philosophical Chemistry in the Scottish Enlightenment* (Edinburgh, 1975), chaps. 1-4.

Cyril Stanley Smith

Remarks on the Discovery of Techniques and on Sources for the Study of Their History

I am not a historian by training, but a metallurgist. In following metallurgy to its limits, however, I have found myself in contact with practically every aspect of mankind's activities. Metallurgy today spreads all the way from public affairs through practical technology to pure science. When one studies its history, as perhaps is true of the history of anything, the extreme complexity of the world becomes evident, as does the fact that human beings can only understand a tiny part of it. Each man must pick out that part which satisfies his particular needs. In looking at the past, it is easy for historians—members of an intellectual establishment—to emphasize those things relating to the development of ideas and social organizations. Personally, however, both as a practical metallurgist and in my attempts to understand the technology of the past, I have learned that innovations in technology have not always been discovered by intellectuals and are not in any way to be understood as "applied science." Although large-scale production is obviously a response to perceived necessity, discovery is usually a product of curiosity, not purpose. Almost all the materials man has used throughout history, until very

This is a transcription of extempore remarks made at the symposium. It was not a prepared paper. For elaboration of the viewpoint expressed, see the following:

Smith, C. S., "The Interpretation of Microstructures of Metallic Artifacts," in W. J. Young, ed., *Application of Science in Examination of Works of Art* (Boston: Museum of Fine Arts, 1967), pp. 20-52.

———, "Art, Technology and Science: Notes on Their Historical Interaction," in Duane Roller, ed., *Perspectives in the History of Science and* recently, first came about as a result of man's sensual reaction: that of a whole man thinking, responding to cultural influences as a social individual, but mainly just enjoying the way materials act and interact to give pleasant texture, form, and color.

All the primary sources for the study of the early history of metallurgical discovery are decorative objects—jewelry, sculpture, or useful objects unnecessarily decorated. Necessity is not the mother of original invention, though it certainly does give birth to development. In fact, discovery from its very nature must at first be illogical, unforeseen, and outside the framework that previously exists. This is as true of a new art style as of a new theory in physics or a new type of material. In moving beyond what is already known and well understood, logical thought is of less value than the complex reaction of the entire human being. An intuitive feeling of relationships and holistic appropriateness is involved, not analytical specification. Of course, most fuzzy ideas or feelings of intuitive understanding vanish as one attempts to reduce them to definiteness, but a few survive to grow to maturity by interacting with a certain ripeness of the total environment. All successful industries had their origin in a vague idea, but they are as forgetful of their origin as raindrops nucleated by an invisible mote floating in the air. The later analysis, enrichment, and exploitation of an idea or process is a rational process, but its discovery is not.

Metallurgy began in the fifth and forth millennia B.C. and reached a very high degree of sophistication in the third millen-

Technology. (Norman, Okla.: University of Oklahoma Press, 1971), pp. 129-65. Preprinted (with some errors) in *Technology and Culture,* 11 (1970), 493-549.

———, "Reflections on Technology and the Decorative Arts . . . ," in I. M. G. Quimby and P. A. Earl, eds., *Technological Innovation and the Decorative Arts* (Charlottesville, Va.: University of Virginia Press, 1974), pp. 1-62.

———, "Metallurgy as a Human Experience," *Metallurgical Transactions,* 6A (1975), 603-23. Reprinted as a separate pamphlet by American Society for Metals (Metals Park, Ohio) and The Metallurgical Society of AIME (New York), 1977.

———, "A Highly Personal View of Science and Its History," *Annals of Science,* 34 (1977), 49-56.

———, "Structural Hierarchy in Science, Art, and History," in J. Wechsler, ed., *On Aesthetics in Science* (Cambridge, Mass.: 1978), pp. 9-53.

nium—much earlier than mechanical engineering of comparable complexity. Why was this? I think it may have been because a man can enjoyably interact with matter using his fingers, eyes, muscles, and mind to make discoveries and be continually enticed to further trials. There is little comparable feeling of satisfaction with simple mechanical devices to suggest further experiments of intermediate complexity—although complex automata are as much fun to play with as the effect of fire on mixed colored minerals.

This meeting is concerned with the modern background of technology. Consider the origins of space flight. The social ground was first prepared by imaginative literature. The spaceship's propulsion derives from fun fireworks, rockets driven by the combustion of gunpowder first used for display, not guns. Its guidance incorporates the gyroscope, discovered and for centuries used only as a toy, the top. In my own field of metallurgy, alloying begins in jewelry and sculpture. All intricate methods of casting and shaping were developed for making works of art, and even welding had its origin in sculpture. There is almost nothing metallurgical prior to 1900 A.D. that began by someone having a theoretical idea and applying it. The common belief that technology is applied science is mainly an anticipation of the future. Philosophers were perhaps responsible for this distortion—it is natural for them to think of an intellectual root for everything—but it did not happen this way. At least in the science of materials, the ideas have followed the observations, and the scientist has had to worry about the explanation of what others have first found empirically.

It must be obvious that I do not believe the scientific method plays a large creative role in the process of discovery in my field —whether in the workshop of the earliest metallurgist, in today's laboratories, or in a museum or library seeking understanding of how a man in the historical past has developed the concepts and practice of science and technology. (Of course, one might choose to define the scientific method as placing oneself in an environment where new things are likely to be found, but it is never very clear where this environment is.) The methodology comes in the validation, extension, and communication of knowledge. Discovery, however, is beyond the intellect. It is an operation of the entire human being.

Discovery without follow-up is useless. Technology is of no importance except as it becomes part of culture and society. Whatever marvels may be discovered, unless they are worked into the social fabric they will not influence human affairs. Historians in general, however, have studied only this stage and have paid little attention to the real beginnings of technological innovation. Personally I am more attracted to tentative beginnings than to massive fruition.

I mentioned earlier that the first uses of most industrially important materials and techniques for working them lay in the decorative arts. From before 1000 B.C. to nearly 1900 A.D. there was not much change in the kinds of materials available for the engineer to choose from, but only in the scale and economy of their production. The decorative arts again played a central role in the development of the newer methods of mass production. The ingenuity preceding the so-called Industrial Revolution was incited by the production of porcelain, the printing of pictures, and the mass production of textiles, dyes, and pigments. (Mechanization did not necessarily result in ugliness. Look at the earliest printing: it has a beauty comparable to the best manuscripts of the day and above that of most.) Lens grinding has its prehistory in gem cutting, and steel ball bearings are made in a machine first devised to make colored glass marbles more exactly spherical. The rolling mill for steel rails and I-beams began as a little device for making the strips of lead for stained-glass windows. It was used on iron—which is harder than lead and must be rolled hot—in making railings and frames with decorative profiles twenty years before Henry Cort incorporated the idea as an essential part of his improved wrought-iron process, which in some ways marks the beginning of the Industrial Revolution. Eric Robinson has shown how important the Birmingham "toy" trade was in developing the attitude and ingenuity of Matthew Boulton. The electrical power industry begins with generators that were first commercially made for silver plating. Over and over again this pattern appears, and it shows, I think, the natural relationship between curiosity and art and discovery. Not only is discovery itself an aesthetic activity, but the aesthetic appeal for a new technology is often what makes possible an enterpreneur's exploitation of it for personal gain.

Now a word about sources for the history of technology, which are, in the main, quite different from the verbal records used by intellectual or social historians. (One might remark parenthetically that verbal records, when formal, are more frequently produced for the purpose of advocacy than for conveying unbiased information!) Technologists have rarely written down what they did, and still more rarely have archivists preserved anything originating in the workshop. Most early references to technology are second- or thirdhand at best. Of course, any surviving record must be exploited for what it will yield, but generally speaking in the technical field one has to use contemporary objects to learn the meaning behind the words. Literary gentlemen usually just do not understand the technology they have written about. There have been encyclopedists galore, but very few technologists themselves recording in detail what they do. China, with all its fine technology and fine literature, seems to have no internal writings on technology by technologists until the present century. In Europe the earliest realistic writer on materials was the monk Theophilus, writing around 1120-1125 A.D. on the crafts needed for the embellishment of the Church. There are collections of chemical recipes from the ninth and tenth centuries which are multiply transcribed and highly corrupt echoes of much earlier experiments and only inadvertently reflect contemporary technology.

The history of material technology, therefore, must depend more upon objects than upon verbal records. The record in the objects is fairly clear, but it is written in a language with which few historians are familiar. The disclosure of its meaning requires laboratory tools. It was beyond the comprehension of anyone before modern science developed methods for studying the composition and structure of materials. In the early part of the nineteenth century the new analytical chemistry was applied to achaeological artifacts and shed a little light on the earliest Bronze Age and ceramic technology. Toward the end of the nineteenth century the science of metallography developed, revealing how the structure of matter varied in response to its chemical, mechanical, and thermal history. By using the optical microscope to study the structure of a piece of metal, it is now possible to reconstruct fairly exactly what treatment it had originally received. In some ways, by reading the details of its microstructure, one can learn more than the man who fabricated the piece of metal could himself have told about temperatures, composition, and treatment. The things that were in his mind, of course, you can't reach, but you can reconstruct and even yourself relive his physical experience in making and shaping that piece of metal. This is sensual history—not intellectual—but it adds to the understanding of man's past in a way that usefully supplements the picture of the cultural environment that is given by the grosser aspects of the artifacts and their relationships with each other. The proper place of technology in human history can never be appreciated unless historians are willing to become as familiar with the material record as they are with the verbal one. This is true right up to the present day. Curators of museums of technology must share with archivists and librarians the necessary task of preserving the record for future new interpretations.

By studying microstructure, and with a little experience in the behavior of metals, one can decide how an object was actually made. To be sure, knowing how a thing was made at a particular time in the past is not technological history, which calls for a knowledge of how the technology interacted with society and social change, and vice versa. Nonetheless, broader questions cannot be studied in the absence of such detailed knowledge any more than words can be read without knowledge of the alphabet. This interaction between levels is, in fact, usually the most interesting and difficult part of understanding any system. Just as crystals, molecules, or matter cannot be understood without realistic knowledge of the atom, the social history of technology is feeble indeed if it lacks some understanding of the human experience of an individual at work.

Analysis is equally useless without synthesis, and synthesis is not simply putting together the parts. It must take into account the real historical accidents of the growth of any complex system. Here, it seems to me, European science has much to learn from Oriental art. My own eyes were opened to the meaning of art in general when I came to study the metallurgy of the Japanese sword. Oriental craftsmen seem to have an intuitive understanding of what the material itself wants to do, and they cooperate with it in shaping the object.

This is particularly apparent in ceramics. Everyone knows the effect of Far Eastern taste on European art in the eighteenth century, the period of *chinoiserie,* but less is heard about the effect of Chinese ceramics on European science. Their arrival, with a quality utterly beyond that of any European ceramics at the time, not only inspired new form and decoration of pots but also stimulated prospecting for minerals, high-temperature research, geological theory, and chemical analysis—all to make more beautiful things. I really believe that the cutting edge of a great deal of technological advance has been just a desire to make things for man to enjoy. Perhaps this is still the primary motive of those who make the greatest discoveries—although the motives of their patrons today are certainly different from those of patrons in earlier times!

E-tu Zen Sun

Chinese History of Technology: Some Points for Comparison with the West

During the past half century there has been a steadily growing interest in the history of technology in China. This heightened awareness of the significance of China's traditional and early modern technology has come about both in China and in the West as a result of several factors. First, the continuing process of economic and social transformation in China since the late nineteenth century has led scholars to inquire into the material base from which that transformation began. Second, there has been a general widening of horizons: specialists in other fields have called for a reappraisal of the validity of widespread clichés about Chinese civilization, utilizing the increased amount of factual evidence now available to scholars. Third, historians have been asking questions regarding China's role in the sum total of human civilization to date. Were there any fundamental differences between the culture that produced Chinese technology and that which has led to European hegemony in science and technology in the last five centuries? Are not all civilizations—as Joseph Needham so vividly puts it—ethnic-cultural rivers flowing into an ecumenical sea of modern science? [1] From this increased interest in the Chinese history of technology has emerged a new field of investigation that promises to yield rich fruit for many more years to come.

We may divide the efforts already made in this field into two main areas of concentration: (1) the reconstruction and explication of traditional techniques and practices, and (2) an examination (still at an early stage of development) of the effect of Western influence on Chinese technology. My remarks will be directed toward a brief summary of the main types of work that have been accomplished and some of the problems that one might encounter in such research.

The Rediscovery and Explication of Ancient Techniques

Essentially, the objective here has been to find modern scientific explanations for the agricultural and industrial techniques that had served the Chinese people so well until the modern era. A journal like *K'o-hsueh* (Science), founded by the first generation of full-fledged modern scientists soon after World War I, from the 1920s on carried articles that analyzed, in modern terms, objects produced by the traditional technology; the objects covered a wide range determined by the authors' own specializations, from ceramics or leather tanning to the chemical composition of Han dynasty coins.[2] This type of analysis heralded one of the major trends in the Chinese history of technology, and the application of modern scientific knowledge to traditional materials has continued since that time. Today the same sort of analysis is most notably done, as is well known, in the field of medicine, where traditional prescriptions and techniques are subjected to chemical and physiological analysis.

Since premodern techniques have continued in use until the present age, the historian of technology has had the opportunity in China to obtain, through field investigation, a firsthand view of the practices in agriculture and the handicraft industry that have survived the centuries. Two works making use of such observation, those by Hommel and T'an, are well known.[3] These are the records of fieldwork done by the authors in the 1930s and 1940s respectively and remain the most useful source materials on Chinese handicraft technology as it was practiced in the twentieth century.

While field investigations of this sort greatly benefit our knowledge of how some traditional techniques looked and worked, the historian also needs to be informed of the origins and evolution of such techniques. Here the researcher faces an abundance of rich source materials which consist of a large variety and amount of written records supplemented with archaeological finds. General availability of the written sources has been gradually increasing over the past few decades, beginning with the reprinting of a few of the best-known classics in this area. In the 1920s the late-Ming (1637) technical encyclopedia, *T'ien-kung k'ai-wu* by Sung Ying-hsing, whom Needham has called "the Diderot of China," was reissued in China after remaining in seeming neglect for over a century, and many more editions of this work have been published since that time.[4] Because it was written at the end of the high point in traditional technology, and because of its comprehensiveness, this particular book has attracted the attention of scholars in different parts of the world, as will be seen below.

Another important classic, *Ch'i-min yao-shu* by Chia Ssu-hsieh (written ca. 540 A.D.), is a record of north China argricultural techniques, from the sowing of grain and the growing of fruits and vegetables to the preservation of foodstuffs, the proper care of farm animals, and proper techniques in some household handicrafts. It also has gone through several modern editions, at least two in China and one in Japan.

Through the centuries there have been an extremely large number of written works in Chinese dealing with nearly all aspects of technology; the only thing needed for the historian is that he know just where to look for the information. Aside from such obvious titles as Chang Hua's *Po-wu chih* (An account of the natural world), a large proportion of the relevant material is stored away in literary works that are not exclusively technological in character. To cite just two examples, Shen Kua, the brilliant Sung-dynasty scholar-official, gives us enlightening passages on such subjects as the making of steel and the prism phenomenon of light in a collection of short essays entitled *Meng-ch'i pi-t'an* (Dream brook notes, published 1086 A.D.); this book has become one of the important sources on eleventh-century science and technology in China. Ch'u Ta-chun (1630-96) gives a famous description of a seventeenth-century blast furnace for iron smelting in Fo-shan (near Canton), but it occupies only one page in his *Kuang-tung hsin-yu* (New discourses from Kuang-tung).

During the past two decades a few notable efforts have been made to translate some of the classics in their entirety. In China, Professor Shih Sheng-han has published two major works based on the *Ch'i-min yao-shu* (CMYS): one is an analysis of the contents of the 150 ancient sources used in the work, and the other a carefully annotated version of the book rendered from classical into contemporary vernacular Chinese.[5] The seventeenth-century *T'ien-kung k'ai-wu* has been translated into two different languages: into Japanese, published together with specialists' essays, by Professor Yabuuchi Kiyoshi and his colleagues at Kyoto University,[6] and into English by E-tu Zen Sun and Sio-chuan Sun, with notes.[7] Also, in the United States, James R. Ware has published a translation of the important fourth-century Taoist work *Pao-p'u tzu* by Ko Hung.[8]

If translations of entire works are rather scarce, surveys and monographic studies are more plentiful and of high quality. Two major research projects that have resulted in multi-volume publications deserve special mention. The first is the general survey of Chinese science and technology put out by Yabuuchi's research group at Kyoto. It consists of a series of volumes arranged chronologically, each containing a number of essays describing the state of technological achievements in specific areas during given periods in Chinese history; the medieval, Sung-Yuan, and Ming-Ch'ing times. The latest volume on Ming-Ch'ing (1368-1912) science and technology is jointly edited by Professor Yabuuchi and Yoshida Mitsukuni and was published before the former's retirement.

The other major survey is, of course, Dr. Needham's magnificent *Science and Civilization in China,* now at approximately the halfway mark with the publication in 1971 of volume 4, part 3. Together with his associates at Cambridge, Needham has produced a work whose wealth of information, breadth of perception, and knowledgeable analysis and reconstruction is unparalleled. I need hardly point out to the reader of the present work the exhaustiveness of his research and the usefulness of the bibliographies in each volume for future scholars. More important, Needham's contribution to the history of civilization goes beyond the narrowly technical, touching upon our understanding of the nature of traditional Chinese culture, and, beyond that, one is given a glimpse of the ways in which influences from different cultures could be, and were, transmitted across vast reaches of time and space and played a role in shaping the culture of a subsequent age. The process of reappraisal is ever present. An example of this is the evidence he offered regarding the use of iron in medieval China: far from being a "wood and bamboo" culture, China had been the world's most advanced producer of iron and steel as well as the largest consumer of these commodities until the seventeenth century.[9]

Chinese science and technology as a serious academic discipline has also made its appearance in the United States. Courses have been offered in this field, for example, at the Massachusetts Institute of Technology, where a new series of specialists' writings, the *M.I.T. East Asian Science Series,* has been launched under the general editorship of Nathan Sivin. Topical in nature, the works offer interpretative insights into various aspects of traditional East Asian science and technology. To date the subjects treated have dealt mainly with Chinese mathematics, medicine, and medical philosophy, but it is the objective of the series to include premodern science and technology of the entire East Asian cultural area—China, Japan, and Korea.

Of the significant works by contemporary Chinese scholars produced in China in recent years, I shall mention only two characteristic ones. They are Ch'en Heng-li's study of the *Book of Agriculture by Shen and Chang,*[10] and Liu Hsien-chou's *Ancient Chinese Agricultural Implements and Techniques.*[11] It seems logical that many of the major works emphasize the agricultural tradition which has determined the economic well-being of the nation. Focus on the history of agricultural technology was, in fact, given official emphasis in 1955 when historians were encouraged by the authorities to "do research in our ancestral land's legacy of agricultural technology."[12] Such research has led to collating of texts and analysis of premodern works.

Let us pause for a moment to consider some of the problems that researchers in this field are likely to encounter. For a scholar who undertakes a translation, the language per se is unlikely to pose problems; what could be difficult at times would be the rendition of technical terms into what one hopes to be the exact (or the most intelligent) Western equivalent and the correct interpretation of certain nontechnical words that would facilitate or obstruct our understanding of the cultural milieu as well as the application of given technological devices. For example, the character *t'ien* could be translated as sky, heaven, or nature according to the context; but being sure of the context presupposes a knowledge of the historical background in religion, philosophy, and social organization at the time concerned.[13]

Before the present century, those Chinese scholars who wrote on technological subjects relied primarily on written records, including inscriptions on oracle bones and bronze vessels as well as books dating from the late-Chou period (ca. fifth century B.C.). Etymological inferences played a crucial part in attempts to establish the technological realities of a previous age. Such was the method employed by Hsu Chung-shu in his "Lei ssu k'ao" (On some ancient Chinese agricultural implements) published in the *Bulletin* of the Institute of History and Philology of the Academia Sinica in 1930. One of his conclusions, reached after extensive study of ancient inscriptions and historical texts and intensive etymological analysis, was that it was erroneous—although it had been accepted for centuries—to regard the *lei-ssu* as two parts of the same plowing implement. Hsu maintains that the term actually designated *lei* and *ssu,* two different kinds of plows.[14] Liu Hsien-chou in his more recent work (which, as he explains in the preface, is the product of some forty years of research and teaching) seems to sidestep this point, but concentrates on the evolutionary sweep of the various types of earth-turning implements and the place of the metal plowshare in ancient Chinese agriculture. Liu's monograph draws more than Hsu's work upon the more recently available archaeological discoveries and epigraphic material. One result, of course, is that we gain a much more concrete image of what the tools looked like and how they must have been used.

If one is puzzled by discrepancies in the diverse written sources concerning a given piece of equipment, the epigraphic representations or concrete remains of that artifact are, if available, properly given the last word. As an example I mention the question of whether farmers actually used three plows pulled by a single ox. The oldest text, in the *History of the Former Han,* was ambiguous, but Chia Ssu-hsieh, the sixth-century author of *Ch'i-min yao-shu,* pointed out that such an arrangement would indeed be absurd—what the text meant was the use of a three-pointed plowshare. That sounded reasonable, but the case was not proven until the frescoes of a Han dynasty tomb in Shansi province came to light in recent years; they depicted without a doubt the use of a three-pointed plow pulled by a single animal.[15] In fact this implement was used, it is at present maintained on the basis of written sources (principally the *Ch'i-min yao-shu*), not only to turn over the earth but also to make furrows for planting the seed. This issue would merit further study.

A study of the evolution of any technological device would inevitably bring one to the question of the relevance of technology to its human environment, and thus into the realm of socioeconomic history. The history of Chinese technology is no exception. Some investigators' tendency toward social and institutional history appears to be more pronounced than that of others; it would appear that placing technological change within its social context is a clearly defined, rising trend among historians in this field. As recent examples of a growing body of studies of this sort one might mention Amano Motonosuke's *History of Chinese Agriculture* and Miyoshita Saburo's treatise on medical services in the Sui and T'ang periods (589-906 A.D.). In China, Ch'en Heng-li's previously mentioned study of the *Book of Agriculture by Shen and Chang* and the many research papers and monographs published in 1962 and 1963 on the life and writings of Hsu Kuang-ch'i, the late-Ming official whose best-known work was entitled *The Complete Book of Agriculture*, show considerable integration of the history of technology with that of the larger socioeconomic framework.[16]

In Ch'en Heng-li's study, for instance, the rice and silk cultures as practiced in Chia-hsing prefecture near Lake T'ai in the early seventeenth century are described meticulously by the original authors. Data from this book of agriculture is taken by Ch'en as the basis of inquiry into the economy of this region in early Ch'ing: agricultural and sericultural techniques, productivity, alternatives in the cropping pattern, man-land relationships, and, finally, what modern planners can learn from this information.

It is well to recall that the writers who have recorded technological tools and practices through the ages, and modern researchers to date, have come mainly from the group who were the patrons and beneficiaries of these techniques and have never been the practitioners themselves.[17] One rather difficult puzzle presented by such a situation is that of determining the exact time and way in which a new technique or tool was invented. To return to the three-pointed plowshare for a moment: it is mentioned in the *History of the Former Han* that a certain prefect Chao Kuo was credited with inventing the triple plow in the first century B.C. while he was an agricultural administrator.[18] Does this prove that he was indeed the originator of the idea that resulted in this improved planting tool? In other words, how many innovations were actually the result of a home-grown process among the artisans and country folk who quietly, over a long period of time, experimented on their own by the trial-and-error method? As a venerable East Anglian farmer explained to a field investigator in 1968 how he came to develop a strain of winter potatoes: "You notice, you pay regard, you say nawthen, and you try it."[19] You say nothing, and you write nothing as well. How often did this take place in history, Chinese and European alike? Pending the unveiling of hitherto unknown sources of information, however, the modern scholar has little choice but to give credence to the written records (which were compiled, let it be recalled, by a literary class in China who placed a high value on authenticity as they knew it) and accept the ascription of inventive ability to such individuals as Chao Kuo and Tu Yu (fl. fourth century A.D.), who was said to have invented the water-powered trip-hammer mechanism for pounding grain, as well as water-powered multiple grindstones.[20]

Studies on the Effect of Western Knowledge

The main question here is how Western elements have interacted with the Chinese environment and become integrated into Chinese practice. Several recent works come to mind, which take as the foci of their inquiry the two periods during which Western activities were prominent in the introduction of technological knowledge into China, the sixteenth and seventeenth centuries when the Jesuits arrived with their scientific knowledge and the nineteenth century when the technologies of industrial Europe (and North America) came hand in hand with expanded commercial, military, and diplomatic interaction with China. For example, Wang P'ing, in her *Introduction of Western Astronomy and Mathematics into China,* gives a factual account of the Jesuits' work,[21] while Needham probes the contrasting "world views" held by the Chinese and the Jesuit astronomers in the seventeenth century. Paradoxically, while the Jesuits possessed a superior technology in making calendars and astronomical instruments, their view of the universe was one that was just then becoming outdated in Europe itself while the Chinese, with an outdated instrument-making capacity, proved to have many elements in their interpretation of the physical properties of the universe that were actually closer to the modern scientific view —such as their belief that the universe was infinite, the stars floating in empty space.[22] Such a historical irony holds fascinating implications. Adrian A. Bennett's study of John Fryer's long career in late-nineteenth-century China as a translator of Western technical works is a recent addition to the history of intercultural contact.[23]

Ultimately, what will be most meaningful to students of Chinese technology will be the elucidation of, first, the nature of the traditional culture and, second, the role of science and technology in the transformation of traditional society in modern times. The reevaluation of Chinese civilization has, in fact, been going on for nearly a century. In the diverse types of scholarly research so briefly sketched above, there were numerous opportunities for the reinterpretation of the total environment surrounding the development and use of a given technology. For Chinese and non-Chinese researchers alike, there is the need to draw upon the fund of previous knowledge; there is also the need to tackle broader questions beyond technology itself. Why had Chinese technology flourished, often ahead of the European cultural area, in the pre-Renaissance eras? What elements in society caused it to halt its creativeness just on the eve of the European age of scientific and technological revolution? Could that halting of inventiveness be explained by economics (e.g., no major innovations were stimulated into being because no new demands were pressing on the producing sector) or was it the result of that particular sociopolitical structure in Chinese society which is called by some historians the agrarian-bureaucratic state? Earlier in the present century there was a spate of furious debate among younger Chinese intellectuals over the future of Chinese civilization. What would be the best course to follow—"modernization" or "total Westernization" or "preservation of the national essence," or perhaps a combination of several factors? Thoughtful observers have been aware of the prolonged, inexorable process of socioeconomic changes that the quest for modernity has brought to China. It is not surprising that a hallmark of modernity everywhere, the technology of the industrial age, should engage historians' attention when they seek some logical explanation for the role of technology in traditional as well as transitional China. The probe is on, and some arresting concepts have already made their appearance. An economic historian with a good grasp of the history of technology has stated, with regard to the lack of technological breakthrough in China after the fourteenth century, that it was a manifestation of the "high-level equilibrium trap," which enabled Chinese society to function at a high preindustrial level by producing increasing quantities of goods and services without an accompanying rise in quality.[24] On the other hand, some more general questions have emerged in the minds of others who have been disquieted by some facets of contemporary Western civilization. How did science and other aspects of culture coexist, in a close-fitting unity, in an earlier age? Since the traditional Chinese culture in this respect showed much similarity to Europe before the Scientific Revolution, might not a study of Chinese science and technology in their social and cultural context offer some clues to problems in the modern West?[25]

In the end, as one looks ahead, perhaps all these differences—in the economic foundations of society, in social and political structure, and in the ways man has explained his own behavior, the phenomena of nature, and his relation with them—have but served to become component parts of the future. Needham's concept of a "grand titration" between the Chinese and European traditions as a way to assess their respective roles in the formation of modern ecumenical science and technology deserves attention. For the historian of technology, to maintain a true historical perspective means to include not only other times, but also other cultures, in one's vision.

Notes

1. This concept underlies Needham's quest. See, for example, the introductory paragraphs of the chapter "Science and Society in East and West," in *The Grand Titration* (Toronto, 1969), p. 191.

2. Wang Chin, "The Scientific Aspects of Ancient Chinese Ceramic Industry," *Science,* VI, no. 9 (Sept., 1921), pp. 869-82; Wang Chin, "The Chemical Composition of the Wu-shu Coins and the Use of Lead, Tin, Zinc, and Lead-Tin Alloy in Ancient Times," *Science,* VIII, no. 8 (Aug., 1933), pp. 839-54; Wang I-chueh, "The Chemical Analysis of a Chinese Gall (Pod) Used in Leather-making," *Science,* VII, no. 6 (June, 1922), pp. 597-601 (all in Chinese).

3. Rudolph Hommel, *China at Work: An Illustrated Record of the Primitive Industries of China's Masses* (Cambridge, Mass., 1937, reissued 1970); T'an Tan-ch'iung, *Chunge-hua min-chien kung-i tu-shuo* (An illustrated description of Chinese folk industries) (Taipei, 1956). Both works cover a wide range of industries investigated, including coal mining, pottery and porcelain production, the manufacture of rattan furniture, and salt making, to name but a few representative areas.

4. In eighteen chapters, *T'ien-kung k'ai-wu* offers detailed descriptions of the techniques practiced in the seventeenth century, principally in the Yangtze River valley, in agriculture, sericulture, mining and metallurgy, the manufacturing of textiles, sugar, salt, paper, weapons, boats and carts, and other objects for daily use.

5. They are *Ts'ung Ch'i-min yao-shu k'an Chung-kuo ku-tai ti nung-yeh k'o hsueh chih-shih* (Ancient Chinese agricultural technology as seen in the CMYS) (Peking, 1957), and *Ch'i-min yao-shu chin shih* (Annotated modern rendition of CMYS) (Peking, 1962).

6. Yabuuchi Kiyoshi, ed., *Tenko kaibutsu no kenkyu* (Kyoto, 1964).

7. Sung Ying-hsing, *T'ien-kung k'ai-wu: Chinese Technology in the Seventeenth Century,* annot. and trans. E-tu Zen Sun and Shiou-chuan Sun (University Park, Pa., and London, 1966).

8. Ko Hung, *Alchemy, Medicine, and Religion in the China of* A.D. *320: The Nei-p'ien of Ko Hung,* trans. and ed. James R. Ware (Cambridge, Mass., 1966).

9. Joseph Needham, *The Development of Iron and Steel Technology in China* (London, 1958, reissued 1964), p. 19. An informative view of the well-developed iron industry in the Sung period is given in Robert Hartwell, "Markets, Technology, and the Structure of Enterprise in the Development of the Eleventh-century Chinese Iron and Steel Industry," *Journal of Economic History,* XXVI, no. 1 (March, 1966), pp. 29-58.

10. Ch'en Heng-li, *Pu Nung-shu yen-chiu* (Shanghai, 1958).

11. Liu Hsien-chou, *Chung-kuo ku-tai nung-yeh ch-chieh fa-ming shih* (Peking, 1963).

12. Shih Sheng-han, *Ts'ung CMYS k'an,* p. 7.

13. Ware, ed., *Alchemy, Medicine,* p. 6, gives an example of this type of problem when the author explains why he has translated *t'ien* as "heaven."

14. An abridged version of this article appears in E-tu Zen Sun and John DeFrancis, eds., *Chinese Social History: Translations of Selected Studies* (Washington, D.C., 1956).

15. Liu Hsien-chou, *Chung-kuo ku-tai,* pp. 31-32.

16. These studies were published to commemorate the four hundredth anniversary of Hsu's birth. Known in the West as one of the high Ming officials in close contact with the Jesuits, Hsu wrote in the fields of agriculture, astronomy, and mathematics. However, *The Complete Book of Agriculture,* published in 1639, shows little European influence.

17. A rare exception might be the autobiography of the painter Ch'i Pai-shih, who began his career in the mid-nineteenth century as an apprentice carpenter; however, the account he gives of his life and work as a carpenter would be of more interest to historical sociologists than to historians of technology. Ch'i Huang, with Chang Tz'u-ch'i, *Pai-shih-lao-jen tzu-chuan* (Peking, 1962).

18. Liu Hsien-chou, *Chung-kuo ku-tai,* p. 31.

19. G. E. Evans, *The Tools of Their Trades: An Oral History of Men at Work, c. 1900* (New York, 1971), p. 60.

20. Liu Hsien-chou, *Chung-kuo ku-tai,* pp. 31-36, 72-78.

21. Wang P'ing, *Hsi-fang li suan hsuch chih shu-ju* (Taipei, 1966).

22. This was a view dating back to the third century A.D.: Joseph Needham, *Chinese Astronomy and the Jesuit Mission: An Encounter of Cultures* (London, 1968), pp. 2-3.

23. Adrian A. Bennett, *John Fryer: The Introduction of Western Science and Technology into Nineteenth-Century China* (Cambridge, Mass., 1970).

24. Mark Elvin, *The Pattern of the Chinese Past* (Stanford, 1973), chap. 17.

25. Shigeru Nakayama and Nathan Sivin, eds., *Chinese Science: Exploration of an Ancient Tradition* (Cambridge, Mass., 1973), pp. xxviii-xxx.

George Bugliarello

The Engineer and the Historian

What relation does the task of the engineer bear to that of the historian? History is written from a platform that changes continuously according to who does the writing and when. Viewed from the platform of the present moment, the past may become crystallized and lose its identity as the changing record of change, but this happens only when the past is seen as immutable or cyclical. As J. H. Plumb put it, "What closed the mind of the Chinese sages to the historical problem was its absence." [1] Therefore, historiography is a never-ending task, and its end product —our sense of history—is endlessly in flux. The corollary is that even if it were possible to conceive of no other changes in human activities and perceptions, science and technology alone insure the existence of history—because they are, intrinsically, change and new explanation.

The problem of technology—the chief challenge to the engineer, the planner, and the designer—is how to make *normative forecasts,* not only of the physical behavior of nature and the man-made, but also of the behavior, attitudes, and development of the human society that interacts with the man-made and is shaped by it. It is evident, for example, that in planning a new town the assessment of future attitudes toward education and socialization is as essential as the forecast of climactic conditions and of the physical behavior and life span of the town's structures.

In the same way changes in the platform from which history is written lead to different perceptions of the past, changes in viewpoints concerning technology represent different platforms from which normative predictions are made and lead inevitably to different projections and different plans. There can be little doubt, for example, that the markedly different views of technology held by classical Chinese society and by Western society were a primary factor in the difference between the technological developments of these two societies.

This does not mean that the engineer disagrees with the position held by many historians that history cannot predict the future. The practice of his own profession has taught him the hard lesson that neither can technology. As Max Nordau said: "One can certainly be assured that mankind will not cease to make discoveries, and that their number and importance will continuously increase. . . . But what these discoveries will be defies completely the imagination of even the most astute researcher." [2] Even if he cannot *predict* the future, however, the engineer cannot escape from the task of projecting plans into it and of including in those plans and operational protocols a conception of future events and human behavior.

The Central Problem of the Definition of Technology

Thus, if one accepts the premise that present-day technology is one of the platforms from which the historian looks at the past, and also the platform from which the engineer projects into the future, the first problem to consider is, what is technology? The answer is neither easy nor univocal. If for our purpose we adopt a very broad definition of technology as the domain of the man-made, two points become important. In the first place, the boundaries of this domain can in effect be defined only by those involved in creating it. Definitions of technology from the outside tend to be narrow (such as the widely held one that technology is applied science), imprecise, or laden with inaccurate value judgments. The definition which holds that technology has as its purpose the benefit of mankind, for instance, would make of military technology a nontechnology. Second, the boundaries of the man-made expand over time. Concepts such as feedback or programming, derived and perfected from the study of the man-made, are brought to bear on an increasingly large cross-section of human knowledge and activities.[3] As the boundaries of technology expand, new concepts and viewpoints emerge, about both technology and other human activities. An example is Jacques Ellul's viewpoint that *la technique,* as a macro-system shaping society, will pervade and eventually determine all human activities.

Two Major Tasks for the Historian

Technology confronts the historian with two major tasks. In the first place, the new concepts and viewpoints that have emerged through technology from the fog of the unknown must be used in providing new visions of the past. This task has a vital bearing on the view society holds of technology, and hence on the place technology holds in our future. The reassessment of history in the light of new concepts arising from technology is occurring today to too limited an extent, if at all. For instance, the insights of communication theory or control theory have not yet been applied with sufficient clarity and detail to the events of our past. Historians have tended to study the concepts that have emerged from science and to neglect the impact of new concepts emerging from technology.

Second, the engineer asks of the historian that he take an intellectual risk and look forward also. We vitally need the historian's comments on the validity of the engineer's perception of the future evolution of technology in society, which the engineer will use as a base for normative forecasts.

This is not to say that a historically accurate perception of the societal reaction to a particular technology will necessarily determine the future of that technology. For instance, a knowledge of past negative reactions to teaching machines does not necessarily tell us that teaching machines should never be developed in the future. It does, however, alert the engineer to factors that, in the past, have affected societal reaction to teaching machines. Unfortunately the dimension of time, both past and future, is almost totally lacking today in our engineering curricula. Rare is the course in structural engineering, for instance, that touches on the history of structures and their societal influence and attempts to project what structures should or could serve the future needs of society. The engineer and the historian share responsibility in equal measure: the engineer for not asking the historian and for not introducing a historical perspective in his designs and his training of other engineers; the historian for not insisting with the engineer that a historical viewpoint be present and for not contributing to the formation of the engineer's plans or being intimately involved in the teaching of engineering courses (as distinct from teaching a survey course on civilization).

The warning of Ortega y Gasset in *History and System* is apropos—even if he does not consider the parallel responsibility of the historian: "My book *The Revolt of the Masses* was written under the haunting impression—in 1928, be it noted, at the climax of prosperity—that this magnificent and miraculous technology of ours was endangered and might crumble between our fingers and vanish faster than anybody imagined. Today I am more than ever frightened. I wish it would dawn upon engineers that, in order to be an engineer it is not enough to be an engineer. While they are minding their own business history may be pulling away the ground from under their feet. Alertness is what we require." [4]

Some Crucial Questions

In this context, one of the most important and urgent questions is that of growth versus nongrowth. It is a question that demands the close interaction of many disciplines and viewpoints and, more centrally than has occurred thus far, of the historian and the engineer. Are there historical examples of nongrowth societies? Can those examples guide us? For instance, what do they tell us about the relation of technology to other societal activities in a nongrowth situation? Does technology in a nongrowth society have distinctive characteristics?

As important as the question of growth, and in many senses intimately related to it, is that of the survival of science and technology. Consider the position science and technology occupy in the consciousness and in the subconscious of our society—in its actions and thoughts, and thus in a sense in its ideologies. Is this position in danger? If the warning of Ortega y Gasset is valid, what could it imply for our future? Are the many examples of high culture overwhelmed by technologically superior barbarians meaningful in this context? Can a society crumble from within because of an imbalance, in one direction or another, between its technology and the rest of its culture?

From a professional viewpoint, these questions do not always have the same import for the historian and for the engineer. Time and human experience—and therefore history—would continue to accumulate even if science and technology as we know them today were to disappear. Conversely, if a new technological barbarism were to prevail, historiography might well vanish as an autonomous activity and become subservient to the pragmatic needs of technology.

I should like to reiterate that neither of these questions is for the historian or the engineer alone. Rather, they are questions in which both the historian and the engineer must become involved and in which both must take risks—professional, intellectual, and even, if necessary, personal.

Some Humbler Tasks

Thus far I have touched on the relationship between the engineer and the historian on a high plane. While concentrating on the central problem of the relationship of the past to the future, however, we cannot neglect some simpler, more humble tasks that also need to be carried out. Society asks of the engineer that even while involved in computers and space probes he should not neglect the more down-to-earth and immediate needs of removing garbage or paving streets. Similarly, society must ask of the historian that side by side with the effort to understand, explain, and synthesize, he also assemble data, documents, and other information. In this context, the engineer and the policy maker demand of the historian that he follow more systematically and more closely the development of new technologies and new technological concepts. The task is too important to be left to chance or to be postponed; it must be systematic and immediate. Today it is already very difficult to trace the history of very recent technologies, for instance computers or bioengineering. Too many of these histories are transmitted orally, often deformed in the process, and rapidly lost. A few years in the development of contemporary technology span the equivalent, in terms of change, of several Chinese or Egyptian dynasties. Thus the accumulation of information of historical value cannot be delayed, lest the information become shrouded in mythology. Part of the reason why this task is urgent is that engineering relies only to a small extent —certainly far less than science—on written communication. Much is communicated by word of mouth, through symposiums, workshops, or meetings around the drafting table.[5]

Often I endeavor to convince several of my faculty colleagues who are historians to look at the history that is indeed unfolding right around them in the science departments and in the colleges of engineering of the university—the prosthetic devices, the new alloys, the new concepts in control theory. Almost invariably I get the reply, "Developments such as these are too recent; it is impossible to have the proper perspective from which history can be written." Possibly so. Still, as we have indicated, the engineer would probably submit that the pace of these developments is so rapid that in waiting for the proper perspective we may lose too much of the source material. Furthermore, the engineer is unconvinced of the traditional view that, while it is warranted to focus much effort on following the vicissitudes of a small town or a minor figure in the fifteenth century, it is unreasonable or undesirable to attempt to chronicle systematically technical events of importance that are occurring right under our noses.

There is another cogent reason to study the present. Harrison Brown, the foreign secretary of the United States National Academy of Sciences, reporting with Theresa Tellez on the academy's foreign programs, has stressed how difficult it has been in many cases to assess their effectiveness, because the memory of past actions and of their results often has been lost.[6] This points to a major and unfortunately widespread deficiency in the increasingly crucial feedback mechanism in our society—the evaluation process. The historian has an important role to play in such a process and should become intimately involved in it. We should insist that no technological project of importance be undertaken unless we are sure that its history, or the information on which its history may be written, will be recorded. The recording could be carried out directly by a professional historian or could be performed under the historian's supervision by the engineer, the scientist, or the manager.

With this concern about the humble but not less vital role of the historian in technology, we have come full circle. If, as we have stressed, the relation of engineering to history is the relation of the future to the past, the evaluation of technological developments arising from historical records becomes an indispensable prelude to planning and forecasting.

Notes

1. J. H. Plumb, ed., *Crisis in the Humanities* (Baltimore, 1964).
2. Max Nordau, *The Interpretation of History,* trans. M. A. Hamilton (London, 1910).
3. See G. Bugliarello, "State Science and Technology Interaction: A Cybernetic View in Science, Technology, and State Government," *Proceedings of Midwest Regional Conference,* ed. Stephen Y. Gage (Arlington Park Towers, Va., 1970).
4. José Ortega y Gasset, *History and Systems,* authorized translation from the Spanish (London, 1932).
5. E.g., G. Bugliarello, "Technological Innovation in Hydraulic Engineering," *Journal of the Hydraulics Division A.S.C.E.,* vol. 98 (May, 1972).
6. Harrison Brown and Theresa Tellez, *International Development Programs of the Office of the Foreign Secretary* (Washington, D.C., 1973).

Harold L. Burstyn

What Can the History of Technology Contribute to Our Understanding?

I

In the fall of 1753, as he traveled on horseback through the back country of Virginia and Pennsylvania, the young George Washington came upon the place where the Allegheny River, flowing from the north, joins the Monongahela, flowing from the south, to form the Ohio, the mightiest of America's westward-running rivers. On that low, marshy, triangular piece of ground, where now stand the handsomest and tallest buildings of the city of Pittsburgh, Washington found only trees. Upstream along the Allegheny, where now a modern industrial city has its factories, hospitals, warehouses, and sewage plants in an unbroken row, Washington could see only the clearing that marked a trading post. The banks of the Monongahela, marked now by blast furnaces and rolling mills on the right bank, open-hearth steel furnaces on the left, was similarly empty of Western civilization except for another trading post.[1]

Stand with me now, high above the central city on the bluff named for Washington,[2] and visualize the wilderness of 200 years ago. The whole of American history is summed up in this contrast of Washington's near-wilderness with the modern city of Pittsburgh on the same site. Indeed, much of the history of the world in the past two centuries is also present here. Though the change from forest to city is perhaps visually starker—wilderness to skyscraper—than in Paris or Peking over the same period, the dramatic conversion of the human habitat from organic to man-made, from natural to artificial, has been the same everywhere.

Like each of us, George Washington had a span of life extremely short compared to the history—even the written history—of the human species. But my intention is not to compare our existential predicament with Washington's. I am concerned neither with Washington the individual nor with the symbolic hero, "the man who never felt a wound, but when it pierced his country, who never groaned, but when fair freedom bled." [3] George Washington, rather than one of his companions, is the focus of my story of proto-Pittsburgh in 1753 primarily because the principal account of the journey is Washington's own.

I want to examine the way historians look at the activities of people in groups. Washington's journey to the site where Pittsburgh now stands brings into sharp focus the task that faces a contemporary historian studying what happened on this spot. The historian has to offer his or her readers a plausible account of how a wilderness became a city in about eight generations, less than four human lifetimes. He or she has to make the whole process unfold with the inevitability that we, who know how the story came out, can see must have been immanent in the choices that people made, whether with freedom or under constraint.

Pittsburgh stands before our eyes, palpable in its reality, yet to write the history of its site from 1753 forward through the eyes of Washington is nearly impossible. What in Washington's lifetime experience could have made it possible for him to foresee what Pittsburgh would be like 200 years later? Washington himself pointed out the strategic importance of the site. Its commercial importance was evidenced by the trading posts that had sprung up close by along both the Allegheny and the Monongahela. Although Washington and his contemporaries may not have known of the mineral riches on which modern Pittsburgh has been built, coal and iron were already eagerly sought after in their time. Nevertheless, I suggest that no forecaster, no master of the laws of probability, no pioneer of extrapolation, could have stood where Washington stood in 1753 and predicted the city of Pittsburgh as it now stands. And yet it is there, and the process by which it developed is the continuous process of human history.

Is the historian's dedication to continuity perhaps outdated? If our world and Washington's are so different that human imagination finds it difficult to portray the growth of one out of the other, should the historian not renounce the attempt? Must one show that the present is plausible when seen through the eyes of the past?

Indeed one must. Because of the power we have seized from nature through our science-based technologies, the future is increasingly ours to make. We need, for this endeavor to fashion our environment to our own specifications, all the assurance the historian can bring us that things do turn out as people order them. In an age increasingly secular, the continuities of human history are all we have to cling to. We need to see that George Washington's vision for the site of Pittsburgh on his first visit in 1753 does indeed encapsulate the city that now stands there. We need to seek evidence that Washington or if not Washington himself then at least some of his contemporaries were thinking along lines that led to our modern material culture.

In my opinion, the historian will find an explanatory framework for the growth of Pittsburgh in the succession of technological developments that have made possible the man-made environment that now confronts us. Unfortunately, however, despite a number of pioneering efforts the history of technology is not well developed. At the present time the history of technology provides a very shaky foundation for the kind of history that can illuminate the lives of our ancestors and show us how to order our own.

My task here is to suggest, from the standpoint of the historian writing about the growth of Pittsburgh or Chicago or any other of the world's great cities, what the history of technology may have to offer our future. To do this, I have chosen four well-defined issues that illuminate the kinds of questions historians ask and to which historians of technology can suggest answers.

I have already stressed one of the historian's essential duties: to make the present appear to be the plausible outcome of the past. The historian has a second essential duty: to be as accurate as one can when one assesses what happened. Unfortunately, as my discussion of cases will show, these two duties can lead in opposite directions. The historian who stresses continuity, as he or she tries to tell a plausible story, can sometimes trivialize important departures from what has gone before. The kind of hindsight a historian needs, as he or she looks through Washington's eyes in 1753, to see the modern city on the triangle of land between the Allegheny and the Monongahela, can lead to diminishing the achievement of those who made the necessary technical innovations: the steamboat, the skyscraper, the open-hearth furnace. Our search for plausibility, for the unfolding of the present mechanized world out of the pastoral world of Washington, can blind us to the difficulty that people find in making new departures. Newtonian mechanics may seem obvious to us, but it took several generations of genius to make it so. The principles of feedback control may have been present in the governor that James Watt put on his early steam engines, but only in our own time has this principle been articulated clearly enough to be generally applied.

An audience that understands modern technology needs no reminder from me of the immense importance of precision. All of us as scholars respect factual accuracy, yet in our search for patterns it is hard to sort out the details that matter from the ones that don't. If the historian can help us by showing how our present world resulted from choices made in the past by people like us, the chief contribution that the historian of technology can make is to insure that the general historian gets the details right. Nothing is easier than to tell a plausible but inaccurate story, and nothing can be more subversive of the lessons that the past can teach us. This need for accuracy arises from the historian's concern for contingencies. It is thus not always appreciated by the philosopher in his quest for necessities. Each has a separate task: the historian searches for the way things were and are; the philosopher, for the way things ought to be.

II

Let me now raise these twin fidelities of the historian—to continuity and to truth, to plausibility and to accuracy—in four concrete situations from the time of George Washington to our own. I turn first to that perennial question, what did the so-called industrial revolution that began in Washington's lifetime owe to the so-called scientific revolution that began two centuries before? What we mean by the industrial revolution is the transition from agricultural to industrial society, and the nature of this transition is easy to define in economic and demographic terms. Before the transition, famine was often a threat, and a Malthusian equilibrium between population and subsistence was always a possibility. Since product per capita increased only slowly if at all, the differential life expectancy enjoyed by the children of the rich over those of the poor meant that, in the aggregate, social mobility was downward. With industrialization came a growth in production greater than any increase in per capita consumption could dispose of: the so-called "take-off into self-sustained growth."

The facts may have proved too refractory to sustain a schematic explanation, and our experience with economic development over the past two decades has blunted our once-optimistic picture of the process. Nevertheless we possess an accurate outline of the chronology of industrialization in Europe and North America. We can even agree that certain innovations in technique were crucial to industrialization. About the relationship between these inventions and the scientific developments that precede them in time, however, there has been a good deal of argument. The older view is given by T. S. Ashton, writing in 1948: "The stream of English scientific thought, issuing from the teaching of Francis Bacon, and enlarged by the genius of Boyle and Newton, was one of the main tributaries of the industrial revolution." [4] With the rise of the history of science, more precise criteria were formulated to assess both the quality of scientific thought and the influence of this thought on technological invention. Bacon was found to be a prophet of the possible fruits of science rather than a scientist as we understand the term, and an examination of the evidence that had led Ashton to his generalization failed to reveal any specific filiation. When David Landes published his monumental study of European industrialization in 1965, he gave wider currency to a new view of the relation between science and technology in George Washington's era:

In spite of some efforts to tie the Industrial Revolution to the Scientific Revolution of the sixteenth and seventeenth centuries, the link would seem to have been an extremely diffuse one: both reflect a heightened interest in natural and material phenomena and a more systematic application of empirical teaching. Indeed, if anything, the growth of scientific knowledge owed much to the concerns and achievements of technology; there was far less flow of ideas or methods the other way; and this was to continue to be the case well into the nineteenth century.[5]

Or, to sum the matter up in L. J. Henderson's famous aphorism: science owes more to the steam engine than the steam engine owes to science.[6]

Two separate lines of thought have converged to form this new view of the independence of science and technology during the first century of industrialization. The first is the result of applying economic analysis to a wide range of quantitative historical data; the outcome has been to stress the nontechnological sources of economic growth: labor mobility, capital supply, the importance of so-called leading sectors. My concern is with the second view, which first denies the importance of technical innovation to industrialization and then goes much further by denying any significant connection between science and technology until the decade 1850-60.

In his article "The Myth of the Technical Revolution," Maurice Daumas argues that the industrial revolution was in no sense technically revolutionary, since the inventions that contributed to it had their origins much earlier.[7] He goes on to maintain that only after 1850 can we discover a science-based technology. It seems to me that Daumas's argument goes too far. In his zeal to deny the old view that eighteenth-century industrialization originated in science-based technology, Daumas has, I believe, offered a myth as false, a dogma as rigid, as the one he would replace. In particular, I wish to question two items of evidence for the new view. Part of the shift from Ashton's position to Landes's is based on interpretations of two crucial inventions. The first is James Watt's development of the separate condenser for the steam engine; the second is the solution of the problem of the longitude. As the latter is less complex, I shall discuss it first.[8]

The hazards of finding one's place on the open ocean are well known. Though latitude can be determined by measuring the altitude above the horizon of the sun by day or the pole star by night, longitude is not so simply found; for lack of it many a ship ran on the rocks. The eighteenth century offered two solutions to the problem of the longitude. One was wholly scientific; the other, wholly technical.

The scientific solution required a theory of the moon's motion adequate to construct tables of its positions in the night sky. By measuring the angular distance between the moon and a selected star and comparing that distance with one for a particular location, the navigator could determine his distance from that location, since the moon moves through the heavens, completing its circuit of the earth in twenty-four hours and fifty minutes rather than in the twenty-four hours taken by the celestial sphere.

The technical method for determining the longitude, first suggested by Gemma Frisius in the sixteenth century, was to build a clock so accurate that, set to the time of a particular location, it would continue to tell the time there even after six weeks or more had gone by. Thus, the navigator had only to compare the local time, obtained by observing the sun, with the time on his accurate clock to determine his distance from the place to whose time the clock had been set. We all know how the uneducated Yorkshire carpenter, John Harrison, won the British Board of Longitude's prize for developing such a clock (later known as a chronometer).

Harrison's triumph has been seen as an important piece of evidence for the independence of technical innovation from science in the eighteenth century.[9] I believe that the history of the technology of navigation does not support this view, or at best supports it only equivocally. Even without examining Harrison's relations with Fellows of the Royal Society as he tried to build his chronometer, we can see flaws in the argument. Harrison's achievement was to show that a chronometer could be built and to develop prototypes of the devices that made it possible, especially the temperature-compensated balance wheel. His clocks were painstakingly handcrafted, one-of-a-kind; they were never copied in sufficient numbers to solve the problem of the longitude in any practical way. Not until forty years after a voyage to the West Indies demonstrated that Harrison had met the requirements of the Act of Parliament for the £20,000 prize could the chronometer be produced in numbers sufficient to make its use possible for more than a few ships at sea. At least another quarter-century elapsed before these expensive instruments became at all widespread. Meanwhile, navigators used the alternative, scientific method—lunar distances, or lunars as they were called—to find the longitude. This method was perfected by the labors of astronomers, chiefly Tobias Mayer, in the same decade that Harrison made his prize-winning chronometer. Of the two solutions to the problem of the longitude, the scientific and the technical, although the latter ultimately triumphed it did not do so until well into the nineteenth century. Not until the twentieth century were the tables necessary to work out lunar distances removed from the annual nautical almanacs. So far as eighteenth-century practice is concerned, science *did* provide a solution to an outstanding technical problem.

What of the second piece of evidence, Watt's separate condenser? Here the historiographic issue has centered on whether Watt's invention followed his learning from Joseph Black of the latter's discovery of the latent heat of steam. That is, did Watt's technological breakthrough depend on his having first learned the scientific principles that allegedly underlie it, or was it purely an empirical discovery? Once again, the issue is not so clear-cut as this oversimplified version makes it. Watt's empiricism was not that of an unlettered mechanic working in some dark corner. Watt was a maker of scientific instruments whose shop was on the premises of the University of Glasgow, and he turned to improving steam engines as the result of having been given the University's model Newcomen engine to repair. His empiricism thus differs little from that of an experimental scientist. While a scientist such as Black experimented to find out how natural processes worked, Watt experimented in order to improve devices. Watt shared with his scientific contemporaries the approach that we may term rational experimentation.

Few nowadays would argue with this characterization of Watt's empiricism. Modern attack on the notion that Watt's technology depended on Black's science rests on two other grounds. The first is chronological: the record appears to show that Watt discovered the properties of steam empirically before he turned to his friend and mentor Black for their explanation. In four references to the sequence of events in the 1760s, Watt is quite explicit that his experiments with steam, during which he noted the phenomenon of latent heat, preceded his conversations with Black about what was happening. In response to Watt's question, Black "explained to me his doctrine of latent heat, which he had taught for some time before this period, (summer 1764)."[10] Moreover, Watt stated later that "this theory . . . did not lead to the improvements I afterwards made in the [steam] engine."[11]

Such forthrightness might be wholly convincing were it not for two difficulties. In the first place, none of the four statements was made earlier than 1796, more than thirty years after the event. At least two, possibly three, of the four come from the early nineteenth century, when Watt (born in 1736) was an old man.[12] Since the occasion for the one or two statements of 1796 was Watt's defense of his patent against infringement, they are likely to overemphasize his originality. The second difficulty with Watt's chronology is that there is earlier evidence that, while by no means conclusive, seems to cast doubt on the sequence of events that has Watt going to Black for an explanation only *after* the experiments were performed.

The accepted dates for these events, also derived from Watt's statements late in life, are 1763 and 1764 for his experiments on steam and 1765 for his invention of the separate condenser. As noted above, Watt dated to the summer of 1764 the explanation to him by Black of the latent heat of steam. Since on Watt's own account his invention of the condenser followed his learning the principle from Black, the latter may be pardoned his belief that his "knowledge . . . contributed, in no inconsiderable degree, to the public good, by suggesting to . . . Mr. Watt . . . his improvements on this powerful [steam] engine."[13] There is further, albeit equivocal, evidence that Watt learned Black's theory before experimenting. First, though Watt's notebook does state that the experiments preceded his learning the theory, the entry in which Watt makes this statement cannot be dated earlier than 1765, when Watt began to write up the experiments he had made some time before.[14] Second, in two letters, one to Magellan in 1780, the other to DeLuc in 1785, Watt dated his learning Black's theory of latent heat "about the year 1763"[15] and "in 1762 or 1763"[16] respectively. Thus, though these two letters contain no evidence about the sequence of events, they appear to falsify the date of "summer 1764" given by Watt to Brewster in 1814. By that time Watt himself was no longer sure of his dates. He knew that Black had taught the principle of latent heat from about 1758, so, after giving Brewster 1764 as the year he had learned it, Watt equivocated: "If I had heard of it [latent heat], I had not attended to it, when I thus stumbled upon one of the material facts by which that beautiful theory is supported." [17]

Even on the most favorable construction, this evidence is ambiguous. The historian is still free to conclude that Watt did discover the latent heat of vaporization before Black taught him what to call the phenomenon. However, the evidence has, I believe, shown something more important: that Black and Watt were in such close communication, when Watt was pursuing his invention, that he must have benefited from Black's scientific knowledge. Furthermore, this benefit can now be shown to be greater than historians of technology have been willing in recent years to admit.

The second argument against Watt's dependence on Black is that the sequence from the principle of latent heat to the separate condenser "makes no sense *in terms of basic physics*." [18] Even if latent heat did not exist, alternate heating and cooling of the same cylinder would still be a wasteful process, and a separate condenser would still be an improvement to a Newcomen engine. Though this argument is correct, it is, as Arthur Donovan shows (in his path-breaking paper in this volume),[19] beside the point. One can admit that Watt discovered latent heat before he learned its principles and still demonstrate that he depended on Black for the invention of the separate condenser.

Twenty-five years ago, in the paper whose argument has dominated recent thinking about the relation between Watt and Black, Donald Fleming noted that Watt was directing his experiments to the phenomenon of back pressure caused by the failure of the cold water injected into the cylinder of a Newcomen engine to condense the steam completely. Fleming noted further that, since condensation increases the vacuum and thus lowers the boiling point of the water, "an astonishingly large injection of cold water is needed to prevent back pressure." [20] Now Donovan has demonstrated that the piece of scientific knowledge Watt required for his invention was the effect of reduced pressure on the boiling point of liquids and that Watt owed his knowledge of this fact to his conversations with Joseph Black (and John Robison). Thus Watt's indebtedness to Black for specific, necessary information is now, I believe, proven. Even those who still believe that Watt experimented before he consulted Black about latent heat will no longer be able to deny that the relationship between the two men shows that science was a source of technology in the eighteenth century. To be fair, most recent students of the subject, regardless of their position on the issues discussed above, have indeed been willing to see Watt as a worker within a scientific tradition.[21]

Having cast some doubt on two principal props of the current orthodox opinion on the relation between science and technology during the early period of industrialization, let me relate this evidence to my general argument. There was, in my opinion, no dramatic technical revolution—neither, as Ashton would have it, shortly after 1760, nor, as Daumas suggests, in the decade between 1850 and 1860. The mid-nineteenth century witnessed the first healthy offspring of the marriage between science and technology, but the marriage had been arranged long before, at least as early as Francis Bacon, and it was consummated no later than the 1760s—Ashton's crucial decade—with the two inventions I mentioned earlier: Watt's separate condenser and the Mayer-Maskelyne tables of the moon's motion that made it possible for the navigator to find his longitude. In other words, science and technology were set on convergent courses during the scientific revolution, which drew much of its inspiration from craft traditions. Their paths became closer and closer as the centuries wore on, until in the middle of the nineteenth century they finally converged with the perfection of the method of rational invention in the chemical and electrical industries. Since then they have together brought forth the artifacts of modernity in bewildering abundance. Though perhaps not so dramatic as the idea of a crucial decade, my suggestion of slow and gradual change in the relations between science and technology or between knowledge and invention is, I believe, more faithful to the historical record. Only when historians of technology provide a detailed outline of this gradual transition, however, will general historians cease to chase the will-o'-the-wisp of abrupt transitions.

Finally, I believe that this gradual transition from separate domains to tight interconnection was a two-way process. Science often found its problems in the sphere of technology: for instance, ballistics in the sixteenth and seventeenth centuries. To these problems science rarely offered solutions, but from them it was led to conquer new intellectual territory: particle mechanics from ballistics, thermodynamics from steam engineering. There is thus no hierarchy of intellectual importance between science and technology. Any notion that there is derives largely from the outmoded social convention that scientific men were gentlemen, since they worked with their heads rather than their hands, while engineers and inventors were not. Technology is as demanding intellectually as science. I believe that the history of technology can demonstrate many examples of people such as James Watt, whose contribution to the contrast between George Washington's Pittsburgh and ours grew from the application of scientific practices, and perhaps scientific principles as well, to the world around us.

III

My first example suggested that a fondness for abrupt change, fed by a lack of the kind of detail that only the historian of technology can provide, has led general historians to stress sharp transition at the expense of continuity. This weakens the plausibility of their accounts of the past two centuries. In my second example I turn from plausibility to the historian's second principle: accuracy.

Eli Whitney's cotton gin, patented in 1794, was the first major invention in American history. By freeing short-staple upland cotton of its seeds, Whitney's gin made cotton production the leading sector of the American economy from 1810 to 1860. For over a century Whitney's reputation as America's first great inventor remained secure in a country noted for its devotion to its heroes. However, with the recent social upheaval in the United States, the ambiguity of Whitney's achievement has become clear. The cotton industry that Whitney's gin made possible fastened chattel slavery on the American South so firmly that its consequences still provide our major social problem a century after Emancipation.

Thus the two modern biographies of Whitney, both written in the 1950s as the civil rights movement gathered momentum,[22] stress a second and potentially even more significant accomplishment: Whitney's development of interchangeable parts, the so-called American system of manufacture, the foundation of modern mass production. If this major social and technical invention—brought to advanced Western Europe from backward America in the middle decades of the nineteenth century and, on both sides of the Atlantic, one of the great contrasts between George Washington's world and ours—if this can be attributed to Whitney, then the ambiguities of the cotton gin need not diminish his heroic stature, since we would owe to him Northern industrialism as well as Southern slavocracy.

The basic facts are not in dispute. In 1798 Eli Whitney contracted with the United States Government to deliver 10,000 muskets in two years. Since he had neither the skilled craftsmen to make them nor a factory in which they could be made, the basis of Whitney's contract was his claim that, rather than make muskets one by one, with a tight fit between two parts insured by their simultaneous manufacture by the same craftsman, Whitney would make each particular part in large quantity and insure their fit by making them to precise dimensions. This use of interchangeable parts, or the "uniformity system" of manufacture as Whitney called it, had earlier been used in the arms industry in France. Its appearance in American manufactures exhibited in Europe during the 1850s, however, led to its reintroduction to Europe in the form in which we now know it, modern mass production.

We have, then, two ends of a historical continuum: Whitney's 1798 contract for the mass production of muskets using interchangeable parts and the American system of manufacture in the 1850s. Does not the principle of continuity, of the plausibility of the present unfolding out of the past, give the historian his framework? Unfortunately, a framework is not a straight line, and Whitney's modern biographers, blinded by the historian's need for continuity, did not pay enough attention to the need for accuracy as well. They failed to understand the gap between Whitney's promise to revolutionize gunmaking and his achievement, between his dream and its realization in the machines of others that were exhibited in the 1850s, two decades after Whitney's death.

As Robert Woodbury has shown in a classic contribution to the history of technology, the American system of manufacture required four technical innovations. The first, the production of parts in bulk, depended on the second, the development of specialized power machinery. In addition, interchangeability of parts, particularly in something possibly lethal to its user like a musket, required two more things: the use of gauges to test the fit of the parts so that each would be exact within the standard of tolerance and standards of precision measurement and their embodiment in working drawings sufficient to insure precise replication. All these criteria were first met in the small-arms industry in the northeast United States by the 1830s, and they then spread from the making of small arms to the making of clocks and locks. However, there is no evidence that Whitney's musket factory was any more advanced than anyone else's around 1800. The fact that he took ten years to complete the contract he had undertaken to complete in two makes it even less likely that he achieved the interchangeability he had promised. The 1798 contract thus cannot have the importance that Whitney's biographers see in it. Rather, Whitney sought from the U.S. government the capital he needed to recoup his failing fortunes, and he obtained that capital by promising to pioneer new methods for manufacturing muskets.[23] His ideas had great appeal in a country where the labor supply was scarce, largely unskilled, and—in a period of little immigration—inelastic.

Whitney's record as an inventor thus stands solely on the cotton gin, and his flirtation with interchangeable parts becomes the first example in American history of what we now call the military-industrial complex. As such, it is of more interest to economic historians than to historians of technology, who must unravel with great patience the detailed development of mass production through precision mechanics elsewhere in the United States and abroad. Yet, precisely because this development is composed of minor improvements, made slowly and haltingly over many years in a number of places, we can never rule out the possibility that Whitney may have contributed to it, in spite of Woodbury's brilliant adverse argument.

This example demonstrates that the historian who lacks a precise understanding of technical detail takes at his peril the truth of the written record. Whitney's correspondence and the memoir by his friend Denison Olmsted are, like Watt's testimony for the infringement cases of 1796, self-serving. They therefore require a cross-examiner's scrutiny. Yet, for all one's passion for accuracy in detail, the records useful to the historian of technology usually lack the completeness of parish registers, voting lists, or business records. Though we have learned to make artifacts, or their traces in drawings, speak to us of their makers, there are often neither artifacts nor drawings left to answer our particular question. Furthermore, if, as I have suggested, the minor improvements may in the long run add up to more than the first device, the historian must pay as close attention to the social conditions favoring adoption of a new invention as he or she does to the details of the invention itself.

IV

This brings me to the third in my series of cases. The most brilliant example I know of a historian making the present plausible from the viewpoint of the past is Geoffrey Barraclough's *An Introduction to Contemporary History*.[24] More than any other work known to me, Barraclough's makes sense of the confusing patterns of world history from the 1890s to the present day. Unlike most historians writing about the period, Barraclough has a sure grasp of both technical and scientific developments and their consequences; indeed, the first thematic chapter in his wide-ranging discussion of the contemporary world and how it got that way is entitled "The Impact of Technical and Scientific Advance."

Having indicated my unreserved admiration for this book, let me draw from it a single sentence whose implications deserve more careful analysis than Barraclough, with his broad brush strokes, could give them. In the chapter just referred to, Barraclough outlines the advances in the preservation of food that resulted when the introduction of aniline dyes made bacteriology possible. He ends his discussion with the sentence: "As a result of Pasteur's researches the pasteurization of milk for general consumption became usual from about 1890." [25]

For Barraclough's purposes this one sentence is sufficient, but it conceals a significant phase of the process by which an important invention was brought into widespread practice. Louis Pasteur's techniques for partially sterilizing fluids grew out of his studies, from the 1850s to the mid-1870s, of the making of wine, vinegar, and beer. By 1877 he had left the field to others, and he died in 1895,[26] about the time when, Barraclough tells us, most milk in American and European cities was being made non-infectious through the use of Pasteur's techniques. The very word *pasteurization* dates from the 1880s.

In fact, the pasteurization of milk caught on very late in France. In England before World War I compulsory pasteurization of milk suffered the same defeats at the hands of the electorate as compulsory fluoridation of water supplies has in our own time. Pasteur may have invented the techniques, but their introduction into commercial milk production required a different kind of pioneering: social inventiveness. The pasteurization of milk began in the Danish cheese industry in the 1860s, a decade after Pasteur's initial publications, when the Danes discovered that their cheese would be more uniform if the milk were sterilized before the cheesemaking cultures were added. Soon they noticed that herds fed on pasteurized milk showed fewer cases of bovine tuberculosis, and in time the law required pasteurization of milk fed to commercial herds, but not to people. Meanwhile, in England, in Germany, and in the United States (in New York City), pioneers of the new germ theory of disease were demonstrating the connection between milk and epidemics of typhoid, scarlet fever, and diphtheria.[27]

Compulsory pasteurization of milk for human consumption began in Germany, where there was no democracy to hinder the growth of a welfare state. In 1892 the New York businessman-philanthropist Nathan Straus, who had been born in Germany in 1848 and had emigrated to Georgia in 1854, made a visit to the old country. Learning of the dramatic decline in infant deaths as a result of the new process, he decided to introduce pasteurized milk into the United States. In fact, many American dairies were already practicing pasteurization, because the growth of cities in the 1880s and 1890s had moved the dairy farms further and further away from their customers, thus leading to greater spoilage in the age of horse-drawn vehicles. However, this pasteurization to improve the keeping qualities of milk was carried out in secret in order not to rouse the consumers' ire, and any improvement in public health that resulted was overlooked.

At the zenith of his considerable philanthropic power, Straus in 1893 overcame popular resistance to pasteurized milk in an ingenious way. When he found that merely offering the milk without cost to the poor was not enough to make them take it, he organized an experiment in which his pasteurized milk was distributed to one block on the Lower East Side of New York while on another block the milk trade was carried on as before. With the help of a scientist from Columbia University, Straus demonstrated, through a careful tabulation of cases of diphtheria and scarlet fever in both blocks, that pasteurized milk prevented disease. In one year Straus's milk depots increased from one to six; the volume of milk they distributed increased nearly ten times. By 1920, when Straus took himself out of milk distribution, he had established 297 depots in 36 cities in the United States and abroad.[28]

The spread of the pasteurization of milk over the world does not seem to us especially rapid. The first compulsory pasteurization law in America was adopted in Chicago in 1908, fifteen years after Straus's demonstration on the Lower East Side. By that time the process itself was forty years old, and its embodiment in legislation almost twenty. Indeed, one wonders how successful Straus's experiment would have been had it been first tried anywhere else. Surely its success had something to do with the nature of Jewish philanthropy. The religious beliefs that Straus and his beneficiaries shared impelled both sides into a communal relationship in which no stubborn peasant ways interfered with the voluntary acceptance of an innovation that could be seen to work. Straus had no difficulty dispensing bread and coal all over the city; for pasteurized milk he had to begin in the ghetto.

The story of how pasteurization of commercial milk came to be adopted after 1890 after a dramatic demonstration of its efficacy demonstrates that the historian must consider not only the technical innovation itself but the social climate within which it is adopted. A machine that carries out an old function in a new way, like the automobile, will be adopted more rapidly than one whose function is entirely new, like a pasteurizer. Hence, the historian who writes about the motorcar can tell a rich story without understanding very much about automotive engineering at all. And if one chooses to stress technology, one will prefer the social technology of Henry Ford's assembly line to the inventions of Charles F. Kettering.

V

This question of the social context of engineering innovation brings me to my final illustration. I must apologize for its brief and sketchy form; it represents work in progress.

Looking into the way in which the British oceanographer Sir John Murray made his considerable fortune, I stumbled upon what seems to me one of the most important and most neglected questions of modern history: how did modern agriculture become so productive? In George Washington's time more than 95 percent of the population worked on the land to provide food and fiber, and even the remaining few, like Washington himself, were almost all engaged in agricultural production or its ancillaries like factoring or blacksmithing. In our day the figures are reversed.

The general explanation to be found in the literature of agricultural history is mechanization—the replacement of men by machines.[29] Though this may be true for the United States up to about 1890, it is certainly of little use in explaining what happened in Britain, where a largely unmechanized agriculture was carried on as early as the 1870s by about 15 percent of the population. There is a better explanation: what F. M. L. Thompson calls "the second agricultural revolution," the shift of farming from an extractive to a manufacturing industry. This shift, which Thompson places between 1815 and 1880 in Britain, saw the replacement of crop rotation by continuous cropping. Soil exhaustion was held at bay, first by changes in the diet of animals to enrich their droppings and then by application of mineral fertilizers directly to the soil.[30] It is these fertilizers, especially guano and its fossilized form which the British called coprolite, that I wish to consider. Though farm machinery was in use in Britain, as in the United States, before 1890, Thompson is persuasive in his argument that only after 1914 was British agriculture highly mechanized. Yet in both Britain and America the importation of minerals for fertilizer began in the 1840s and has a continuous history.

The growth of the fertilizer industry followed shortly after the beginnings of agricultural chemistry, which was firmly established by Justus von Liebig at the University of Giessen. Though he minimized the need for nitrogen in the soil, Liebig's stress on potassium and phosphorus stimulated efforts to provide both elements to growing plants.[31] Phosphorus came first from bones, usually ground into dust, but John Bennett Lawes established, at his experimental farm at Rothamsted in England, that phosphorus in bone meal was readily available only when it was first treated with sulfuric acid. Since the treatment of bone meal had preceded his experiments, Lawes turned in 1847 to coprolite to make his patent manure, or superphosphate.

My question is, what is the nature of the guano imported into the United States and Britain beginning in the 1840s, and what is its relationship to superphosphate? Though I think I see a tentative outline taking shape, the literature of agricultural history is, so far as a preliminary survey shows, innocent of the distinction I want to make. The problem arises out of a simple chemical fact: nitrates are far more soluble than phosphates. The result is that when birds or other animals deposit their dung it is generally converted by rainfall into a substance rich in phosphoric acid.[32] That is, in a humid climate the nitrates are leached away. If the deposit is underlain by limestone, the phosphate replaces the carbonate in the rock to form tri-calcium phosphate. On a few small Pacific islands, large deposits of up to 85 percent tri-calcium phosphate are known, and it was the discovery of the mineral composition of one of these islands—Christmas Island in the Indian Ocean—that brought wealth to Sir John Murray.[33] The reason Murray became wealthy was that, beginning in the 1890s, world demand soared for phosphate rock to be converted to superphosphate fertilizer. In spite of the opening of vast deposits in North Africa and in Florida, where phosphates are still mined, Murray's small island began to pay its way, and it still does.[34]

Prospecting for mineable quantities of guano and related substances on Pacific islands did not begin in the 1890s. American seafarers had begun doing so in the 1840s, but none of them ever found an island such as Murray's, where the surface deposits are underlain by vast quantities of phosphatized limestone. Two other such islands were, however, found by a New Zealander, Sir Albert Ellis, in the early 1900s: Ocean and Nauru islands.[35] (The former is now mined out.) My effort is directed to determining the relationship between the phosphatized limestone from these islands which I know to have been made into superphosphate, the guano that Ellis and his American predecessors scraped off many other islands in small but marketable quantities, and the raw material of the same name that was turned into superphosphate in Woods Hole, Massachusetts, and elsewhere, also about 1900.

Moreover, though I believe that all the raw materials I have just mentioned are basically of the same composition, it is necessary to relate them to the Peruvian and Chilean guano that was spread directly on the soil in both Britain and America beginning about 1840. I believe that my knowledge of chemistry and meteorology can direct me to a distinction between mid-Pacific and South American guanos. If my hunch is borne out there are two types of guano, depending upon the climate of the region from which they come. Guano from a humid climate is phosphatic and thus not very useful for fertilizer until it has been converted to superphosphate. Guano from an arid climate, such as the west coast of South America, is on the other hand still rich in its original nitrates. There is no rainfall there to leach them out. This South American guano, then, would be a powerful fertilizer without treatment, and its phosphorus content would be partially available as is the phosphate from untreated bone meal.

I must apologize for presenting suggestions so tentative, but I do so to make a point. Only when historians begin to appreciate these kinds of details—the difference between nitrogenous and phosphatic fertilizer, for example—can they write history at once plausible enough to give us confidence and accurate enough to claim the respect we reserve for truly factual accounts. The story of fertilizers, smelly and unromantic though it may be, is of enormous importance to our understanding of the differences between George Washington's world and our own; it has, I believe, yet to be written.

VI

These four examples—the relationship of science and technology in the industrial revolution, the role of Eli Whitney in developing the American system of manufacture, the nature of the innovation that brought about the pasteurization of milk, and the role of fertilizer in modern agriculture—provide a sample of the issues historians face for which historians of technology can provide guidance. Up to now the guidance has been weak. Still, progress is being made. Astonishing in both its comprehensiveness and its readability, David Landes's *The Unbound Prometheus* [36] gives a detailed and lively account of the debt we owe to the changes in technology from George Washington's time to our own. Yet Landes's subtitle, *Technological Change and Industrial Development in Western Europe from 1750 to the Present,* shows that his brilliant account has been framed to answer the questions posed by economic development. His is a world of aggregates. Moreover, Landes has drawn his structure from the history of nation-states. Both technical development and economic change are expressed through national economics: Britain pioneers, Germany catches up, the United States overtakes both of them. Can this view offer insight into a world where economic development and multinational corporations may make nation-states obsolete?

My own suggestion for the direction in which historians should move grows out of a sense of the growing discrepancy between the life each of us leads as an individual and the institutional structures that we create in groups, be they corporations, social classes, or nation-states. Landes, though individuals appear from time to time in his pages, has spoken to the history of institutions or aggregates. Can there be a history in which this distance between individual and institution is bridged? Can history be written so that the contrast I have posed between George Washington's journey on horseback and our own by automobile or airplane becomes more than a rhetorical device? Can the history of technology provide a description of the changes in our material culture over the past two centuries so compelling that the impact on the lives of individuals becomes apparent to all who write history? Without such a history to span the gap between the person and his or her environment, I am afraid that the confidence to face the unforeseeable future, that confidence to the strengthening of which historians must be dedicated if they are to deserve an audience, will be lacking, and we will turn increasingly to those techniques of reducing human beings to indistinguishable and interchangeable components of aggregates that are (correctly, in my view) perceived to be inhumane. Only with a history that shows us, from its study of the past, how as individuals we choose our future can we humans gain the confidence we need to achieve all that we are capable of achieving through our institutions.

Notes

1. The *Journal of Major George Washington* (London, 1754). A facsimile of this rare pamphlet is in Hugh Cleland, *George Washington in the Ohio Valley* (Pittsburgh, 1955). Douglas Southall Freeman, *George Washington,* 7v. (New York, 1948-57), v. 1, chap. 8.

2. There is no evidence that George Washington ever set foot on Mount Washington, the bluff on the left bank of the Monongahela from which may be seen the most spectacular views of the city of Pittsburgh.

3. Daniel Webster (4 July 1800), as quoted by Irving H. Bartlett, "Daniel Webster as a Symbolic Hero," *New England Quarterly* 45 (1972): 496. I am indebted to Dr. Bartlett for first calling my attention, over dinner atop Mount Washington, to the contrast beween Washington's view and our own.

4. T. S. Ashton, *The Industrial Revolution, 1760-1830* (New York, 1964), pp. 12-13.

5. D. S. Landes, "Technological Change and Industrial Development in Western Europe, 1750-1914," in M. M. Postan and H. J. Habakkus, eds., *Cambridge Economic History of Europe,* 6v. (Cambridge, 1952-65), v. 6, 293-94. See n. 36 below.

6. Though I do not know where in Henderson's writings this often-quoted sentence appears, he probably got it from the following: "I am still doubtful whether scientific thought has, at the end of our century, as yet balanced the debt which it owes to practical inventors. It is instructive . . . to consider how much . . . science has learnt from the steam-engine, and to reflect whether from all the theoretical insight gained any really radical improvement of the steam-engine . . . has resulted" (J. T. Merz, *A History of European Thought in the Nineteenth Century* [Edinburgh, 1896], v. 1, 331). Henderson's copy of Merz is in the Harvard College Library.

7. M. Daumas, "Le mythe de la révolution technique," *Revue d'histoire des sciences,* 16 (1963): 291-302.

8. L. A. Brown, *The Story of Maps* (Boston, 1949), chap. 8; E. G. R. Taylor, *The Haven-Finding Art* (London, 1956), chap. 11.

9. R. T. Gould, *The Marine Chronometer* (London, 1923); H. Quill, *John Harrison* (New York, 1966).

10. Watt's note to David Brewster on John Robison's dissertation on steam engines, J. P. Muirhead, *The Origins and Progress of the Mechanical Inventions of James Watt,* 3v. (London, 1854), v. 1, lxxiv.

11. Watt to Brewster (May 1814), Ibid., v. 2, 354.

12. In addition to the two statements cited above, one from 1814 and the other from about that date, they are "A Plain Story," Watt's narrative for the patent litigation of 1796 (Ibid., v. 1, lxxxi), and Watt's holograph note on John Robison's narrative for the 1796 litigation (Ibid., v. 1, xlix, and E. Robinson and A. E. Musson, eds., *James Watt and the Steam Revolution* [New York, 1969], 27). Though Robison's narrative can be dated, Watt's note on it cannot, so far as I know, and I should like to argue that Watt's note is as likely to have been written in 1805, when he was preparing his recollections of Robison on the latter's death, as in 1796.

13. J. Black, *Lectures on the Elements of Chemistry,* ed. J. Robison, 2 v. (Edinburgh, 1803), v. 1, iii.

14. E. Robinson and D. McKie, eds., *Partners in Science. Letters of James Watt and Joseph Black* (London, 1970), part 2. The relevant entries by Watt in his notebook are on p. 439.

15. Watt to J. H. Magellan (20 March 1780), Ibid., p. 85.

16. Watt to J. A. DeLuc (28 May 1785), J. A. DeLuc, *Idées sur la météorologie,* 2v. (Paris, 1787), v. 1, 504-5 (my translation). The English original has not to my knowledge been published.

17. Muirhead, *Inventions of Watt,* v. 1, lxxiv.

18. D. S. L. Cardwell, *From Watt to Clausius* (Ithaca, N.Y., 1971), p. 42.

19. Arthur Donovan, "Toward a Social History of Technological Ideas: Joseph Black, James Watt, and the Separate Condenser," pp. 19-30.

20. Donald Fleming, "Latent Heat and the Invention of the Watt Engine," *Isis,* 43 (1952): 5.

21. Ibid., pp. 3-5; R. E. Schofield, *The Lunar Society of Birmingham* (Oxford, 1963); A. E. Musson and E. Robinson, *Science and Technology in the Industrial Revolution* (Manchester, 1969). For the contrary argument that science did not lead to technology so early see A. R. Hall, "Engineering and the Scientific Revolution," *Technology and Culture,* 2 (1961): 333-41.

22. J. Mirsky and A. Nevins, *The World of Eli Whitney* (New York, 1952); C. McL. Green, *Eli Whitney and the Birth of American Technology* (Boston, 1956).

23. R. Woodbury, "The Legend of Eli Whitney and Interchangeable Parts," *Technology and Culture,* 1 (1960): 235-53. For further discussion see E. A. Battison, "Eli Whitney and the Milling Machine," *Smithsonian Journal of History,* 1 (1966): 9-34, and "A New Look at the Whitney Milling Machine," *Technology and Culture,* 14 (1973): 592-98; also see M. R. Smith, *Harpers Ferry Armory and the New Technology* (Ithaca, N.Y., 1977).

24. G. Barraclough, *An Introduction to Contemporary History* (London, 1964, 1967).

25. Ibid., p. 49.

26. R. J. Dubos, *Louis Pasteur. Free Lance of Science* (Boston, 1950).
27. E. E. Morison, *Men, Machines, and Modern Times* (Cambridge, Mass., 1966), 211-13; S. C. Prescott and M. P. Horwood, *Sedgwick's Principles of Sanitary Science and Public Health* (New York, 1935), pp. 218-22, 238-42; C. E. North, "Milk and Its Relation to Public Health," in M. P. Ravenel, ed., *A Half Century of Public Health* (New York, 1921), pp. 236-89.
28. D. deS. Pool, "Nathan Straus (1848-1931)," *Dictionary of American Biography,* v. 18, 129-30; obituary in *New York Times,* 12 January 1931, p. 5.
29. G. Borgstrom, "Food and Agriculture in the Nineteenth Century," in M. Kranzberg and C. Pursell, eds., *Technology in Western Civilization,* 2v. (New York, 1967), v. 1, 408-12; G. E. Fussell, "Agriculture: Techniques of Farming," and "Growth of Food Production," in C. Singer et al., eds., *A History of Technology,* 5v. (London, 1954-58), v. 4, 13-43, and v. 5, 1-25; F. W. Kohlmeyer and F. L. Herum, "Science and Engineering in Agriculture: A Historical Perspective," *Technology and Culture,* 2 (1961): 368-80.
30. F. M. L. Thompson, "The Second Agricultural Revolution, 1815-1880," *Economic History Review,* ser. 2, v. 21 (1968): 62-77.
31. A. J. Ihde, *The Development of Modern Chemistry* (New York, 1964), pp. 260-64, 321-26; M. Rossiter, *The Emergence of Agricultural Science* (New Haven, 1975), chaps 2, 3.
32. G. E. Hutchinson, *The Biogeochemistry of Vertebrate Excretion* (New York, 1950).
33. H. L. Burstyn, "Science Pays Off: Sir John Murray and the Christmas Island Phosphate Industry, 1886-1914," *Social Studies of Science,* v. 5 (1975): 5-34.
34. M. Williams, *Three Islands* (Melbourne, 1971).
35. A. F. Ellis, *Ocean Island and Nauru* (Sydney, 1935); Ellis, *Adventuring in Coral Seas* (Sydney, 1936).
36. D. S. Landes, *The Unbound Prometheus* (Cambridge, 1969), an expanded version of the long article cited in n. 5 above; the quotation referenced in n. 5 appears in the book on p. 61.

Note added in proof: Further research has validated my 1973 conjecture (see Part V above) that there are two types of guano. Although historians do not seem to be aware of the distinction, users of fertilizer were. The *précis* of evidence in Lawes's suit to protect his superphosphate patent (1853) states: "It is especially those kinds [of guano] which are poor in ammonia, and therefore of little value when applied as manure in their natural state, that are particularly rich in phosphate of lime: these are well adapted for decomposing with sulphuric acid" (quoted by A. N. Gray, *Phosphates and Superphosphates,* 2nd ed. [London, 1944], p. 11). See H. Voss, "The Phosphate Deposits on Islands in the Pacific Ocean: Their Origin and Importance for Agriculture," in W. Ramsay and W. MacNab, eds., *Seventh International Congress of Applied Chemistry* (London, 1910), Section VII, pp. 196-202.

Nathan Rosenberg

Technology, Economy, and Values

Introduction

For an economist to discuss values is not quite as inappropriate as we may suppose, conditioned as we are by Oscar Wilde's remark that a cynic is a person who knows the price of everything and the value of nothing. An economist, after all, is a cynic with credentials.

First of all, economists do have a very strong interest in certain kinds of values, those which have significance for the choices which individuals make in the marketplace. In particular, the economist is intensely concerned with each person's process of valuation as it reveals itself in his market behavior. This does not, to be sure, exhaust the many possible meanings of "value." The notion that economists thought that it did is presumably what underlies Oscar Wilde's observation. In the larger view, economics as a central preoccupation is concerned with the *choices*—with both the *logic* of choice and the *consequences* of choice—which individuals reveal in their market behavior. Economic values are, in this sense, a species of a much larger genus, but it is as true of noneconomic values as of economic ones that their content is revealed in the act of choice.

There is a second important reason why economics can contribute to a better understanding of problems concerning values. Although it is intuitively clear to many observers that changes in technology are going to be a prime mover of changes in values, the nature of that relationship usually remains obscure. I will argue in this paper that important insights into this relationship can be achieved by examining the impact of technology upon the performance of the economy and the effectiveness of certain economic institutions. Our central focus will be on the mediating role of the marketplace. In this region economic analysis is an indispensable guide. Economic analysis provides a conceptualization of economic activity which makes it possible to trace the repercussions of a given change upon the functioning of a complex, interdependent system. It will be a major theme of this paper that technology is changing the nature of these interdependencies and that it is, in particular, introducing new paths along which people interact with, and affect, one another's welfare. In order to appreciate the impact of technology upon values it will be necessary to examine the new network of interrelationships which is in the process of being established by technology.

From a narrowly economic perspective, the consequences of technological change may be described as (1) a growth in the productivity of the economic system, i.e., producing a larger volume of output from the same quantity of resource inputs, and (2) the introduction of entirely new commodities and services, plus qualitative alterations in old ones. Thus technological change presents us with immediate results which have both quantitative and qualitative dimensions. The growing productivity of our economic system, toward which technological change has made such a decisive contribution, has in turn generated a whole vast chain reaction of consequences, partly because of the different ways in which people at higher levels of income have chosen to dispose of their incomes. Thus, higher levels of per capita income in the past 150 years have been associated with changes in the composition of demand and therefore final output. Perhaps the most spectacular and far-reaching is the decline in the relative importance of the agricultural sector, as the percentage of consumer expenditures on food products declines; [1] closely associated is the rising importance of that highly heterogeneous collection of activities rather indiscriminately lumped together in the "services sector." [2] The impact of technology, then, lies not only behind the changing productivity of our economic inputs but behind the drastic changes in the *composition* of output and the shifting composition and allocation of inputs (such as changes in the industrial and occupational composition of the labor force). Moreover, the imperatives of technological change have generated spatial shifts in the location of economic activity and have thus contributed in a massive way to the whole complex of phenomena associated with urbanization. Indeed, as I will argue, some of the most important consequences of technological change and their implications for values revolve around the peculiarities of human interactions in a predominantly urban environment. Finally, the long-term downward decline in the length of the work week and the corresponding importance of leisure-time activities have been the joint product of growing productivity combined with a set of tastes which has treated leisure time as a superior good.

Effects of Technology Depend upon Values

This is a convenient point to state a major theme of this paper. Those who study the relationship between technology and values have devoted much attention to the impact of changes in technology upon changes in values. I will treat this relationship at some length. However, I would also like to emphasize that the kinds of changes which technology produces in a society will also depend very heavily upon that society's value structure. What we *do* with the fruits of technology will, inevitably, depend on what we *value*. The same technological changes may therefore produce very different consequences in societies with different value structures (or in the same society at different periods of time). Thus, it is noteworthy that there has been no decline in the number of hours worked per week in Japan in the years since World War II, in spite of a remarkable rate of growth of per capita incomes.[3] Evidently, the response to rising productivity in that country has been dominated by a higher valuation of an increased output of goods and services than of the alternative of increased leisure time.

Another important theme also concerns the dependence of technological change upon a prior system of values. Technological change is basically the result of certain *problem-solving* activities and is an activity to which we can allocate a larger or smaller portion of society's resources. Furthermore, within any given total we possess a wide range of choices as to the *kinds* of problems we will undertake to solve. For if, as Whitehead once stated, "The greatest invention of the nineteenth century was the invention of the method of invention," [4] then it follows that the *consequences* of technological change will depend, to an important degree, on the direction in which we propel our inventive activity.

This argument, in turn, has important implications for the relationship between technology and values, because it underlines the role of choice and therefore also the role of values in determining the direction of technological change. Technological change is often discussed as if its rate and direction were somehow predetermined and as if it were something to which individuals and society could make only rather passive adaptations. Actually, these things are largely the outcome of a social process in which individuals and larger collectivities make choices determining the allocation of scarce resources, and these allocations inevitably reflect the prevailing system of values. For example, the kinds of technological changes which we have been getting in recent years have been very much influenced by the predominant role played by the space-military complex in research and development activity. The relative technological stagnation of some sectors of our economy is, to some degree at least, the result of a failure to allocate more resources to research in these areas. If we are dissatisfied with the results (and, of course, not everyone is dissatisfied) it is foolish to blame some impersonal, uncontrollable force called "technology" rather than the values and social structure which are responsible for the past allocation of resources.[5]

Our growing technological skills are, in fact, dramatically intensifying the importance of valuational activities for the simple reason that they are endowing us with the capacity to do things which were formerly not possible. We have far more choices to make, choices which technology has made not only possible but unavoidable and which will, inevitably, have far-reaching repercussions. As recent breakthroughs in medical technology have made obvious, these choices will be no less than matters of life and death. Indeed, deliberate social choices have had such a significance for some time. Medical technology has abolished poliomyelitis by immunological techniques and has developed similar controls over a whole range of childhood diseases, but it has made comparatively slow progress with the chronic, debilitating diseases which afflict the aged. In part, this reflects the past allocation of medical research resources. Given the whole wide range of human afflictions, and in view of the high opportunity cost of these resources, how do we go about deciding how they shall "best" be distributed? [6]

It is estimated, for example, that the establishment of a network of kidney dialysis centers could save the lives of 6,000 people who each year die a slow, painful death due to poisoning from kidney failure. The cost, however, was $28,000 per patient per year in 1968.[7] Is this a good investment? The point is that we now possess the awesome power of deciding who shall be saved.

Although the further development of the technology of death control involves new and painful choices, it suggests a closely related problem of even more explosive potential. While the technology of death control is almost universally welcomed and highly valued, the technology of birth control is not. This asymmetry in human values underlies the ominously high rates of population growth in many parts of the underdeveloped world. Pilot programs which have made the latest birth control techniques available at no cost to poor Indian villagers have had only negligible effects in reducing the birth rate. Followup studies have suggested that the reason is, essentially, that the villagers continue to want large families.[8] The problem does not have a purely technological solution: for such technology to be effective, people must attach a high value to the *goal* which the technology makes attainable or otherwise the technology remains irrelevant. Clearly, modern birth control technology permits us to limit family size with a high degree of effectiveness. The effect of the technology, then, will depend on the size to which we decide to limit our families. The consequences of the new capacity are not a technological matter but, rather, a valuational one.

The point at issue is a general one. We cannot predict the consequences of a change in technology without understanding the values of a society. In the extreme case the technology may be irrelevant because society rejects the goals which the technology makes possible. In other cases the effects will depend upon a whole range of values: work-leisure preferences, attachment to residential location, willingness to acquire new skills or to discard old forms of work organization which are no longer appropriate, willingness to assume risks, and consumer tastes. A mass-produced gun or a ready-made suit of clothes differs from its custom-made predecessor. More significance was attached to these differences in England than in the United States in the nineteenth century, and these taste differences had a great deal to do with differences in the rates of development of the technology of mass production and its consequences in these two countries.[9]

There is a closely related point to be made here. Social goals and individual aspirations are both intimately connected with our capacity to realize them. In this sense, one of the most important consequences of the rising productivity which flows from technological change is our growing realization that poverty and its attendant suffering and indignities can be eliminated in the American economy. The elimination of poverty has, in fact, recently become an integral part of our public dialogue. No less interesting from the longer-term perspective is that the *definition* of poverty has itself undergone drastic upward revision in the past several decades.[10]

In a sense, our growing technological capacity is creating a powerful challenge by producing what amounts to a national "moment of truth." Although the abolition of poverty has long been part of our national rhetoric, it is only recently that a vague, utopian possibility has been transformed into something within the reach of our present productive capacities. In the past, poverty and its various manifestations have been tolerated on the grounds of unavoidability. Today it is becoming clear that the abolition of poverty is something concerning which we now have a genuine choice. It seems apparent that a growing awareness of these new possibilities has a good deal to do with the rising tensions in our society. The real challenge which technology currently presents to our long-expressed egalitarian values is that it has deprived us of any technological excuse for failing to fulfill long-professed ideals. Technological change and rising productivity have been responsible for placing the elimination of poverty high on our national agenda. By thus confronting us with social options which were once inconceivable, it has forced certain questions to the forefront of our attention. By presenting us with the possibility of new choices, technology is forcing us to reexamine and to clarify our values.

The consequences of technological change over the past several decades have not only bright egalitarian ideals within the prospect of possible attainment; they have also worked, in a variety of ways, toward the reinforcement and the diffusion of democratic values. The biases of modern technology have, by reducing the demand for unskilled labor, brought an upgrading of labor skills and with it a reduction in wage inequalities. The higher educational levels required by modern technology have been a pervasive force for a general rise in educational standards, and this in return has been important in accounting for the high degree of social, occupational, and geographical mobility which has been a notable feature of the American labor force. The demand for personnel possessing skills, talent, and the capacity to accept responsibility results in the establishment of universalistic criteria in employment practices based upon objective evaluations of performance and proficiency. Modern technology, with its vast expansion of white-collar and service-sector employment opportunities and labor-saving household appliances, and in combination with the technology of birth control, has played a strategic role in the social as well as political enfranchisement of women. Finally, the strong bias in modern technology has been toward the elimination of the design and production of highly differentiated products catering to the individualized tastes of small elite groups and toward the production of large quantities of highly standardized products for mass consumption.

The Role of Externalities

I turn now to a diverse set of considerations which have as their common denominator the impact of technological change upon the effectiveness of existing institutions for decision making and resource allocation as they are presently organized around the market mechanism. There are elements of both paradox and dialectic here, since the growing limitations upon the effectiveness of market institutions are, in large measure, a product of their *past* effectiveness. Capitalist institutions and a dynamic technology have been, after all, enormously successful engines of growth which have produced the present high state of affluence of western societies. Moreover, although it would take us much too far afield to discuss, this historical conjuncture of capitalist institutions with technological dynamism was no mere historical coincidence.[11] The question which must now be posed is this: having attained our present affluence, and given some of the characteristics of modern technology, what are the implications for a highly individualistic value system and the closely associated ideology of the marketplace? I will argue that they are far-reaching, as they have rested not only upon the moral propriety of the rewards distributed by the market mechanism but also upon the social optimality of the resource allocation achieved by this mechanism.[12]

First, with the growth in per capita incomes which has owed so much to past technological innovations there has been a shift in the composition of demand in favor of goods and services which, by wide general consensus, the market has never provided effectively, and in the provision of which the government has usually played a significant role. This would include such services as education, health, and transportation. The role of government in providing such services has reflected a mixture of considerations. In part it has involved equity rather than efficiency, a determination that some services should be provided regardless of an individual's ability to pay. In other cases the manner in which certain services are to be provided cannot be resolved without dealing simultaneously with broader issues, for instance the impact of mass transportation systems in urban environments upon land use for residential, industrial, and recreational purposes.

Transportation systems shape the patterns of land use in urban environments in a decisive way. Even where the form of transportation is entirely privately owned—as in the case of the automobile—there is a high degree of complementarity between its use and certain facilities which cannot be effectively provided by the private sector, among them road networks and traffic control facilities. Because of these complementarities between the public and private sectors, the profusion of automobiles flowing out of Detroit for several decades has, in fact, been a potent force leading to the growth in government expenditures.

In the increasingly urbanized environment which has, in the past, been such a characteristic feature of modern technology, many further services are either provided in some public form or subjected to public regulation because competition is either impossible or inefficient. This would include sewage and garbage disposal, utilities (gas, water, power), often recreational facilities, and other amenities.[13] Systematic shifts in the composition of consumer demand, on the one hand, and the rising share of the government sector, on the other, are generally observed phenomena which appear to be deeply rooted in a relatively stable structure of human tastes.[14] Indeed, a large portion of the history of industrialization and economic change could be written around the direct economic consequences flowing from these shifts.

One of the most distinctive characteristics of modern technology is that it has been associated with new kinds of human interaction. The nature of these interactions is highly significant both for economic analysis and for social policy because they restrict the area in which the market mechanism, even under the most favorable competitive conditions, may be said to bring about an optimal allocation of resources. These new interactions may flow directly from peculiarities of modern technology, from social organizations related to technology and particularly to urbanization, or they may be of a long-existing kind which has recently acquired much greater relative importance.

The market mechanism is an efficient allocator of resources, and the private pursuit of wealth maximizes the economic welfare of all members of society (for a given distribution of income) when there are no significant external effects. If, however, on the one hand my actions impose costs on other persons for which I am not required to provide compensation or, on the other, my actions confer benefits upon other persons for which I cannot receive remuneration, then purely self-seeking behavior as mediated through the signals of the marketplace may lead to inefficient patterns of resource use. In the first case, too much of society's resources will be employed, and in the second case an insufficient amount. Modern technology-cum-urbanization has multiplied many times the frequency and the significance of such interactions. Furthermore, there is a large class of cases of goods and services whose consumption is necessarily shared, where *more* of something for X does not mean that there is *less* available for Y or Z. Here too a reliance upon purely private cost-benefit calculations will lead to an insufficient volume of society's resources being devoted to the activity.

Consider the case of the supersonic jet as an example of a negative externality. Once the jet is built and placed in operation, I cannot choose whether or not I want to listen to the sonic boom. It is a cost which is imposed on me whether or not I make use of air travel. When some people travel at supersonic speed they produce intensely unpleasant experiences for people who are *not* traveling. I do not have a choice in the matter, short of withdrawing impractically far from civilization. So long as Boston's Logan Airport continues to operate, Walden Pond will not evenly remotely serve as an adequate retreat.[15] Similarly, when large numbers of people opt to travel by car they produce smog and traffic congestion even for those who choose to walk or use public transport. When the teenager in the next apartment decides to play her latest rock records, I find myself a distraught member of a captive audience.

It is a distinctive feature of a technologically dynamic society that more and more of the choices and actions of individual consumers directly affect the welfare of others. This they do by altering the nature of the environment for other members of the community. The automobile and the supersonic jet plane are only the most notorious examples. We would have to include all cases involving the joint use of certain facilities from, quite literally, the air we breathe in varying degrees of impurity and the rivers which we pollute to the common use of public highways where A's reckless driving or negligence poses a threat not only to his own life but to that of B and C driving on the same road. How my neighbor disposes of his leaves in the fall or clears his sidewalk of snow in the winter is of significance to me. Similarly, whether in a supermarket, a subway, a movie house, library, or laundromat, our behavior as consumers impinges in important ways upon the environment or "living space" of others, and thereby affects their welfare.

The beer bottle which bears the ostensibly reassuring message "No deposit, no return" neatly symbolizes many of a growing class of externalities. The cost of disposing of the bottle is not borne directly by either the brewery or the consumer, but the replacement of bottles on which a deposit was paid by cans and nonreturnable bottles which can be conveniently discarded has significantly added to the volume of trash which has to be collected. The disposal of solid waste in fact poses a problem of increasingly serious proportions in affluent societies. It is estimated that, in New York City alone, some thirty thousand automobiles are simply abandoned by their owners every year, leaving the disposal problem to the municipal authorities. The underlying economic forces, of course, are the high price of labor and low prices for scrap materials which are fast rendering the traditional activities of the junkman uneconomic. The end result, however, is that new and costly burdens are added to the services performed by the public sector. To consider another example, the use of tungsten carbide studs in automobile tires has been shown to provide significantly improved traction when driving on ice and hardpacked snow. As a result the sale of such studs has been booming in the past few years. While this undoubtedly represents a desirable innovation from a safety point of view, the properties of tungsten carbide which result in increased traction also result in considerable destruction to road surfaces, a cost which will again be borne by the public sector.[16]

It seems to be a technological condition of modern life, therefore, that we engage in more numerous activities which generate costs not borne directly by those who generate them. The converse, however, is equally true. Members of urban communities not only generate costs for which they are not directly responsible but also generate benefits to other members of the community for which they receive no compensation. My neighbor's rose garden and azalea bushes cost me nothing yet constitute a considerable source of pleasure to me. When a new electronics firm arrives in a community, it may train workers in computer programming skills which can be subsequently exploited by neighboring firms, since they may not have to incur the costs of such training. More generally, all expenditures made by my neighbors on the education of their children accrue as a benefit to me since such expenditures improve my environment by raising the level of cultural and political sophistication in the community in which my family lives.

My welfare, in other words, is at least partly a function of other people's activities, tastes, and consumption expenditures.[17] While this has always been true in some respects, the thrust of modern technology and of its implications for social organization is to increase sharply the number and importance of such interdependencies. While they might have warranted an occasional footnote in a Jeffersonian society of small-holding farmers, they constitute some of the most vital issues of the densely populated, late-twentieth-century world. One may even conjecture that our growing national concern with poverty reflects more than just a sense of equity or human compassion. It reflects also the discomfort and even danger which the existence of poor people poses to those more comfortably situated. My welfare may be affected not only by the well-known disutility of envy but also by the poverty of others, whether this takes the form of social disturbances generated by urban slums, driving hazards compounded by motorists who neglect to replace wornout tires or brake linings, or the hazards of living in a society where emotionally disturbed people are too poor to afford the costs of psychiatric care.

The plausibility of this argument seems to be supported by the fact that the recent increasing concern over the plight of the poor has coincided with a general improvement in their material well-being. In fact, the poverty line has been redrawn at a level which, in constant prices, is over 50 percent above the one which it displaced.[18] For this reason, even a continued rise in the real incomes of the poor will not necessarily reduce public concern with the problem of poverty.

The Role of Public Goods

The market mechanism, then, is an imperfect device for allocating resources under circumstances where externalities are significant. Since such external effects involve interdependencies which are not appropriately incorporated in the decision-making calculus of firms and households, reliance on market evaluations will not lead to a best possible use of scarce resources—that is, we will have an excess of some goods and services and an insufficiency of others. There is a related class of goods that cause markets to function inadequately and have come to be called "public goods."[19] The distinctive characteristic of such goods is that they are, in some significant sense, indivisible: they must be provided on an all-or-none basis, and therefore consumed in such a way that more for me does not mean less for you. Consider a flood control project. Such a project benefits indiscriminately all those who were formerly subject to the occasional rampaging waters of an uncontrolled river. Indeed, there is no way in which a resident of the affected region could "opt out" of the benefits of such a project. Similarly, the application of oil or DDT to a swamp which served as a breeding ground for malaria-bearing mosquitos benefits indiscriminately all residents in the vicinity. A meteorological forecast, once completed, can be made available to additional "users" who benefit from it (assuming it to be correct!) at no extra cost whatever.

Indeed, we may go a step further and say that knowledge itself is the ideal public good. Once a new bit of knowledge has been produced, it may be made available to any number of additional users at no extra cost whatever (aside from the costs of transmittal). The knowledge has to be *produced* (its research costs incurred) only once. Goods and services of this kind are supplied on an inadequate basis when society relies exclusively on private profit calculations. Because the benefit is indiscriminate, it will be worthwhile for individuals to dissimulate, to understate their true preferences, in order to shift the cost of these services to others. There is, in other words, what Samuelson has called a "free loader" problem. The market mechanism provides no effective way of determining the optimal amount of such commodities, or even of commodities which, while not purely public, are tinged with varying degrees of publicness. If our society got only as much public health service, flood control, and knowledge as profit calculations would induce from private individuals, we would be seriously deficient in all of them.

The technological progressiveness of the American economy is increasingly dependent upon a distinctive input, knowledge, which possesses very special characteristics. In fact, it is now clear that in our productive enterprises we are pressing against the limits of available knowledge in a growing number of areas. Knowledge, of course, has always been a critical factor of production. In the past, however, economic activity was conducted, in most areas, well inside the limits of our knowledge frontier. The Industrial Revolution, involving the exploitation of new sources of fuel and power and their application to machine production, was conducted for the most part upon a knowledge base of long standing. Technical advance took place primarily on an empirical, trial-and-error basis,[20] which demanded a high degree of skill, imagination, and mechanical ingenuity. Economic progress did not press too hard against the limits of knowledge until the late nineteenth century and, even then, only very selectively. Since then there has been a rapid growth in knowledge-based industries, industries which are dependent for further advances upon the expansion of the available knowledge stock. This has been so in chemicals and synthetics, in electronics, aeronautics, nuclear energy, metallurgy and materials generally, agriculture, and a growing host of others. Perhaps plastics may be taken as the prototype of the new relationship. Plastics, a material of extraordinary versatility, was entirely a product of inorganic chemistry. Technical progress in its exploitation has always been closely linked with the advance of scientific knowledge. It is now a large industry which could not conceivably have been developed along the lines of the casual empiricism of an early-nineteenth-century iron producer.

With Mendelejeff's formulation of his Periodic Table of the Elements in 1869 it became possible not only to explain the physical properties of the known elements but even to predict the properties of elements which were unknown at the time—specifically, gallium, germanium, and scandium. This kind of knowledge has given us a control over materials and their properties which enables us to manipulate and recombine their characteristics by a new order of magnitude. At the heart of much contemporary technological change is this recently acquired scientific base which makes it possible to alter and modify single characteristics of materials in response to new demands. In agriculture, for example, scientific knowledge in genetics and biochemistry has recently made it possible to develop new disease-resistant, high-yielding seed varieties which can be appropriately modified to suit local ecological requirements. The resulting growth in agricultural productivity (the "Green Revolution") may well turn out to be the most significant innovation for the human enterprise of the twentieth century.

The production of knowledge has now become a major industry. In this industry the research costs are extremely high, the output is most uncertain and often long deferred, and its eventual applications impossible to predict. Yet its long-term benefits, if the past is any guide, are likely to be massive. Since the private returns to a firm engaged in knowledge production are likely to be much smaller than the social returns, such activities are neglected by profit-seeking firms.[21]

Some Policy and Institutional Implications

The market, then, is an inadequate organizational device for allocating resources in a world where externalities are important and where goods of varying degrees of publicness, including knowledge, play a major role. Such externalities and publicness increasingly characterize urban-industrial societies. Where do we go from here? Several observations are in order.

First, evidence that private profit calculations perform badly in a certain kind of environment is not a sufficient case for transferring a responsibility to the public sector. This sector is subject to problems of its own, too notorious and too often rehearsed to require repetition. Clearly the "imperfectness" of the market does not *ipso facto* establish the "perfectness" of government. We may reject as inadequate for the late twentieth century the eighteenth-century view, as expressed in Pope's couplet: "Thus God and Nature linked the general frame, / And bade Self-love and Social be the same. . . ." It is nevertheless a total *non sequitur* to feel that this requires us to embrace the polar opposite of *dirigisme*. Technological change is forcing upon us the need to rethink our conceptions of economic and political organization. We seem to have entered a world unsatisfying to the ideological purists, one where we shall have to make pragmatic choices between imperfect markets and imperfect governments. It is apparent that, in reality, we have already been exploring this complex intermediate realm for some time. Not only do we have public production of the TVA variety but we also have public regulation and control of private utilities, public subsidy of private enterprise, public highways financed on a user cost basis, public and private educational institutions living in some reasonable coexistence, public support of medical research which is simultaneously financed partly from private sources, and public sponsorship of specialized activities contracted to private organizations—the Atomic Energy Commission and even the RAND Corporation. One of the most striking institutional developments in recent years has been the growth of nonprofit enterprises. Needless to add, even this list is merely suggestive rather than exhaustive.

As a practical matter, we have passed well beyond the naive and fallacious view which looks upon the private and public sectors as natural antagonists and interprets gains for one sector as losses for the other. From a conceptual point of view it is important that we recognize a continuum of possible institutional arrangements rather than a sharp dichotomy,[22] and also that we recognize the growing complementarity in the functions performed by the public and private spheres. Many of the activities now performed in the public sector are indispensable to the operation of the private sector and would have to be provided by them were they not provided by government. One may legitimately debate whether the public sector is efficient in its provision of roads, hospitals, schools, garbage disposal, fire and police protection, and so on. It is clear, however, that these services are not expendable and would have to be provided by private enterprise in the absence of the present public arrangements. Given our preference for the automobile as a form of transport, there can be no question of the need for the complementary services of highways or, for that matter, ambulances and hospitals. Given the dependence of our technology upon the production of knowledge, the massive support which this activity currently receives from the federal treasury may well represent an investment of resources second to no other in the size of its social rate of return.

Furthermore, we are only beginning to raise certain important questions concerning the relationship between technological changes and their implications for social, political, and economic change. The impact of such changes upon the workings of our institutions is, at best, only dimly perceived. How does technological change structure the positive and negative externalities? To what extent has our recent preoccupation with the negative externalities of production reflected a particular stage of technology—specifically, one which relies very heavily upon the exploitation of fossil fuels? The technology of horsepower, waterwheels, windmills, and sailing ships differed radically in the externalities which it generated from the technology of gasoline-powered automobiles, steam engines, diesel engines, and power-generating stations. How would the problems of urban and industrial pollution change in a world which relied upon atomic power or solar radiation rather than fossil fuels? Clearly the external effects are very different. We need to sort out, furthermore, the extent to which some phenomena are the result of technology and the extent to which they reflect particular social institutions. The recent heated controversy in the Soviet Union over the impending pollution of Lake Baikal by the effluvia of a new paper mill strongly suggests that certain painful social choices may be independent of economic systems, in spite of the efforts of several decades of Marxist critics to pin the tail of pollution and environmental destruction upon a capitalist donkey.[23] It seems obvious that we have a long way to go before we can claim an adequate understanding of the precise ways in which specific technologies shape social choices and institutions.

All this raises also the question of the adequacy of our present conceptual framework and measurement devices for examining the incidence of technological change. It seems to be one of the very peculiar things about modern man that he has not adapted himself to a sophisticated way of evaluating the consequences of a technology which is highly sophisticated. He has so far devised no effective ways for introducing some of the costs of technological change into his decision-making calculus. In part, this has been because many of the costs impose themselves in unconventional ways, at least from the point of view of a society which has been trained to think of costs solely after the manner of a business accountant. Where the costs have imposed themselves in an indirect or remote way—involving not a current dollar outlay but a deterioration in some aspect of the environment—there has been a general reluctance to take them into our calculations. This neglect is aggravated by the difficulty or impossibility of expressing these costs in quantitative terms. The bias against costs which appear to be remote in time is reinforced by the propensity for treating as unreal all costs which cannot be expressed in terms of some obvious money equivalent. Furthermore, there may be a systematic bias in our perception of environmental decay. Such decay, although it is cumulative, is often so gradual as to make it very difficult to perceive. The human response to various forms of environmental decay might well be more drastic if the evidence of the decay were more easily perceived. Similar difficulties appear when we attempt to make decisions concerning expenditures which will raise the quality of life in an urban environment—where, for example, the benefit is "merely" aesthetic. Similarly, although we are learning to treat educational expenditures as an investment in human capital, we are less successful in relating such expenditures to improvements in the quality of our lives. What we urgently need is some set of social indicators responsive to the many dimensions of human well-being which escape what Pigou called the "measuring rod of money." Much recent work in such areas as criminology and medicine suggests that some such indices are not only feasible but might serve as valuable supplements to more conventional monetary calculations.[24]

Ecological Interactions

The necessity of learning to make new kinds of value judgments is very much intensified by the growing power of our technology to produce ecological changes on an unprecendented scale. In fact, what is needed in order to make such judgments is a new paradigm which will encompass the man–technology–natural environment interdependencies more effectively than any currently available. One possibility is that in addition to the more traditional models employed by the economist we might develop models which incorporate a sufficient number of ecological features to compel a more explicit recognition of the impact of our actions upon the quality and the stability of our natural environment. There is no real conflict between maximization models and ecological models. Maximization models provide a guide to the behavior of individual firms and to problems of resource allocation more generally; ecological models, on the other hand, may relate the individual person or firm and larger social groups to relevant and limited aspects of the natural environment and therefore may be expected to yield insights about how human behavior interacts with and modifies this environment. Such models, therefore, should be particularly useful wherever the environment plays a creative role in the productive process—as in agriculture, forestry and fishing [25]—or wherever the productive process has substantial side effects upon this environment—as in air and water pollution resulting from industrial activity or in the construction of a dam which drastically alters the surrounding ecosystem.[26] In addition, of course, the informational inputs required for such models necessarily involve a substantial crossing of traditional disciplinary boundaries. Intelligent decision making, however, will require that we collect and exploit such information concerning the nature of interrelationships which, until very recently, have been virtually ignored. Such relationships can no longer be ignored. Even now, every ecologist has his own favorite catalogue of the horrors of environmental destruction which man has perpetrated due to neglect, indifference, or failure to anticipate the biochemical side effects of some new agent or technique—whether it be insecticides, fertilizers, refuse disposal, or radiation.[27] Without question we need to devote more resources than we do at present to exploring and attempting to anticipate the consequences of these interactions.

There is no excuse for making decisions which may have important ecological consequences without the best possible information about the likely nature of these consequences. Once having done this, however, we are going to have to move beyond the simplistic conservationist view which regards the identification of one of these processes as a sufficient reason for terminating the productive activity which causes it. We must face the fact that it is quite impossible to utilize any sophisticated technology—or, for that matter, many primitive ones—without engaging in innumerable actions which alter some important dimension of the environment.[28] The costs of alternative approaches will often prove to be extremely high, even to a society at our present state of affluence (and effluents!). We all want unpolluted rivers and streams just as we all lament the death of Lake Erie, but are we prepared to pay the huge costs of alternative methods of waste disposal? This, surely, is the critical question. After all, if such destruction were merely wanton, if it could be terminated at no cost whatever, it would be done immediately. The relevant question is how much we are prepared to pay to purify the Potomac or the Hudson or the Charles. How much urban renewal, medical research, foreign aid, and vocational rehabilitation are we prepared to forgo so that our rivers may run pure once again? How much, in other words, do we *value* such purity? Such scrutiny of our values is as unavoidable as it is painful.

We can, with our *present* technology, substantially reduce the amount of air pollution emitted by the electric utilities in big cities either by removing the offending substances prior to combustion or by burning fuels which are more expensive but contain smaller quantities of offensive substances. In this respect, fuel oil is preferable to bituminous coal, and natural gas is preferable to both. A substantial portion of our present air pollution, in other words, is not an inevitable product of our present technology but rather the product of a decision to utilize low-cost fuel supplies.[29]

Furthermore, techniques are presently available for achieving a high degree of purity of our water systems. One rough estimate places the cost of returning effluents to water courses in a state of "pristine purity" at around $20 billion. The author of this estimate also states his certainty "that most people would find such an undertaking ridiculous." [30] While he may well be correct in this judgment, it is hardly self-evident. It is certainly correct, however, that the decisions confronting us here are not of an all-or-none variety. There exists a gradation of degrees of water purity, greater than the low state which we presently tolerate, available at progressively higher cost all the way to the upper-bound estimate for pristine purity. Here again, choice is inevitable.

The painfulness of our choices may be seen even more forcefully in a different context. The public has been made increasingly familiar in the past few years—especially since the publication of Rachel Carson's *Silent Spring*—with some of the deplorable side effects of insecticides, herbicides, and commercial fertilizers, yet these have played a major role in the growth of farm output in the United States in the past thirty years. The use of such inputs was instrumental in the production of the large wheat surpluses which have been recently shipped to India during years of crop failure in that country. When this evidence is placed in the balance, the moral self-righteousness of some of the opponents of the use of such inputs may be seen in a distinctly different light. What, after all, is the final value judgment placed upon the use of a chemical agent which damages birds and wildlife at home and at the same time mitigates mass starvation in India.[31]

In some measure, technology itself may be invoked to reduce the painfulness of the choices which confront us. Our sophisticated technology may be put to work at the task of reducing the destructiveness of some of the obnoxious side effects of modern technology, or even to providing new uses for waste materials. It is really an old story that new technologies create problems which require offsetting or remedial actions within the technological sphere, just as, throughout history, new offensive weapons have generated responses in the form of improvements in the technology of defense. We can, by appropriate incentive schemes and organizational changes, reduce the time lag between awareness of a problem and its eventual technological solution. In fact, in some areas progress is being made. Many agricultural chemicals of recent development decompose into harmless substances more rapidly than did their predecessors. Modern detergents are incapable of rising up into billowing mountains of foam as they once did. The tamer detergent formula, however, took some fifteen years of diligent and expensive chemical research to develop. Furthermore, although detergents are now biodegradable, their high phosphate content continues to act as a fertilizing agent encouraging the growth of algae and fostering algae pollution in rivers and lakes.

It will not do, then, to state that many of our modern ills are simply a product of modern technology. This is at best a half-truth. The other half of the truth is that these ills are a product of human choices. We have *chosen* not to reduce the unpleasant side effects of our technology in at least two important ways: first, we have failed to devote more of our resources to seeking out new techniques for reducing these side effects, and, second, within the spectrum of alternatives offered by the present state of our technology we often select alternatives which are less costly in terms of money outlays but which generate higher levels of pollution—as in the case of our choice of fuels. From this perspective, environmental pollution is simply the result of a decision to adopt a less costly method of production.

Some Implications and Conclusions

Modern technology, then, is compelling us to make certain choices, and it is the unpleasant duty of the economist to continue to remind all who will listen that everything has a price tag, that there is no such thing as a free lunch. Furthermore, the new interdependencies of modern technology compel us to make many decisions in a collective way. Just as I cannot go out and purchase my own share of flood control, traffic regulation, or national defense, so I cannot purchase my own share of pure air or unpolluted rivers. These goals can be achieved only by decisions and actions which must be pursued in a communal way. In this sense they have the characteristics of public goods. Moreover, many of these decisions will have to involve new forms of cooperation at higher levels of government. The municipal government of Cleveland cannot, by its own action, solve the problems of a polluted Lake Erie, nor will combined action by the states of Michigan and Ohio be sufficient. A satisfactory solution will require agreements among many municipal governments, the governments of several states, and the government of Canada as well. Clearly the general problem here is that of developing new institutional mechanisms which will mobilize the active cooperation of all the interested and affected parties.

It should also be obvious that a highly individualistic system of values, particularly as such a system pertains to the rights of property, to systems of contractual relationships, and to notions of unqualified consumer sovereignty, will not be a useful guide in dealing with many of the intractable problems which confront us.[32] Indeed, the meaning of property and the rights of property become exceedingly difficult to define in conventional terms. The growing economic role of knowledge in a technologically dynamic society has created out of such knowledge a new form of property which is only very partially appropriable by private individuals. Since such knowledge, as we have argued, partakes of the nature of a public good, it is fortunate that our patent laws are of very limited effectiveness in allowing individuals to appropriate such property for their private use. Any legal system which places restrictions upon the use of existing knowledge must lead to an under-utilization of such knowledge and therefore to an inefficient use of this particular form of property. On the other hand, the absence of such restrictions creates the obvious dilemma that it weakens the private incentive to undertake the production of knowledge in the first place.

Of course this limited effectiveness of market forces in achieving a sufficient allocation of resources to knowledge production has been recognized, at least implicitly, for a long time. Financing research activity, especially basic research, has been a responsibility of public agencies for many years.[33] Total federal expenditures for research and development rose from $3.1 billion in fiscal year 1954 to $5.0 billion in 1958, $7.7 billion in 1960, and $15.4 billion in 1966.[34] As a percentage of Gross National Product such support has risen from a few tenths of 1 percent before World War II to almost 3 percent today. Moreover, some 60 percent of all research and development activity is now financed through the public sector.

The growth in public support of knowledge production, especially at the level of basic scientific research where private incentives are weakest, poses important questions of public policy. To a substantial degree, such research has become socialized, and both the size and the direction of research efforts are increasingly the product of allocative decisions made within the public sphere. Much of the technological change in the past quarter century has been the result of intensive public efforts to pursue goals formulated in relation to the needs of national security—jet propulsion, computer technology and electronics more generally, and atomic energy. Indeed, in recent years almost 90 percent of federal research and development spending has been accounted for by three agencies: Department of Defense, National Space and Aeronautics Administration, and Atomic Energy Commission. Our growing sense of failure in some areas of our national life surely reflects partly a collective failure to allocate research resources to areas which have been relatively neglected—education, environmental pollution, urban housing, and public transportation. It is obviously becoming a matter of growing importance that we develop institutional mechanisms for formulating and for achieving goals in areas of our national life, other than defense, where the market mechanism has proven to be an unsatisfactory instrument. These mechanisms include the not-for-profit organizations, which perform much of the research financed by governmental agencies. Such organizations are playing and undoubtedly will continue to play an important role in the production and distribution of knowledge. More generally, there is an urgent need for developing new forms of collaboration between the public and private sectors which will direct our technological potential more forcefully toward the solution of nonmilitary problems.

Another major area of property rights where our individualistic value system seems to require modification is related to the externalities discussed earlier. Technological change and the closely associated phenomenon of urbanization create new forms of interaction through which individuals impose uncompensated costs and benefits upon third parties. Property rights, in general, may be regarded as the set of socially sanctioned rules which regulate these rights to enjoy benefits and to impose costs.[35] Such rules work efficiently in regulating wide areas of human behavior and interaction. However, our preoccupation with property rights as they pertain to *private* property has now led to serious misuse of property which is not privately owned—that is, common property resources. The extent of the pollution of our two most important common property resources, air and water, is a consequence of the fact that our present system of property rights imposes few restrictions upon the freedom of individuals to use common property resources in this way. Pollution costs have been ignored by individual decision makers because such costs have been external to their private accounting systems. It is hardly surprising, therefore, that waste disposal facilities which are "free" to the individual user will be intensively utilized. When total population sizes and urban concentrations were much smaller and when our technology was more primitive, such unrestricted access to common property did not matter too much. Now a combination of larger and more densely concentrated populations together with a far more powerful technology is producing environmental damage which is widely regarded as intolerable. Our common property air and water resources are, quite simply, overloaded as a result of unrestricted rights of access.[36]

Clearly what is required is a modification of our present system of property rights which will compel individuals to internalize some portion of the costs which they now impose upon their external environment. This can be achieved in a variety of ways: regulation, subsidy, or a system of charges for the right to pollute the common property resource in question. Whichever devices are adopted, they must involve a recognition of a community interest in certain portions of the natural environment and the legitimacy of a new set of property rights regulating the access of individuals to this environment.[37]

In the simple, rustic environment of early-nineteenth-century America, where the costs and benefits of individual actions had far fewer public effects than at present, an individualistic system of values was a reasonably satisfactory guide to private and social actions. Now, largely because of technological change, individual actions have such extensive influence on the welfare of other people and on the environment itself that an individualistic calculus and an individualistic value system can no longer suffice in their present forms.

Notes

1. The proportion of the American labor force in agriculture declined from 63 percent in 1840 to 8 percent in 1960. See Stanley Lebergott, "Labor Force and Employment, 1800-1960," in Dorothy Brady, ed., *Output, Employment, and Productivity in the United States after 1800* (New York, 1966), p. 119.

2. "The services sector comprises a variety of economic activities, ranging from professional pursuits demanding high skill and large investment in training to domestic service and other unskilled personal services; from activities with large capital investment, such as residential housing, to those requiring no material capital; from pursuits closely connected with the private market, such as trade, banking, and related financial and business services, to government activities, including defense, in which market considerations are limited. They have one basic feature in common: none of the activities represents in any significant way the production of commodities; each renders a product that is intangible and not easily embodied in a lasting and measurable form. . . ." Simon Kuznets, *Modern Economic Growth* (New Haven, 1966), pp. 143-44.

3. Saburo Okita, *Causes and Problems of Rapid Growth in Postwar*

Japan and Their Implications for Newly Developing Countries, Japan Economic Research Center, Center Paper No. 6 (March, 1967), p. 36. The response of working hours to a rise in income is a complex issue which cannot be pursued here. Although the long-term downward decline in the length of the work week has been sharp and unmistakable in the United States, the evidence of the last twenty-five years or so does not point unambiguously in the direction of a continuation of this trend. It is not at all clear to what extent shorter hours of employment currently reflect worker preferences as compared to union bargaining tactics. The growing prevalence of "moonlighting" and the evident widespread willingness to work overtime, even at standard hourly rates, suggests that many people would prefer longer hours and higher money incomes. All this, plus the substantial increase in female participation in the labor force, strongly suggests that we are still far from having completely internalized the values of a leisure time–oriented society. We may be already approaching the plateau beyond which leisure time begins to become an "inferior good"—a likelihood which may be increased by the high money cost attached to many leisure-time activities, such as foreign travel.

4. A. N. Whitehead, *Science and the Modern World* (New York, 1947), p. 141.

5. A point frequently made is that the military orientation of our federally sponsored research effort does not really matter so much because of the technological "spillovers" to the civilian sector. Such spillovers represent a highly inefficient way of solving problems in the civilian sector, quite aside from the fact that they leave untouched vast areas of high-priority social concern. Expenditures on military-oriented research can hardly be justified economically on these grounds alone. The social opportunities costs of such military research are extremely high, because it is absorbing precisely those scarce human and intellectual resources which might otherwise be directed toward the solution of social problems.

6. For example, in an institutional context, how should the National Institutes of Health expend its large annual budget of federal funds?

7. *New York Times,* May 26, 1968, p. 64.

8. J. B. Wyon and J. E. Gordon, "The Khanna Study," *Harvard Medical Alumni Bulletin,* Spring, 1967.

9. See Nathan Rosenberg, ed., *The American System of Manufactures* (Edinburgh, 1969); Ed Ames and Nathan Rosenberg, "The Enfield Arsenal in Theory and History," *Economic Journal,* Dec., 1968.

10. See Eugene Smolensky, "The Past and Present Poor," in *The Concept of Poverty* (Washington, D.C., 1965); and Eugene Smolensky, "Investment in the Education of the Poor: A Pessimistic Report," *American Economic Review Papers and Proceedings,* May, 1966. See also Robert Lampman, "The Low Income Population and Economic Growth," Study Paper number 12 (prepared in connection with the Study of Employment, Growth, and Price Levels for consideration by the Joint Economic Committee, Congress of the U.S., 86th Congress, 1st Session), (Washington, D.C., 1959).

11. For a most forceful presentation of this point of view, see the sweeping historical survey in the first few pages of *The Communist Manifesto.*

12. The case for the market has always rested upon a far more complex

foundation of institutional presuppositions than its more vociferous ideologues have been prepared to recognize. See, for example, Nathan Rosenberg, "Some Institutional Aspects of the *Wealth of Nations*," *Journal of Political Economy*, Dec., 1960.

13. Each family *could* undertake to assure the purity of its own water supply. It is, however, much more efficient to filter and chlorinate our water supply centrally. The form which the technology takes also has important implications for public control. It is much more difficult to institute controls over the air pollution generated by London's three million domestic hearths than it is to control pollution created by the centralized power plants.

14. See Kuznets, *Modern Economic Growth*, chap. 5. Here again the mechanism is a dual one. Demand and resource use have altered over time not only because of the role of technology in raising incomes, as a result of which consumers shift their expenditure patterns. Technology has also developed the new products and services upon which households are so anxious to spend their rising incomes. If detailed budget studies were available for 1800 as well as for 1968, they would certainly show that a very high fraction of present consumer expenditures are for a range of goods and services which simply did not exist in 1800.

Moreover, modern technology seems to possess a distinct and systematic antiagriculture bias. Although there have been massive improvements in the productivity of resources employed in agriculture, there has been little development of new agricultural products, while new products have emerged in profusion from the manufacturing sector. Thus, the nature of technological change has been a major influence in accounting for changes in industrial structure.

Furthermore, centralizing effects due to the economics of large-scale production require an offsetting growth in certain services. The decline of household baking and the rise of large commercial bakeries necessarily involves an associated growth in transportation and retailing services.

15. See the excellent cost-benefit analysis of the projected supersonic transport in Stephen Enke, "Government-Industry Development of a Commercial Supersonic Transport," *American Economic Review Papers and Proceedings*, May, 1967. Enke points out: "About 85 percent of U.S. residents have never flown, those who do fly do not always take long-haul flights, and perhaps less than 5 percent of all Americans will ever fly SST's at their higher fares. Private, nonexpense account, long-haul passengers will mostly continue to fly subsonically. . . . Further, American SST passengers will tend to travel to and from a few areas, such as New York, Chicago, Los Angeles, San Francisco, Seattle, Washington, D.C., and Miami. Americans living elsewhere may never use an SST except on international flights. But 100 million Americans may find themselves subjected daily to sonic booms if overland SST flights are permitted" (pp. 78-79).

16. Some states have prohibited the use of such studs by law.

17. Tastes should be defined here to include preferences with respect to family size. Indeed, other people's tastes concerning family size may be the most important externality of all. The most significant decisions which "the rest of the world" makes affecting my welfare are the decisions which, collectively, determine the size of the total population of which I find myself a member.

18. Smolensky, "The Past and Present Poor." Elsewhere Smolensky argues: "The disutility that arises from the persistence of poverty is not confined to the poor, and the effective demand for federal antipoverty programs does not emanate solely, nor perhaps even primarily, from a concern for equity. The primary objective may very well be efficiency in the Pareto sense. If this be so, a poverty line which considers only the needs of the poor may not be the most pertinent guide to federal policy. Perhaps the most relevant requirement of a poverty line is that it serve as an index of the disutility to the community of the persistence of poverty." "Investment in the Education of the Poor," p. 371.

19. Paul Samuelson, "The Pure Theory of Public Expenditure," *Review of Economics and Statistics,* Nov., 1954. This subject has been brilliantly analyzed and summarized in Francis M. Bator, *The Question of Government Spending* (New York, 1960).

20. This was notoriously so in the critical case of metallurgy. It seems fair to say that no one really understood what was being done in iron and steel production until metallurgy began to be placed on a scientific foundation in the last third of the nineteenth century.

21. For an interesting attempt to measure the social rate of return in a single, important instance, see Zvi Griliches, "Research Costs and Social Returns: Hybrid Corn and Related Innovations," *Journal of Political Economy,* Oct., 1958. Griliches attempted to estimate the social rate of return over the period 1910-55 to the public and private resources that had been invested in the development of hybrid corn. His deliberately over-conservative methods yielded an estimate on the order of 700 percent.

22. This continuum is thoroughly explored in Robert Dahl and Charles Lindblom, *Politics, Economics, and Welfare* (New York, 1953).

23. A nice example of such an effort is provided by J. D. Bernal in his *Science in History* (London, 1954), pp. 676-77. After referring to the impact of "modern mechanical agriculture and lumbering" in "ruining a dangerously large proportion of the soil of the planet and of changing climate unfavourably to almost all forms of life," he goes on to state: "This large-scale intensive devastation has nothing to do with the inherent wickedness or stupidity of man, or of his unrestrained desire to propagate, as many publicists want people to think; it is simply due to the essentially predatory nature of capitalism, now spread as imperialism over so large an area of the world. The destruction of the soil has been enormously accelerated in the last fifty years by the methods characteristic of ruthless capitalist exploitation for immediate profit." And then, in striking contrast: "In the part of the world now saved from the operation of the free market and the monopoly trust . . . more especially in the Soviet Union, where alone there has been time for long-term projects to mature, the improvement of soils and the reclaiming of deserts have been going on for twenty years."

24. For the biases implicit in our present national income accounts and for an illuminating treatment of how these accounts lead us to overstate the net output of goods and services in an industrial-urban society, see Simon Kuznets, *Economic Change* (New York, 1953), chap. 6. The matter of a new set of social indicators was explored in the May and November,

1967, issues of the *Annals of the American Academy of Political and Social Science,* Bertram Gross, ed., entitled "Social Goals and Indicators for American Society."

25. See H. S. Gordon, "The Economies of a Common-Property Resource: The Fishery," *Journal of Political Economy,* Apr., 1954.

26. "The ecological approach attempts to achieve a more exact specification of the relations between selected human activities, biological transactions, and physical processes by including them within a single analytical system, an *ecosystem.* In ecology generally, an ecosystem consists of a biotic community of interrelated organisms together with their common habitat and. . . . thus emphasizes the material interdependencies among the group of organisms which form a community and the relevant physical features of the setting in which they are found. Clifford Geertz, *Agricultural Involution* (Berkeley, Calif., 1966), p. 3.

27. For a sobering recital of such horror stories, see "Development in the Poor Nations: How to Avoid Fouling the Nest," *Science,* March 7, 1969, pp. 1046-48.

28. One of the most drastic of ecological disturbances is the conversion of forest and plain to cropland. For a systematic treatment of the effect of westward expansion upon wildlife in America, see Peter Mathiessen, *Wildlife in America* (New York, 1959); see also John V. Krutilla, "Some Environmental Effects of Economic Development," *Daedalus,* Fall, 1967, esp. pp. 1060-62.

Lynn White, Jr., has argued: "Ever since man became a numerous species he has affected his environment notably. The hypothesis that his fire-drive method of hunting created the world's great grasslands and helped to exterminate the monster mammals of the Pleistocene from much of the globe is plausible, if not proved. For six millennia at least, the banks of the lower Nile have been a human artifact rather than the swampy African jungle which nature, apart from man, would have made it. The Aswan Dam, flooding 5,000 square miles, is only the latest stage in a long process. In many regions terracing or irrigation, overgrazing, the cutting of forests by Romans to build ships to fight Carthaginians or by Crusaders to solve the logistics problems of their expeditions, have profoundly changed some ecologies." "The Historical Roots of Our Ecologic Crisis," *Science,* March 10, 1967, p. 1203. White's main thesis in this paper is that the present ecologic crisis has its historical roots in the exploitative attitude toward nature fostered by western Christianity, which he calls "the most anthropocentric religion the world has seen."

29. "In most large cities . . . the electric utilities consume up to half of all fuel burned. Most utilities have made reasonable efforts to reduce the emission of soot and fly ash. . . . Utilities, however, are still under pressure, both from the public and from supervision agencies, to use the cheapest fuels available. . . . New York . . . utilities burn large volumes of residual fuel oil imported from abroad, which happens to contain between 2.5 and 3 percent of sulfur, compared with only about 1.7 percent for domestic fuel oil. . . . Recent studies show that the level of sulfur dioxide in New York City air is almost twice that found in other large cities." Abel Wolman, "The Metabolism of Cities," *Scientific American,* Sept., 1965, pp. 187-88.

30. Allen V. Kneese, "Research Goals and Progress toward Them," in Henry Jarrett, ed., *Environmental Quality in a Growing Economy* (Baltimore, 1966), p. 71.

31. The issue may, of course, be more complex than the manner in which it is formulated here. If the use of such chemicals has a long-run cumulative effect in reducing soil fertility, say through ecological disturbances or by upsetting bacteriological activities, the situation then becomes different. That is to say, higher productivity today may be being purchased at the price of lower productivity at some future date.

32. On contractual relationships in government procurement, Price points out: "The contractual relation is not the traditional market affair: the contract is not let on competitive bids, the product cannot be specified, the price is not fixed, the government supplies much of the plant and capital, and the government may determine or approve the letting of subcontracts, the salaries of key executives, and a host of other managerial matters. A sizable proportion of the government (and nation's) business is done this way." Don K. Price, *The Scientific Estate* (Cambridge, Mass., 1965), pp. 37-38.

33. A useful description of the role of the federal government before World War II appears in A. Hunter Dupree, *Science in the Federal Government* (Cambridge, Mass., 1957).

34. *Federal Funds for Research Development and Other Scientific Activities, Fiscal Years 1964, 1965, and 1968*, vol. XIV, National Science Foundation Publication No. NSF 65-19, July, 1965.

35. "Harming a competitor by producing superior products may be permitted, while shooting him may not. A man may be permitted to benefit himself by shooting an intruder but be prohibited from selling below a price floor. It is clear, then, that property rights specify how persons may be benefited and harmed, and, therefore, who must pay whom to modify the actions taken by persons." Harold Demsetz, "Toward a Theory of Property Rights," *American Economic Review Papers and Proceedings*, May, 1967, p. 347.

36. "Consider . . . the earthy and undramatic problems of waste disposal in urban centers and of pollution in general. These clearly require policy decisions at the societal level. But it was not always so. Families in sparsely settled regions were once free to dispose of their waste on the basis of self-interest; they did not adversely affect the environment of others. The habit so formed persisted when cities came into being. Progressive cities then proscribed use of the streets as sewers, and substituted direct discharge into rivers, lakes, and oceans. In terms of local interest this is reasonable—sewage treatment is expensive. But it is becoming anachronistic for large communities to determine their methods of disposal independently. The problem has been greatly augmented by the growth of industrialization. The wastes of modern industry have reached huge volume. In terms of individual interest, industrial firms—paper mills, chemical plants, refineries, and so on—find economy in downstream disposal of untreated waste and heat, but this reduces and often destroys the value of the streams for individual use, recreation, and other purposes. In terms of their interest, it is equally reasonable for great metropolitan centers to claim pre-emptive rights to regional waters for waste disposal, as when Chicago

reduces water levels in the Great Lakes to flush its sewage down the Mississippi River." Harold Barnett and Chandler Morse, *Scarcity and Growth* (Baltimore, 1963), pp. 254-55.

37. The policy alternatives for dealing with the problem of pollution of common-property resources are treated in an absorbing little book by J. H. Dales, *Pollution, Property and Prices* (Toronto, 1968). Dales favors a system of fees for the right to discharge waste materials into common-property resources.

Steve M. Slaby

What Should We Ask of the History of Technology?

The question discussed in this paper is not "What do we ask of the history of technology" but "What should we ask of the history of technology?" What, too, should we ask of the people who research, record, and interpret this history?

Since I believe interdisciplinary or cross processes constantly occur in the world, I will not single out a particular group of scholars or specific disciplines to damn for causing the ills of today's technological world. On the contrary, all intellectuals, scholars, and academics, in my opinion, bear a special responsibility for today's society. We therefore have a special obligation to attempt to understand one another if we hope to deal realistically with the momentous problems that face humanity today, including the problem of human survival. Any criticisms, therefore, are made in a spirit of learning, and of judging no one. We are all in the same boat.

In reviewing a number of books on the history of American Indians, Michael Rogin said, "History written from present concerns can distort the past, but it can . . . help make the past accessible." Professor Rogin also quoted Andrew Jackson who said, "The past we cannot recall . . . but the future we can provide for." The underlying philosophy in effect says that the history of past wrongs is detached from present solutions and has a limited influence on the ability of human beings to determine their future.[1]

There is a tendency for us to look at history—if we look at it at all—as a force detached from us as individuals and detached from the mass of individuals who make up our human society. Yet it is the masses of individuals in a society who actually make history.

History is the past—but simultaneously it is being created now. If we could learn from our own past and present history, our history could become a catalyst to promote justice and nonviolent social change as well as a potent power to help develop and preserve humane values while discarding and destroying barbaric ones.

As recorded by professional historians with claims to objectivity, history seems to have been written for other historians. For them, the invention of esoteric phraseology wins prizes. Actually, some of the lessons of history may be comprehended by a sensitive but politically powerless academic elite. However, since the work of historians does not reach the people, the masses who create this history do not understand their own creations and continue through life, generation after generation, not learning from past mistakes and past wrongs. Always we are starting from scratch.

Why do we learn too little from history? Perhaps one reason may be that no major attempts are being made, by most of those who record and teach history, to make it relevant and understandable to people at an early stage of their lives. What passes for history in many of our schools tends to be a form of nationalistic and ideological pap to promote the glorification and mystification of a particular political and economic order. This has resulted in what Cedric Belfrage (the cofounder of the *National Guardian*) calls "enshrined history"—a rewrite of what has actually occurred, supplied by a school of historical revisionists. In his opinion, today's confusion is due to not really knowing what happened in the past.[2]

The present form of historical analysis perpetuates the status quo of political forces and is not, in itself, an innovating force for positive social change and justice. Consequently the future, in terms of political power and social injustice, will continue to reflect the present.

The Indochina war is an excellent example. "Peace with honor" gave us continued killing and destruction at unimagined costs in moral, spiritual, human, and material treasure. The domestic political scandals that envelop us today also reflect this same tragic historical amnesia. We have learned little from the distant past, the recent past, and the daily present.

Among the many reasons for this, we must include the inadequacies of traditional history and its lack of communication with the public. History should have an orientation which not only looks at the micro and macro aspects of technological development, but which also considers, as Professor Howard Zinn of Boston University has stated, "the relationship of government to corporate power and of both to movements of social protest." [3] The importance of the relationship of technology and technological development to corporate power and national interests should need no elaboration.

Contemporary history of technology tends to read as a form of propaganda. The public, including many intellectuals, is being brainwashed and is brainwashing itself into believing that the continued unlimited application of undirected technology ultimately is going to solve most human problems. John McDermott described this as technology being the "opiate of the intellectuals." [4]

The history of technology should depict past experiences in a way that gives people a meaningful perspective on the impact that technology and other forces have on their daily lives. As a catalyst it can trigger social and political changes and inventions needed to create conditions for improving the quality of life and protecting the natural environment now so that there can be a future. Traditionally, the history of technology has glorified the leaders of technology, nations, armies, corporations, and religion. What were the thoughts and emotions of the unsung and unrecorded millions who made past history? Why have historians largely neglected them?

Historians of technology as well as scholars in all the other disciplines must become critical participants in social movements, according to Zinn, if we are to cease our near-parasitic status in society: "We have been honored, flattered, even paid, for producing the largest number of inconsequential studies in the history of civilization. . . . We publish while others perish." [5] To deal with events today intelligently and perceptively as they continuously become part of contemporary history, it is imperative that we have access to the documentation in which contemporary events are recorded. The press offers one form of this documentation but it is woefully inadequate since much of its information is secondhand. Much of the recorded primary source documentation which is not made available to the contemporary historian or the public is buried in government archives for up to one hundred years. National archives of this kind are the mausoleums of history. We cannot help repeating past wrongs with this arrangement, based on the value judgments of special-power interests who suppress history to protect themselves from the people. When the documentation and records are not exhumed from the grave until a hundred years after the fact, then "present concerns can distort the past,"[6] since such hidden records of the past have little value in guiding humanity to a better future. What would be the effect of technological innovation if new inventions were locked up in archives and not released for one hundred years? Perhaps technological development would be on a par with social development. However, the fact is that technological innovations normally are rushed to the people at a rate which surpasses their capacity to absorb or reject them.

A macrohistorical point of view surveys the continuous struggle and development of human beings through the ages. However, the microhistorical events of technology—inventions and their application—have imposed themselves on historians and on all of us at such a rapid rate, especially during this century, that we have not been able to cope with the influence this technological growth and change have had on social change. Technological innovation has outstripped social innovation, resulting in continuous wars, massive dislocations of people, and destruction of individual cultures which are rapidly being replaced with a mass culture. This technological force keeps imposing technical structures on the social and cultural life of people. This is leading not only toward more tyranny by technology but also toward the dehumanization of people. Herbert Marcuse points out: "Technological transformation is at the same time political transformation, but the political change would turn into qualitative social change only to the degree to which it would alter the direction of technical progress—that is, develop a new technology. For the established technology has become an instrument of destructive politics."[7]

A good many historians of technology tend to look upon technology as a collection of machines when in fact it is vastly larger in scope. Jacques Ellul contends that technique forms the basis of technology in its broadest meaning. Technique, according to Ellul, "refers to any complex of standardized means for attaining a predetermined result. Thus, it converts spontaneous and unreflective behavior into behavior that is deliberate and rationalized. . . . [A technical civilization] is a civilization committed to a quest for continually improved means to carelessly examined ends. . . . Technique transforms ends into means. What was once prized in its own right now becomes worthwhile only if it helps achieve something else. And, conversely, technique turns means into ends. Know-how takes on an ultimate value." [8]

The history of technology should reflect the intimate relationship of technology to political power through which technology itself can become a powerful political force for constructive social change. In Herbert Marcuse's view, "When technics becomes the universal form of material production, it circumscribes an entire culture; it projects a historical totality—a 'world.' " [9] Who is in control of this "historical totality?" According to Swedish writer Harry Martinson: "What is essential in technology's very being is . . . our . . . spiritual ability to instill in technology a soul and in this way to make the dream of man come true. That dream of man is something entirely different from the dream of technology. . . . It is the dream of man's ultimate discovery of himself and of his place in the world beyond all concealing disguises. Technology as such can also become a kind of iron curtain, which, with its perpetual illusionary facade, obliterates all spiritual vision." [10]

The history of technology can either further perpetuate Martinson's "illusionary facade" or—on the contrary—it can lead to "man's ultimate discovery of himself." The history of technology can become surrounded by this "iron curtain," however; there are signs that this is happening. In the process the ultimate of follies is being committed. Some researchers are attempting to develop a technology of history with which history is to be quantified and computerized. Games theory mathematical techniques are being applied to test the validity of different historical theories as they relate to international relations. This same trend exists in certain quarters of political science and the other social sciences. History cannot be separated into neat sanitized packets.

The history of technology is an integral part of the general history of human beings since all human events interact upon one another, directly and subtly, and include the creative thoughts and emotions of the people who make the history. To play mathematical games with history is to dehumanize history. Where does this lead us as human beings?

Roger Garaudy, professor of philosophy and aesthetics at the Sorbonne, has referred to the "contingency of historical hegemonies":

> History is always written by the vanquisher, and moreover from his point of view; and that view is almost always that his victory was an historical necessity. But this is obviously untrue. . . . what would have happened if Hitler had succeeded, as we feared, in using the atomic bomb *before* his opponents? I emphasize this contingency of historical hegemonies not only to suggest a better reading of the past, but to show that at every moment a *plurality* of possible ways are conceivable and realizable, that future is not a scenario already written without us, that we are not puppets, but beings fully responsible for our own history and for our own future. In order to fertilize the future, we must fertilize history.[11]

What should we ask of the history of technology? Let us ask that it become a political force in order to enlighten the present, so that we can thereby "fertilize the future." Let the history of technology cease being a passive observer and an elitist servant to those who control the centers of power in society. Let it become an active participant in the ongoing life process of the mass of people who make history. Let all of us become personally involved in a value-based social movement whose ultimate value is reverence for all living things.

Thomas Jefferson felt that "the study of history in particular, would teach the young to judge the actions and designs of men . . . to know ambition under every disguise it may assume, and knowing it, to defeat its views."[12]

The young are the future—and yet are present with us today! The history of technology should prepare them to understand that technology and technological change profoundly influence social change and, furthermore, that technology is not the answer to all human problems. Technology must be directed.

In order for technology to be directed, it cannot be assumed to be value free or neutral. Technology contains its own value order as it relates to living things. The historians of technology, as well as all of us who operate under different labels, are part and parcel of history in the making. Therefore we cannot continue to ignore the values inherent in our own particular disciplines by hiding behind the curtain of "objectivity." Martin Nolan defined objectivity as "a code word for playing it safe, covering up, and superficiality." This kind of objectivity "let the most unexplained war in history go on without challenge until one and a half million people were killed." [13] Howard Zinn wrote that "there is no one objective description. This search for a nonexistent objectivity has led us, ironically, into a particularly retrogressive subjectivity, that of the bystander." [14] *Let us cease being bystanders!*

Epilogue: December 10, 1973

Experiences in academia, industry, consulting, and politics have convinced the author that the dedicated, innovative people in these institutions have not managed to exercise any major influence on decision making in their own institutions or on setting priorities for the nation. They are politically powerless and impotent.

To cope with this reality of powerlessness, mass academia perpetrates a largely self-imposed myth. This is the myth of the *independence* of scholars in our institutions of higher learning. Heavy reliance upon federal government grants, foundation support, and other external sources of money has created the illusion of total freedom of inquiry for those who receive such funds. In reality, this freedom is severely limited, is not spontaneous, and involves serious moral compromises.

In this system, one is judged by his peers—those in society and in academia who have the most to gain from the status quo. The innovator is ridiculed or at best not rewarded or respected for the expression of freedom, since innovative ideas often conflict with, and therefore challenge, the existing order and value system. Therefore, to advance in the hierarchy of established scholars requires special kinds of compromises on the part of innovators, compromises which abate the free spirit and the free intellect.

The process corrupts the utopian concept of academic freedom and has contributed to the development of academic elitism and arrogance.

Critical analysis of the society in which we and our institutions exist is frowned upon. For example, how long has it taken historians to write an accurate history of the American Indians and the Black African people in our country for the general populace and especially for the youth in the schools—a history which exposes the realities of the lives and deaths of these people? How many Marxist scholars are to be found in our colleges and universities today? How long did it take before professional societies admitted to, much less dealt with, the carnage that is still occurring in Indochina? How many of these professional societies actually confronted this issue at any level? What about the subservient status of women and the poor in our society? Where are the historians and philosophers on this issue?

What kinds of conscious and unconscious compromises are made by academics to remain in the good graces of their sources of funding? A few examples can perhaps put this question in perspective. In 1961, at a parents' convocation at Michigan State University, then-President John A. Hannah remarked, "Our colleges and universities must be regarded as bastions of our defense, as essential to the preservation of our country and our way of life as supersonic bombers, nuclear-powered submarines, and intercontinental ballistic missiles." [15] The interesting sequel to this remark was the revelation, in 1966, of Michigan State University's involvement in the internal affairs of Vietnam, through the Vietnam Project at Michigan State University and the Michigan State University Group in Vietnam.[16] This official university project, under contract to Saigon and Washington, was to be responsible "for the proper functioning of Diem's civil service and his police network, the shaping up of the 50,000 man . . . militia, and the supplying of guns and ammunition for the city police, the civil guard, the palace police and the dreaded Sûreté." [17]

In 1968 and 1969, about sixty-two officially listed research projects at Princeton University were being funded by the Department of Defense, supported by the Air Force, Army, and Navy. Titles that sound like typical academic basic research topics were listed in the fields of astrophysical sciences, biology, chemistry, geological and geophysical sciences, mathematics, statistics, physics, plasma physics, psychology, aerospace and mechanical sciences, electrical engineering, and economics and for the Center of International Studies. The Army, for example, supported research such as "Physiological Mechanisms of Leaf Abscission and Senescence" and "The Genesis of Civil Violence." The Air Force funded research on "Fundamental Concepts in Theoretical Physics" and "Perception of Time-Varying Stimuli." The Navy supported a "Workshop on the Social Basis of Stable Rule." [18]

In 1970-71, Pentagon-funded projects at Princeton included "Experimental and Theoretical Investigations of Gravitation" (Navy), "A Study of Network and System Sensitivity of Signal Processors for Use in Air Force Systems Such As Command and Control Systems" (Air Force), and "Economic Planning (Korea)" (Air Force).[19] Such projects have a huge dollar value. The Massachusetts Institute of Technology, Stanford University, and the California Institute of Technology all received funding above that of Princeton University, whose government funding in 1970-71 was $25,419,170.

The Defense Department has established "military research centers at selected universities and has enlisted the help of university administrators in the creation of organizations such as the Institute of Defense Analyses." [20] The secret and highly classified Communications Research Division of the Institute for Defense Analyses, for example, has been located on the Princeton University campus since 1962.[21] The director and assistant director of the Communications Research Division also simultaneously hold visiting lecturer positions on the Princeton University faculty. However, I.D.A. has been the cause of much controversy and dissension at Princeton since 1967—so much that I.D.A. is moving off the campus into the town, where citizen action shows that it is not particularly welcome either.

University administrations' uncritical, political ties with government showed that the Pentagon "could count on the enthusiastic cooperation of many scientists and administrators . . . [because] for the first time in American history, scientists and academicians had come to enjoy positions of considerable prestige and influence in Washington. . . . The establishment of large research organizations had freed many professors from conventional academic procedure and permitted them to pursue laboratory studies without being accountable to their colleagues in tradition-minded university departments."[22] As a result, "the universities developed a deep dependence on the government which caused them to anticipate what the government wanted from them and brought them to believe the government's interests were the same as their own."[23]

This intimate university-government relationship has created conditions in which the concept of *independent* universities and scholars has become a self-deluding myth. A statement made by Lt. General A. W. Betts, Army Chief of Research and Development, in 1970 leaves no doubt as to his views concerning the dependence of academia upon government support and approval of its activities. General Betts, speaking at the ROTC awards banquet at the Massachusetts Institute of Technology, was partly reacting to those "misguided individuals [who] contend that such research [for the military] interferes with the teaching process": "The new knowledge gained from research invigorates teaching. . . . Money available for salaries permits a university to extend its teaching staff. . . . With strong support of research, the university is able to attract and hold higher quality scientists. . . . The image of the university that results from its research accomplishments is often a large measure of the reason for the high esteem in which the university is held. Also, research projects provide the necessary focus for graduate student training. . . . And lastly, support of research provides funds to pay for student employment."[24] *Ordnance* commented: "There is a great deal of evidence that universities would find it difficult, if not impossible, to stay in the forefront of technology without access to the technological advances that come from defense science. Some of our most vital weapons systems, upon which depends free-world survival in these troubled times, were made possible by research conducted by American educational institutions. The universities sacrificed nothing in the process; instead they gained much from it."[25]

Relatively few academics challenge the collusive nature of the relationship between institutions of higher learning and the government. As *Ramparts* magazine noted: "We lack historical perspective. We have been conditioned by our social science training not to ask the normative question, we possess neither the inclination nor the means with which to question and judge our foreign policy. We have only the capacity to be experts and technicians to serve that policy. . . . Our failure in Vietnam was not one of technical expertise, but rather of historical wisdom." [26]

What the illusion or myth of academic independence has done is to have created and nurtured the development of establishment scholars, including establishment historians, philosophers, and technologists. These particular scholars serve the interests of those who fund their research and therefore are indebted to those who possess the controlling power in society. The result has been that very few scholars have raised critical questions to challenge the nature and effect of the American political, social, and economic system. Gunnar Myrdal wrote:

> I have often reflected on how much of our intellectual exercises in the social sciences is escapism: turning away attention from what is highly important towards something which is "safe" in the political surrounding of one's own country. In such a situation those who are rebellious are given an axe to grind, which often is equally nonconducive to truly unbiased research.
>
> A few years ago I visited a great university in the United States which was starting a separate department for peace research, and would award doctoral degrees in that subject. When I looked at the curriculum of the department, I found attention given to very abstract and abstruse topics and to exercises in mathematical model-building of all sorts. But no place was awarded for what people are really anxious about and discussing in the United Nations and elsewhere: the raging wars in the third world, often spurred by the superpowers, the accelerating arms race, the continued qualitative improvement of weapons for mass murders, including the most cruel and even illegal ones, and the failure to reach effective disarmament agreements.
>
> When I explained my astonishment about this to the dean of the new department, he answered that research on these more immediately practical and political international issues could well be done in Stockholm but not in the United States. I, of course, replied that if I were a professor at an American university, I would certainly not feel motivated to avoid these important issues in my research. Well, he replied, in that case *you would get no money for research, neither*

from government sources nor the foundations which are all eager to conform to what are considered to be national policy interests.[27]

The record of academics and scholars involved in the Indochina war reflects the degree to which these establishment scholars have been willing to compromise the high standards and concept of open intellectual inquiry to which academia theoretically aspires. Noam Chomsky observed that "by any standard, the man [Kissinger] is one of the great mass murderers of the modern period."[28] Chomsky's characterization of Kissinger will offend many people, but history is reflected in the following data: between January 1, 1969, and June 30, 1972, under the Nixon-Kissinger administration 4.5 million Indochinese civilians were killed, wounded, or made homeless, 1.5 million soldiers on all sides were killed or wounded, 40,000 South Vietnamese civilians were executed without trial under the Phoenix program, 3.7 million tons of bombs were dropped (equivalent to two tons every sixty seconds), 13 million bomb craters were created, 20,000 Americans were killed, and 110,000 were wounded.[29]

However, we cannot place all the blame for this killing and destruction on a few people. Where was Congress during this period of history? Where were the corporations, the universities, and the objective and unbiased scholars and faculty in these universities? Except for a few, most scholars and academicians found a safe refuge in directing their research upon problems that did not question the policies of their government during this period of history. Where were the citizens of the United States? Where are we today with a proposed military budget approximating 100 billion dollars in 1975?

Since we exist in the present, history can only have real meaning for us as it can help us as individuals and a society to understand and cope with the contemporary period—as *applied history*. At present, history in general has little meaning for us as a people since it remains an entity detached from our present frame of mind. This detachment has been very instrumental in leading us to the precarious state—a state of crisis—in which we find ourselves today as alienated individuals who have lost our historical roots and found a "country [that] is in terrible disarray."[30]

If the history of technology is concerned mainly with the niceties of particular theories of the past, it can have no application to what is happening in the present nor can it have any major influence on what will happen in the future. That kind of history exists as an intellectual artifact to be rehashed in professional journals and forums while contemporary life continues on its way, moving on to the future in its highly irrational manner. History, rather than being a great and influential teacher, becomes a junk heap of human experiences and ideas.

An analogous situation exists in the area of technology. (Let us assume for the moment that technology can be observed as an independent force in human affairs.) Since academic and practicing technologists are intimately related to the power centers in the society, they promote technology as the ultimate remedy for solving the problems created by technology. This philosophy neglects the social and moral aspects of the impact of technology and therefore intellectually and politically detaches technology from the rest of the components which make up the total society.

An obvious difference between traditional historians and philosophers and technologists is that technologists have a profound effect on what is happening today and what will be happening in the future. In fact, it seems as if many historians and philosophers justify the works and effects developed by technologists rather than generating a critical analysis of technology. Such analysis from both historical and contemporary perspectives is continually needed if the technologists and the consumers of their products are to be influenced to examine their own ideologies and compromises critically. Technologists have largely ignored history in the larger philosophical and human sense, while intensively using history in the narrow technological-scientific sense to develop new technologies.

The technological process generally is cumulative. Past knowledge is intimately related to present knowledge, which is pragmatically applied to develop innovative technologies and new knowledge. Even theoretical and abstract formulations have bases in the past. If this were not so, each generation of scientists and technologists would have to reinvent what was created earlier. For example, in 1838 George Boole invented Boolean algebra. In 1938, one hundred years later, Boolean algebra found its application in modern electronic machine language.

If people in general could make intensive use of past and contemporary human history to the same extent that technologists use their specific history, then repeatable human disasters might be lessened and even eliminated, leading to a higher quality of human life the world over.

There seems to be an irreconcilable gap between the application of general human history as a learning and teaching tool and the application of the specific history of technology to develop new tools and inventions out of past ones at an ever-increasing rate. A dichotomy results. Technological innovation, based on past accomplishments and failures, has resulted in rampant technology, misdirected and misused, but nevertheless ever growing in content and ever innovative. The intellectual content and basic understanding of this technology, however, are primarily restricted to professional technologists.

The overall history of human beings, from which the history of technology in reality cannot and should not be detached, is much too neglected except by professional historians and by society's power centers which selectively and parochially apply, manipulate, or cover it up. The applied history of human beings trails far behind the applied history of technology, with the result that very few new social and political inventions are developed over the generations to deal with the changes that technology and other forces impose upon people. Parochial application of selective history cannot contribute to controlling galloping technology, the values inherent in it, the values it destroys, and the powers it sustains.

Technology is moving on while history is made but its lessons are not effectively taught to the people. Both technology and history seem to possess self-destructive qualities simultaneously with utopian qualities. Communication between historians and technologists is tenuous at best. Communication between these intellectuals and the masses of people is elusive, mysterious, and deceptive. The intellectuals speak to each other in terms that do not convey the essence of their thoughts, ideas, and feelings, but rather in academic rhetoric with which they cannot truly learn from one another and which cannot help ordinary people to comprehend and learn from them.

Academicians are comparatively few in number and politically almost powerless. However, as a group they possess immense intellectual powers and as part of the masses of people—not as intellectual elite—they hold the potential of humane political power. Their actual powerlessness will continue as long as academicians and scholars cater to—and thus are the servants of —the social institutions holding *status quo* power. Only if academicians and scholars can disengage themselves from this servant status is there a chance that they can acquire that independence which is crucial to true free inquiry, critical analysis, and wisdom. Only through this real freedom and sense of mutual responsibility to humanity, rather than to a national state, can academicians be transformed from servants and bystanders to active participants in the making of history. For this to happen, each of us must begin to make a searching reevaluation of our present status and role in society. We must try to develop a form of applied history that has meaning to the masses of people who continue to create their costly history.

Notes

1. Michael Rogin, *New York Times Book Review,* Dec. 24, 1972, p. 4.
2. Lecture by Cedric Belfrage, "The American Inquisition," Princeton University, May 7, 1973.
3. Howard Zinn, *The Politics of History* (Boston: Beacon Press, 1970), p. 100.
4. John McDermott, "Technology: The Opiate of the Intellectuals," *New York Review of Books,* July 31, 1969, p. 24.
5. Zinn, *Politics of History,* pp. 52, 5.
6. Rogin, *New York Times Book Review,* p. 7.
7. Herbert Marcuse, *One Dimensional Man* (Boston: Beacon Press, 1964), p. 227.
8. Jacques Ellul, *The Technological Society,* introduction by Robert Merton, trans. John Wilkinson (New York: Random House, Vintage Books, 1964), p. xi.
9. Marcuse, *One Dimensional Man,* p. 154.
10. Leif Sjöberg, "Harry Martinson: Writer in Quest of Harmony," *American Scandinavian Review,* Winter, 1972-73, p. 366.
11. "Roger Garaudy and His Epic of Mankind," Center for the Study of Democratic Institutions, *Center Report* (Santa Barbara, Calif., 1973), pp. 13-14.
12. Christopher Lasch, "Inequality and Education," *New York Review of Books,* May 17, 1973, p. 24.
13. Martin Nolan, "A Code Word for Objectivity," *Village Voice,* Apr. 29, 1971.

14. Zinn, *Politics of History,* p. 40.
15. *The University-Military-Police Complex* (New York: North American Congress on Latin America, Inc., 1970), p. 2.
16. *Ramparts,* Apr., 1966, pp. 11-14.
17. Ibid., p. 14.
18. Princeton University Office of Research and Project Administration, *Summary: Department of Defense Research July 1, 1968-June 30, 1969 Appendix A.*
19. Ibid.
20. Michael Klare, "The Military Research Network," *The Nation,* Oct. 12, 1970, pp. 327-32.
21. The Institute of Defense Analyses (I.D.A.) was created in 1956 as a nonprofit corporation by M.I.T., Stanford University, Tulane, Case Institute, and the California Institute of Technology. Princeton joined the corporation in 1962 as did the Universities of Michigan, California, Illinois, Chicago, Columbia, and Penn State. (See *The University-Military-Police Complex* for a more complete picture of I.D.A. and its relation to institutions of higher learning.)
22. Klare, "Military Research," p. 328.
23. Ibid., p. 330.
24. *Ordnance,* Nov.-Dec., 1970, p. 245.
25. Ibid., p. 245.
26. *Ramparts,* p. 13.
27. Gunnar Myrdal, "Peace Research and the Peace Movement," Center for the Study of Democratic Institutions, *Center Report* (Santa Barbara, Calif.: June, 1974), pp. 5-6.
28. Noam Chomsky, "Watergate: A Skeptical View," *New York Review of Books,* Sept. 20, 1973, p. 8.
29. *Six Million Victims* (New York: Indochina Peace Campaign, 1972). Primary sources for this document, prepared by Project Air War, came from the Pentagon Information Office and the U. S. Senate Subcommittee on Refugees.
30. Margaret Mead, testimony before the U. S. Senate Subcommittee on Labor and Public Welfare conducting hearings on "American Families: Trends and Pressures," *New York Times,* Sept. 26, 1973, p. 19.

David Joravsky

What Do We Ask of the History of Technology?

Too often the historian of ideas goes looking for the influence of great ideas in the historical process and winds up discovering the far greater influence of petty ideas, the cliché-ridden thought patterns of the masters of wealth and power. In the twentieth century great ideas have become the toys of sequestered intellectuals, while the world at large is ruled by devotion to wealth and power and the kind of technology that serves wealth and power.

You may say that this is nothing new, that the world has always been this way. I don't think so. The value of great ideas has become so dubious that creation of them has virtually ceased. In our time there are many fine minds but no great creative thinkers. The neural circuits still discharge as intelligently as ever: see how technologists solve the problems fed to them by the masters of wealth and power. But the creative intellect, when it functions at all, expresses an overwhelming sense of its own futility.

In the nineteenth century passion for great ideas was still strong, but the premonition of its end can be found in the widespread efforts to reduce thought to some nonmental process, to show that it is a function of social or biological processes. Marx, to take the most famous of the social reductionists, made a great idea of his disdain for the self-sufficiency of ideas and strove to show that changing ideas are a product and servant of changes in socioeconomic formations, with changing technology acting as the ultimate determinant of the process. Marx wedded this potentially anti-intellectual doctrine to passionate faith in the creation of a just and free society by working-class revolution.

His followers have divorced these two parts of his thought. Those who share Marx's faith in justice and freedom are impotent dreamers in intellectual ghettos; those who have achieved political power turn that faith into a ritualistic incantation while they cleave with genuine passion to the belief that ideas are justified only if they help to increase wealth and power. That is the essence of Stalin's major contribution to Marxist thought, his reinterpretation of the principle that practice—or praxis—is the criterion of truth. Students of Soviet history, appalled by this degeneration of Marxism into a weird cousin of the businessman's ethos, frequently seek reassurance in the history of technology, trying to find in technology forces that must push the masters of wealth and power toward respect for justice and freedom—even for the creative intellect.

Let me show you how this happened in my own case, how an investigation of Soviet intellectual history led—to my surprise—into a study of technology as the main determinant of the ideas I was investigating and ultimately led me to hope that the power of technology to shape ideas might become greater than it has been so far in the Soviet Union. The hope, as you will see, is rather shamefaced, for it involves derogation of the intellect.

I began with a study of Soviet philosophical controversies in the 1920s. I discovered an interesting contest of positivist and metaphysical interpretations of Marxism, oriented, on the practical side, toward the new rulers' problem of winning the allegiance of "bourgeois specialists" (the Bolshevik name for the inherited class of scientists, technologists, and the like) while creating a proper Weltanschauung for a new class of "red specialists."

That practical orientation did not harm the philosophical discussions of the twenties; if anything, it gave them a breadth, a resonance of social significance, reminiscent of such discussions in the nineteenth century. But suddenly, at the end of 1929, Stalin demanded a new kind of practicality from intellectuals. The presumption of a long-term utility of autonomous intellectual endeavor was denounced as a bourgeois delusion, an excuse for failing to provide immediate aid to the feverish construction of industrial socialism. The old Marxist dictum that practice (or praxis) is the criterion of truth was reinterpreted in a very narrow, authoritarian way.

Since the supreme form of practice was political rule, it followed that political rulers were the supreme readers of the lessons of practice, the ultimate judges of truth in all lesser realms. The bosses did not choose to exercise their wisdom with equal vigor in all fields. At one extreme, in the humanities and the social sciences, they quickly and almost completely crushed the autonomy of intellectuals. At another extreme, they barely touched mathematics, while physics and chemistry experienced only minor turmoil on their philosophical frontiers. I discovered that the Stalinist bosses, for all their authoritarian bluster, were rather reluctant to tell natural scientists and technologists what truths were dictated to their disciplines by practice.

Biology and agronomy were exceptions to the rule. In those disciplines the Stalinist bosses did intervene, in the mid-thirties, forcefully promoting their reading of the lessons of agricultural practice. Thirty years later, in the mid-sixties, they finally gave up such efforts, restoring to biologists and agronomists the autonomy enjoyed by other natural scientists and technologists. To explain that anomalous campaign, I was obliged to study the bosses' experience with agriculture.

At the beginning of the long process it seemed possible to destroy the autonomy of peasants while preserving that of scientists and technologists. Leading biologists and agronomists endorsed the Stalinist conviction that the peasants' antiquated social organization had to be abolished—not only because it obstructed the greatest possible extraction of produce from them, but also because it obstructed the modernization of agricultural techniques. Indeed, the two considerations were closely linked. Rapid collectivization, it was thought, would rapidly improve the techniques of production, thereby enabling everyone to have his penny and his pie. The state could increase its exactions from the peasants while leaving them sufficient residual income to make them appreciate the advantages of the new order. The Stalinists perceived modern agronomy and biology, coordinated on the pattern of the U.S.D.A., as an essential element in forceful collectivization. Anyone looking for evidence that modern science and technology *can* serve as instruments of oppression and exploitation will find abundant evidence in the collectivization drive of 1929-33.

But very quickly a rift opened between the political bosses, on the one hand, and the agronomists and biologists, on the other. The Stalinists were bitterly disappointed. They had achieved a great increase in the state's share of the peasants' produce, but they had not achieved a great increase in the peasants' productivity. Quite the contrary: productivity fell sharply in the early years of collectivized farming, and then, returning to the primitive level of traditional peasant agriculture, tended to stagnate there. The Bolsheviks turned against modern agronomy and biology because practice had proved them useless. Bolshevik support went to cranks like Lysenko and Williams (or Vil'iams, the son of an American railroad engineer who came to Russia in the mid-nineteenth century to help build the first major line). They promised spectacular increases in yields to be achieved by simple nostrums; the only significant cost of the increases was to be the largely unpaid labor of peasants and the energy of the straw bosses who mobilized that labor. Genuine agronomists and biologists were subjected to the power of such cranks, thus suffering a loss of autonomy as the peasants had, and for the same reason: they had disappointed the Stalinist leaders. Anyone looking for evidence that modern science and technology *cannot* serve as instruments of oppression and exploitation will find abundant evidence in the protracted aftermath of the collectivization drive, the Stalinist turn from genuine agronomy and biology to crackpot agronomy and pseudobiology.

Should we conclude that modern science and technology are neutral in their social uses, like the proverbial ax that can be used by forester or murderer—or thrown away in frustrated rage by an incompetent user? I think not. Soviet experience seems to me to support the view that modern science and technology are incompatible with the kind of oppression and exploitation that Stalin devised. Note how quickly the Stalinists grew disillusioned with genuine agronomy and biology, and then note how long it took them—thirty years—to grow disillusioned with crackpot technology and pseudoscience, in spite of their constant failure to produce the promised results. Modern technology required two things that the Stalinist bosses were reluctant to concede: a major increase in the peasants' share of agricultural produce and a major increase in the autonomy allowed to peasants, agronomists, and biologists. (Also involved in the Stalinists' protracted clinging to crackpot agronomy and pseudobiology was the usual reluctance of political leaders to confess they had made an enormous blunder.)

I am suggesting, as I have put it elsewhere, that "there is something in plants, soil, farmers, agronomists, and scientists that requires rationality and genuine self-criticism of those who would manage them. My hope is that this inherent necessity is limitless in scope, that it will make itself felt in areas of Soviet life where it now seems utopian to expect it" (Karcz, ed., *Soviet and East European Agriculture,* p. 172).

This kind of hope is supported by other case studies of Soviet intellectual history. Pavlovian psychology, to take the case I am currently investigating, is widely supposed to have provided Stalinist bosses with methods of conditioning their subjects. Not only is this myth completely untrue, but the real history of Soviet psychology follows the pattern of agronomy and biology. For a comparatively short time—in the twenties and early thirties—Bolshevik leaders favored the vigorous development of experimental psychology with the expectation that it would be of great practical use in education, occupational placement, propaganda, medicine, and engineering. They were soon disappointed, however—most bitterly so in education, where IQ tests (the most important technical device that has issued from experimental psychology) predicted that the children of workers and peasants would not do so well in school as the children of the professional and white-collar class.

In the mid-thirties IQ tests were scrapped, and experimental psychology was denounced as bourgeois pseudoscience. A genuine science of psychology was supposed to be developed from the intuitions of teachers like Makarenko, who had allegedly discovered in practice how the least promising of lower-class children, delinquents, could be turned into educated, dedicated participants in socialist construction. The Pavlovian school—or rather, ossified segments of the Pavlovian school—were endorsed for their supposed contributions to neurophysiology and medicine.

In these fields the Bolshevik retreat to standard science and technology has been slower and less complete than in agronomy and biology, if only because the uses of experimental psychology in the educational systems and the medical practice of Western countries are nowhere near as obvious as the uses of agronomy and biology in Western agriculture. Nevertheless, the post-Stalin leadership has moved a considerable way toward the restoration of autonomy to experimental psychologists and neurophysiologists. They have been impelled in part by the inadequacy of intuition and bluster in modern education and medicine, in part by recognition that experimental psychology may not do much good in those fields, but it doesn't seem to do much harm either.

The most well known case of modern technology and science pushing Soviet leaders toward decentralization of authority is, of course, industrial management and planning. Back in the twenties Soviet economists made exciting breakthroughs in planning, only to be cast aside when the Stalinist leaders plunged toward instant industrialization. The necessity of retreat toward rationality, of restoring some autonomy to economists, was already felt during Stalin's lifetime. Since his death there has been a fairly vigorous rebirth of economics, impelled by accumulating evidence of gross inefficiencies in the extremely centralized, authoritarian system of planning and management that the Stalinists created. Thoroughgoing reforms have been stalled by the reluctance of the central authorities to make the choice that leading Soviet economists are urging upon them: a choice between centralized power, which entails an increasing economic slowdown, and decentralization, which promises technological improvement and economic growth.

I am somewhat ashamed to offer this analysis, for it seems to measure the value of ideas by their contribution to economic growth. I long to agree with Lev Ventsov, a Soviet dissident who recently expressed his passionate belief that "culture can be not only a means but also an end." He declared his faith in "the self-sufficient value *(samotsennost')* of the spirit and in the spirituality of life," his belief "that social circumstances come and go, but the achievements of free human creativity remain, that crude physical power is powerless against truth, justice, and beauty" (quoted, from an unpublished *samizdat* manuscript, in Roy

Medvedev, *Kniga o sotsialisticheskoi demokratii,* p. 85). That faith is a reproach to those who regard the intellect either as an instrument for building wealth and power or as a harmless toy of the intelligentsia.

When the intellectual historian is drawn into the history of technology as the ultimate determinant of a society's shifting patterns of thought, he is willy-nilly sharing in that society's contempt for the self-sufficient value of the intellect. It is small reassurance to say that I can hardly do otherwise, for that is the way the people whose history I am analyzing think. Nor is it great reassurance to find in technology grounds for hope that the Soviet masters of wealth and power may come to resemble our own: that is, they may be learning the possibility of dropping thought control because free thought in a well-developed industrial society serves wealth and power and is otherwise impotent.

Heather Lechtman and Arthur Steinberg

The History of Technology: An Anthropological Point of View

I think it is fair to say that most anthropologists, at least in the United States, look upon culture as their master concept.[1]

Introduction: An Anthropological Framework

While acknowledging the concept of culture as perhaps anthropology's most important and, indeed, most useful contribution to the understanding of people as social beings, anthropologists have interpreted the concept in a wide variety of ways. Moreover, the discipline often seeks to review and reevaluate this notion, this tool that has proved so fundamental and generative to the unfolding of the "science of man." [2] In spite of the array of definitions of culture, however, the argument has been made and made well that "if we put aside what anthropologists assert to be the essential nature of culture . . . and instead look at how they actually work with the concept in the context of their empirical research, we will find a greater degree of agreement than the bewildering array of definitions . . . might suggest." [3] Culture turns out to be a powerful, integrative formulation through which we can describe and understand the patterned and interrelated traditions within human communities. Certainly we would all agree culture is that uniquely human mechanism by which people have adapted to the circumstances of the world around them. Culture, in this sense, can be thought of as an adaptive strategy, a strategy not only for survival but also for growth and elaboration.

On the other hand, when an anthropologist talks about any specific culture, he or she is generally referring to a distinct group of people that exists in a particular environment, is organized in some structured pattern of social arrangements, and shares a common set of values. Any group of people that can be said to participate in a common culture thus participates in a particular concatenation of these three realms—environment, social structure, ideology—that is characteristic of and peculiar to the group and serves to distinguish the group from any other.

Moreover, these realms are systemically interrelated within any given culture; that is, the interaction of one realm with another shapes a series of cultural phenomena that constitutes a system whose parts affect and influence each other. According to such a holistic view, as Steward has been careful to point out, all aspects of culture are functionally interdependent but the degree and nature of interdependency are not the same with all features.[4] In fact, the interrelations of these realms in specific historical contexts have led to a wide variety of sociocultural forms. Any such system is, of course, capable of changing, and as it changes the patterns of interrelation alter.

The fact that anthropologists tend to view cultures systemically derives in part from their insistence on understanding them as wholes and characterizing the synthesis of their parts or subsystems. It has often proved convenient to identify and study individual subsystems, but a unique thrust of anthropology lies in its ability to see and interpret the processes of interaction and synthesis.

We are concerned here with that subsystem of culture known as technology, the specific patterns by which it unfolds, and its interrelations with other cultural subsystems. Technology is a subsystem of culture just as religion, art, mythology, social structure, and values are. As Robert Merrill says in his article in the *International Encyclopedia of the Social Sciences:* "Technologies are the cultural traditions developed in human communities for dealing with the physical and biological environment. . . . [they] are important not only because they affect social life but also because they constitute a major body of cultural phenomena in their own right."[5] Technologies are thus part and parcel of the mainstream of cultural inclinations and are irrevocably bound to the social setting in which they arise.

One of the issues we see as uppermost in understanding the nature of technology within cultural settings is the integration of any technology or system of technologies with the cultural matrix in which it is manifest. The question of integration is concerned with the extent to which any technology has a life and a style of its own, bringing with it a set of inherent properties or characteristics that are inescapable and independent of specific cultural milieus. Some of these characteristics may be grounded in the physical properties of matter itself; others may relate to the energy sources necessary for the technical manipulations taking place. On the social structural level, there is the difficult but crucial question of whether specific technologies require and predict certain organizational forms or whether the social organization of technology derives from that of other cultural subsystems, such as kinship or religion. Within the range of expressions any technology may take, how much of a given manifestation is predicted by its materials and energy systems and how much by organizational modes, value systems, mythologies, or science?

The issue, then, arises from an attempt to isolate the deterministic aspects of technologies—if in fact they exist—to describe the inescapable components that appear to be built into any given technological system, and to evaluate the way in which they have been culturally informed. We would like to ascertain the extent to which technologies have internal forces that drive them and the extent to which culture is determinative in shaping their content and structure. Where does the balance lie between cultural values and cultural choices on the one hand and the realities of the physical world on the other? How can we describe the cultural or the noncultural component of technology? What are the internal consistencies of a technology, and what is its cultural style?

We can evaluate the relations between the imperatives of technology and those of culture only by understanding thoroughly a given technology and a given culture. Both are certainly within our capabilities. As anthropologists and archaeologists of technology, we must know technologies from the inside just as we have traditionally attempted to know cultures from the inside. However, we must always work within the framework and orientation that is essentially anthropological; that is, from a definition of technology as an institution of culture.[6]

Let us be quite clear that in examining those elements of technology that may appear to be culture independent we are not subscribing to the currently common view of technology and society as separate aspects of human life that affect one another, but that behave quite independently of each other, each with a will and a life of its own. There is no point in talking about technology *and* culture or about technology *and* society as many people and some academic programs currently do. We have been brainwashed into conceiving of modern technology almost as a willful force that compels us into courses of action and can later be blamed and used as a scapegoat when those actions backfire or prove disastrous. It is difficult to be objective about our own system, particularly since it is so complex, but, as Merrill argues: "Deliberate technological change has been institutionalized in Western Societies for some time. Most modern technologies include not only traditions for making and doing things but also traditions for 'advancing the state of the art,' for producing new knowledge, processes, and products. [Thus] modern technologies are culture-producing as well as culture-using sociocultural systems."[7]

The rapid technological changes we are currently living through are not the result of a technology run wild—free of the control of men, wreaking havoc in a random way with our cities and our countryside, to say nothing of annihilating subject peoples—but rather an institutionalized form within a much larger cultural system. To the extent that industrialism and capitalism have become worldwide, overriding geographic and ethnic barriers, contemporary industrial technology tends to look alike to us wherever we find it—to partake of a monolithic quality, a quality of separation from the rest of society—and to be invested with a personality and a will. The fear and anxiety people feel and express about technology today, although real, are of course misplaced, but it is convenient to blame technology rather than to see it as symptomatic and expressive of value systems and orientations that characterize our world view.

In asking what the cultural component of technology is, we are also asking what technology can tell us about culture. We must be concerned not only with the bodies of skill and knowledge Merrill speaks of, not only with the materials, processes, and products of technology, but also with what technologies express. If we claim that technologies are totally integrated systems manifesting cultural choices and values, what is the nature of that manifestation and how can we "read" it?

We all recognize that human beings have developed many modes of cultural expression and communication. Language predominates among these, and as such has traditionally come under close scrutiny by anthropologists, most recently in an attempt to relate modes of cognition with linguistic modes of expression.[8] But visual art, music, dance, costume, and gesture are also important and fundamental communicative systems. Some ideas, some orientations, can only be expressed nonverbally, and it remains for the new field of aesthetic anthropology [9] to demonstrate the congruence among such systems within a given culture as well as their relationships to fundamental patterns of cognition, learning, and performance. We would argue that technologies are also particular sorts of cultural phenomena that reflect cultural preoccupations and express them in the very style of the technology itself. Our responsibility is to find means by which the form of that expression can be recognized, and then to describe and interpret technological style.[10]

Material Culture—Methods of Investigation

At the moment, material culture as a category of phenomena is unaccounted for. It is scattered between interior decorators, advertising firms, and historians of technology. But when one considers how little we know about how material culture articulates with other cultural subsystems, one begins to see the potential. There exists a completely empty niche, and it is neither small nor irrelevant.

Should archaeology become the science of material objects or technology, many of the aims, problems, methods, and data of the field would be completely transformed.[11]

If there is a sense in which we are asking a great deal of technology, a sense in which we are trying to mine it for all it's worth, this stems in part from our situation as archaeologists. For us, culture has been reduced for the most part to the material remains of people's existence. Our data are in the nature of artifacts, whether these be pots or roads, sculpture or irrigation canals. The one element common to all of these material things, however, is that they are all products of technological processes. Each is the result of some human technology, each involved work of some kind or play of some kind. Since certain aspects of any technology remain an inextricable part of its products, appropriate study of artifactual materials starts us on our way to a fuller understanding of early technologies and, thereby, of the societies that utilized them. Even if there is little else that we can reconstruct about a culture from examination of any one of its artifacts, we *can* say something about the technology that produced that artifact—often a great deal. Every object retains, at least in part, the history of its own manufacture.

Lynn White has elegantly chided fellow historians about methods and sources of information appropriate to their work, yet rarely used by them: "If historians are to attempt to write the history of mankind, and not simply the history of mankind as it was viewed by the small and specialized segments of our race which have had the habit of scribbling, they must take a fresh view of the records, ask new questions of them, and use all the resources of archaeology, iconography, and etymology to find answers when no answers can be discovered in contemporary writings." [12] We agree. As impoverished as archaeology must always remain, by the very nature of its sources of data, it has nevertheless failed to exploit that single area in which it is unquestionably rich and through which it may be led to a deep understanding of the technological aspect of man.

There is only one way of extracting an object's technological history from it, and that is by appropriate laboratory studies, where we assume that the laboratory includes the excavation site and the science lab. One of the reasons we have such an insubstantial fund of information about basic technological processes is that few archaeologists have ever cared to excavate a site primarily for the technological activity it might reveal.

The idea of excavating those areas of settlements or cities where workshops may be found or where craftsmen may have lived has not been a motivating factor in the selection of sites; nor have many archaeologists bothered to investigate such areas thoroughly at sites which are already under investigation.[13] Religious and ceremonial complexes, monumental architecture, and cemeteries are still favorite digging grounds; except for occasional surveys or excavations at pyrotechnological sites, no one goes looking for technology.[14] Because it is crucial to understand any system of technology within a specific ecological setting, we must increase our efforts to excavate appropriate sites.

The study of the development of metallurgy, for example, can no longer rely so exclusively upon analysis of the products of metallurgy, i.e., metal artifacts. We must systematically look for mines, ore-processing areas, furnaces, slag heaps, cupellation hearths, and so forth, and we must aim to describe such complexes in terms of the particular characteristics of the environments as well as to seek in them information about the social organization they entailed. The laboratory for technological analysis must, in short, be moved into the field.

Another factor in our paucity of information about early technological processes has been the lack of sufficient laboratory facilities to undertake the necessary analytic research. There are few scientists with sufficient sophistication in archaeology to be able to interpret the data they obtain experimentally and far fewer archaeologists sufficiently sophisticated in the rigors of laboratory regimes to undertake the necessary analytic studies. To understand a technological process, one must understand it in its entirety, including the most technical or scientific aspects of it. One then begins to appreciate the inputs of materials and energy and to evaluate the latitude for error, the range within which substitutions or alterations are allowable, the mechanisms of quality control, and so forth. As Brooke Hindle has so forcefully argued, "The greatest need is to stand at the center of the technology—on the inside looking out."[15]

Scale in the Study of Technology

Within the framework we have outlined, studies of technology must be conducted on at least two levels: general, large-scale, historical developments on the one hand, and microstudies of the cultural manifestations of certain technologies on the other.

As an instance of the way technologies have been used to characterize or even explain historical change at the most general level, consider that archaeologists have always referred to specific time periods by the natural or man-made materials they see as indicators of major changes in man's utilization of his environment: stone, ceramic, copper, bronze, and iron ages or eras. But how meaningful are these designations? For example, metallographic examinations of Luristan iron swords from first-millenium B.C. Iran have shown that these swords are not purposefully carburized along their edges to harden them; that they were less useful for cutting or stabbing than similarly shaped bronze weapons; that, moreover, these swords imitated the shapes of cast bronze daggers made before the introduction of iron, their hilts sometimes sheathed with bronze to make them look like the copper alloy. Most significant, however, is the fact that they were fabricated from many individual, separately forged pieces of iron which were assembled by techniques characteristic of bronze joining. Luristan smiths did not exploit the particular properties of iron which permit versatility in hammer welding it, thus missing one of the most highly developable aspects of the metal.[16] To what extent, then, is the Luristan culture an Iron Age one? Is it reasonable to use a technological classification for a culture which, technologically at least, barely took advantage of the very technology we attribute to it?

To take another example, there appears to be a direct correlation between the development of food production and sedentary life and the first uses of clay,[17] so that to distinguish between ceramic and aceramic or preceramic cultures is a crucial distinction. Do we imply similar vast cultural alterations in the change from the Bronze to the Iron Age? Is there anything in the technologies of these two metals and in the consequences of their use that could compare with the marked evolution apparent in the shift from a nonceramic to a ceramic complex? Such classification systems as these appear to be internally inconsistent and are based upon technologies whose utilization and cultural impact varied markedly in time, in place, and in emphasis. Pleiner, for example, has shown that although iron was used in Greece from the twelfth century B.C. on, and used fairly extensively from the eighth century on, one cannot speak of a "full-fledged iron age" until sometime in the fifth century when iron was employed in virtually every aspect of Greek technology.[18] If we are to use such systems, we must invest them with a rigor based on scientific evaluation of the technologies we are referring to in the particular settings in which they occurred. Instead, archaeologists have tended to equate the mere presence of a material as sufficient grounds for classifying a culture within such general schemes. We must understand individual technological processes and their cultural impact before their bearing on cultural development can be assessed and certainly before they can be used as indices of such development.

Another general scheme of classification relying upon technological change has been used to demarcate stages of cultural evolution with the focus on the food-procurement technologies of foraging, herding, gardening, and irrigated farming. Here again, however, it is subtle interactions of a number of different technologies with different internal structures and forms of human organization that are the forces of change—not merely the food-procurement technologies alone. Accordingly, we might suggest that the gradual change from food-collecting to food-producing systems in settled villages in the Zagros mountains between the tenth and seventh millennia B.C. should be studied not only from the floral and faunal remains which serve as the primary data indicating the actual change in food-procurement systems, but also from the evidence of other associated technologies. The observable developments in the stone industries that add pecking and grinding of new tools and vessels to the old flaking tradition, developments in the manufacture of containers (where basketry and wood seem to precede ceramics), changes in the permanence of living structures (with a concomitant change in building materials and techniques), and general changes in settlement patterns all add up to profound technological changes which accompany—whether in a cause or effect relation we cannot say—this evolutionary sequence.[19]

A similar nexus of technological change in food production, stone-tool industries, and containers is apparent in the Tehuacan sequences[20] and apparently also in the Oaxaca assemblages,[21] and this set of interrelations could be probed and developed much more intensively than it has been in the past. More striking still is the technological complex surrounding the development of urbanization in southern Mesopotamia in the fourth millennium: this includes irrigated agriculture, large-scale ceramic industries, metallurgy based on extensive trade, an important textile industry, monumental building with imported stone, the development of a writing system, long-range communications networks via the irrigation waterways, and so on. All those other technologies should be focused on now as carefully as scholars have looked at the technology of food production in the past, for it is in the interaction of these various technologies that we can learn much more about cultural processes such as task specialization, social differentiation, shifting settlement patterns, labor organization, long-distance trade, colonization, imperialism, and diffusion.

In this connection we should note that while the stress on food procurement in the anthropological and archaeological literature is, at least in part, a result of a preoccupation with simple societies, our own studies tend to be concerned instead with complex, urban societies in which it appears to be precisely occupational (i.e., technological) specialization that lies at the root of social differentiation and stratification. As a result, our understanding of the dynamics of the "urban revolution" in the ancient Near East or Far East or of the "industrial revolution" in western Europe or the United States must, in fact, be built on a close knowledge of the cultural role of a number of basic technologies. For the former we have already noted several technologies dealing with the exploitation of new materials, whereas in the latter case the concentration must be instead on a host of new machine and power technologies, all of which are concerned with the exploitation of new and more efficient sources of energy. The understanding and evaluation of these major changes in man's development are then to be studied as the interactions of these technologies with each other and with other cultural institutions.

Approaches to the Study of the Archaeology of Technology

The preceding discussion of large-scale technological developments inevitably points to the fact that our ability to understand such changes in depth will rely largely upon a prior set of many smaller, specific studies of individual technologies and groups of technologies. These kinds of data still remain to be collected both in the field and in the laboratory. The following are several approaches we consider profitable both for collecting such data and for interpreting them.

A Single Technology in One Culture Area

Clearly one fundamental mode of specific study of a technology is that which looks at its development within a single culture area, documenting the various changes that occur over a certain period of time and attempting to show how these relate to other cultural changes. Though such a study may begin with the laboratory examination of artifacts, accompanied by archaeological investigations of relevant "technological sites," it will inevitably raise basic questions about technology as a complex cultural process. Some of these questions will be concerned with the processes of invention and innovation, with those factors that make technological change acceptable to a culture at a given time, with the climate of receptivity to technological change. How is it, for example, that certain technological complexes are incorporated wholesale into a culture while only portions of others are used? Or what is it in a culture that causes certain techniques to be perpetuated even when the materials usual to those techniques and employed in the fabrication of objects changes, so that the techniques are no longer the natural ones for the new material yet they go on (as in the case cited above of the Luristan iron swords)?

Still other questions will be concerned with the internal variations in a technology as we see it in operation at different levels or in different zones of a large, complex culture area. How, in fact, is a technology disseminated from a center to outlying areas? What is different about a technology practiced in a city from the same technology in a rural environment; how different are the technologies of imperial workshops and provincial ones? Are these not different in more than just scale and level of production?

Of course, once we deal with complex societies, the number of interacting technologies is greatly increased, and our problems of reconstruction become equally complex. We have to look at the rates at which technologies change within a society and how those different rates affect one another. We are not even aware of the extent to which the techniques of individual technological systems influence each other within a specific culture setting. To what extent do they develop separately according to their own sets of properties or, alternatively, along strongly culture-bound lines? For example, is there any way in which fundamental changes in metal technology—as, perhaps, the change from forging to casting a metal or the change from an exclusively copper-oriented metallurgy to a copper-and-iron metallurgy—affect ceramic production or weaving or domestic building? These questions need not be far-fetched, especially when we notice the extent to which stylistic and iconographic borrowing among crafts has been so common.

Comparing Several Technologies within One Culture Area

An exciting comparative study might involve a consideration of the relations between the fundamental pyrotechnical arts—ceramics and metallurgy. Did the first smelting of copper appear where high temperatures were commonplace in the potters' kilns? How much do the metallurgists' clay moulds owe to contemporary potters' skills? Is there any relation between the development of the alloying of metals and a wider use of colored minerals in decorating pottery? Are the differences between the developments of metallurgy in the Americas and in the Near East related to the associated different ceramic traditions?

How much contact, if any, has there ever been between potters and smiths, and what has been their relative role or position within particular societies? Certainly they have expressed in their works common stylistic preferences, common iconographic schema, common decorative elements. But, aside from sharing in these formal aspects of style which were part of their cultural ambience, do we also find that they influenced one another more directly in the purely technical sense? For instance, is there any evidence that the desire to execute an object in metal so that it looked like objects previously executed only in clay produced certain constraints in the metal technology, so that, rather than developing entirely along lines that might have been most "natural" to it, metal was made to conform to criteria related to the technical and stylistic norms of a totally different material?

Here, of course, we are back again to questions about the "style" of technology. Whether artisans work a material according to techniques and properties natural to that material or whether they use techniques basic to other materials or a technology that imitates another are important considerations. Problems of the full mastery of materials and processes set against the necessity to make things that look like something else or that function like something else or are thought to be something else are fundamental to understanding the cultural aspect of technology.

The Transfer of Technology

Finally, we must address ourselves to the mechanisms of diffusion of any technology we may study. The question of technology transfer has been nicely treated by Brooke Hindle in a recent paper [22] in which he distinguishes between the importation of "mechanics" from England and Europe to develop new technologies in nineteenth-century America and their ultimate achievement of those goals, as opposed to the importation solely of machines and processes into a technological environment that was simply not able to cope with them on its own, though it tried with far less success to adapt itself. Or, from ethnographic contexts we can see that itinerant craftsmen may often have been important to the spread of technologies. A fine example of a revolutionary technological innovation carried by itinerant craftsmen, and then soundly established in other cultures that were fully prepared for it, is the rapid dissemination of moveable type by itinerant German printers in the Italian city-states of the fifteenth century. Matson [23] has shown in his studies in Afghanistan that pottery styles and techniques are carried about by itinerant potters who, by bringing their technology with them, can either satisfy local tastes with local wares or, conversely, can disseminate one kind of ware over vast regions. Thus, though the products of the technology may appear different in different regions, they are still the work of the same small group of craftsmen, and it is not at all clear to what extent the technological expertise itself is actually disseminated. A similarly ambiguous situation prevails in parts of Crete where, a generation ago, potters moved from village to village setting up shop to produce the local wares, while now one kind of ware is produced in a village such as Thrapsanon and distributed by truck quite widely over the island, making for a uniform pottery distribution, but hardly a spread of technology. We must pay close attention to such variations in the archaeological record and be prepared to understand them in terms of such mechanisms of diffusion.

Cross-cultural Comparisons

Problems of technological transfer or spread point out the importance of a cross-cultural approach to the study of technology. By documenting carefully the technical aspects of a given technology in specific culture settings and noting how similar or how different those aspects are in each we may begin to shed some light on that vexing question of the extent to which a technology's development is determined by the physical properties of materials and processes rather than primarily by cultural forces. A study that would be interesting in this regard involves the particularly fascinating technological complex associated with wrought iron, cast iron, and steel. A comparative investigation of the development of ironworking in the Far East and in the West is absolutely necessary if we are to form any coherent explanation of the motivating factors that led to the casting of iron very early in China, where it was perfected and maintained in addition to forging, while in the West forging iron was the only process utilized until the fifteenth century A.D., when casting was introduced.

Was there, in fact, a choice to be made in each case between casting and forging or were other factors so pressing that the choice never really existed? What were the kinds of objects first made of cast iron by the Chinese? Do they bear any relation to the objects they were so skilled in casting in bronze? Is there any sense in which the iron technology can be understood, at least initially, as a carry-over from a highly refined and status-bearing bronze manufacturing industry? What can technical examination of cast-iron objects tell us about such cultural factors that may have been operating? Certainly it will give us a good idea of the capabilities of the oriental foundrymen and the state of their art. They were able to produce temperatures in their furnaces high enough to melt iron and cast it, whereas in the West, even though the high temperature necessary for the casting of iron could be attained, the much-lower-temperature forging process became entrenched as a matter of choice and never really yielded until the modern period. Joseph Needham's [24] masterful work *The Development of Iron and Steel Technology in China,* which is so far the definitive treatment of this subject, deals mainly with the cultural and historical developments of the technology; and even though he chronicles the technological changes that took place, he does not consider the technical factors that influenced those changes.

On the other hand, bronze casting was also an important tradition in the Near East; yet there ironworking seemed to emerge technically quite independent from casting, though it imitated in forged iron many of the cast-bronze forms. Conversely, it is during the first millennium B.C., when iron forging came into its own, that the forging of bronze suddenly became popular too. This was especially the case in an area stretching from northern Iran through eastern Turkey and across northern Syria to the Levant—very likely the area in which ironworking was also developed. In unravelling these technological and cultural connections, will further excavations and laboratory studies point to technological "determinism" or to cultural predilection as the prime operator in each case?

Ethnographic Parallels

Sometimes archaeologists are fortunate in being able to supplement their excavation and laboratory research with ethnographic or ethnohistorical evidence. While everyone is aware of the caution that must be maintained in any recourse to analogy between contemporary and ancient situations of whatever kind, it is often the case that the very conversatism of technology, its strong dependence upon tradition and the social institutions devised to maintain it almost statically, have sometimes resulted in the preservation of certain technical systems over incredibly long periods of time. In such circumstances, evaluation of the ethnographic situation can be extremely helpful in suggesting hypotheses, for instance about the social organization of technology, that must then be looked for in the archaeological record.

For example, one such study, which was conducted among the Diola potters of southern Senegal by Olga Linares,[25] was valuable because by observing manufacturing techniques she was able to look for evidence of them in the archaeological artifacts and could get a reasonable idea of the per capita production of a single woman. Moreover, she was also able to distinguish between the ways in which pottery knowledge was transmitted among the women of an isolated village where marriage was exogamous as opposed to a large, accessible one in which marriage was primarily endogamous. From these marriage patterns she could determine differences in the way standard forms and techniques were diffused from these two areas, one by circulating the women, the other by circulating the pots. And, of course, she could establish the fact that the pots were made by women on a seasonal schedule—and all that that entails in terms of social consequences.

Another recent study, one which attempts to use the descriptive terminology for various clays currently used by potters in the Yucatan to help reconstruct Mayan sources and uses of clay for pottery and building, combines ethnographic, linguistic, and historical data.[26] The increasing interest of archaeologists in the value of ethnographic data as possible aids in solving archaeological problems in technology is illustrated by a number of articles in *World Archaeology*.[27] In these, archaeologists have tried to create an environment in which to interpret the role of ancient craftsmen. Only careful excavations will allow us to test the reliability of such interpretations.

The Individual as Technologist

The most elusive aspect in the technology-culture relation is the role of the individual as craftsman, technologist, inventor—call him what you will. Cyril Stanley Smith has singled out as one of the important reasons for studying the technology of works of art the pleasure a viewer derives from understanding the workman's experience.[28] The extent to which we can ever really recall that experience, however, especially by working solely with archaeological artifacts as opposed to artifacts that can still be studied within an ethnographic setting, is very much in question. We may be able to achieve some of that sense by observing a man or woman at work in his or her own environment,[29] but can we really expect to understand, by reproducing in the environment of a laboratory something similar to the product of an individual's skill, anything of the internal, psychological experience of a craftsman as he works or even about the innovative process? It is rather doubtful that we can.

On the other hand, the relation between an individual and the materials he works with is a one-to-one relation in many cases. A single pot is generally made by a single man or woman, although obtaining the clay and preparing it for use may have been the task of someone else, as may the eventual firing of the pot and its final decoration. Nevertheless, as there is no intermediary between the individual and the artifact during the actual process of manufacture, we can certainly gain some insights into individual skill and craftsmanship—the level of control of the properties of the material, the extent to which its potential is exploited, the sense of play with the medium itself, the meticulousness and attention to detail or the casual or even shoddy quality of the work. Some of this, of course, depends not so much on the care and capability of the individual craftsman as on the function of the object and on who uses it.

We do not expect to find the same care given to the execution of common ware pottery for hearth cooking as we do to a ceramic pot for ceremonial use. But, as Deetz has pointed out: "It is unlikely that more than one individual was responsible, in a direct sense, for the attribute configuration of any given artifact of the type commonly encountered in archaeological analysis. For this reason, patterning of attributes at the level of individual activity, is archaeology's only case of perfect association. Neither rodent activity, incorrect excavation procedure, nor improper laboratory sorting can destroy the association of cord-impressed decoration and lip thickening on a rim sherd." [30] This perfect association also holds with respect to the skill, the artistic capability, the sensitivity, and the sensual involvement of the man or woman whose work we can still examine, and some of these qualities are demonstrable in the material evidence.

If the affective and psychological nature of the individual and his experience can never be drawn from the physical result of that experience, perhaps the best we can do is to feel our own personal sense of elation, excitement, or whatever in an intimate confrontation with the artifact, having fully understood what made it what it now is. There is a sense in which all of archaeology is vicarious, after all. Historians are much luckier in that regard, for the written word does carry emotional and intellectual content. Brooke Hindle, in describing "the exhilaration of early American technology" expresses precisely what it is that historians may achieve and that archaeology probably never will:

> The men who constructed the elements of early American technology were themselves a part of it. Their words reflect an emotional satisfaction in reaching forward with each improvement. It is doubtful that the central role of technology in American history will ever be apprehended until this drive is appreciated. Its recognition alone will constitute a response to recent critiques emphasizing the evil results of technology—a response in the form of an explanation, not necessarily a denial. . . .
>
> Most important of all is the enthusiasm of the early American craftsman, mechanic, and engineer. Unless the historian can catch some of that spirit he will render a better service by studying in some other field.[31]

The archaeologist cannot move to another field, but there is recourse to ethnography. We need only mention Ruth Bunzel's *The Pueblo Potter* as an example of a sensitive study from the inside which succeeds in describing a technological activity from the point of view of the individual—her mental and physical processes—within a particular culture setting. Bunzel's work is an indication of how fruitful such an approach can be. The extent to which we can draw analogies from such results to the conditions we study archaeologically remains moot.[32]

In fully subscribing to Cyril Smith's definition of technology as "the integrated work of man's hand, eye, and mind," we recognize any artifact as the product of that integration and expect that we must be able to say something about that synthesis or at least about some of the elements comprising it by careful technical study of artifactual material. As far as understanding art is concerned, Steinberg has recently remarked that "if enough such studies could be amassed, we might well achieve a marvelous sense of the creative and reproductive processes in ancient art, which we have not been able to attain thus far through stylistic studies alone."[33] We may also begin to assess Smith's contention that "esthetically motivated curiosity, or perhaps just play, seems to have been the most important stimulus to discovery."[34] It is worth quoting some of his argument here, for he arrives at this conclusion in part through his familiarity with archaeological materials:

> Practically everything about metals and alloys that could have been discovered with the use of recognizable minerals and charcoal fires was discovered and put to some use at least a millennium before the philosophers of classical Greece began to point the way toward an explanation of them. It was not intellectual knowledge, for it was sensually acquired, but it produced a range of materials that continued to serve almost all of man's needs in warfare, art, and engineering until the end of the 19th century A.D. It is of basic significance for human history that, from the cave paintings on, almost all inorganic materials and treatments of them to modify their structure and properties appear first in decorative objects rather than in tools or weapons necessary for survival.[35]

This is a hypothesis which is certainly capable of being tested and which, in its examination, will bring us as close to the individual and his response to his material environment as we are ever likely to get. The element of play in that response is an important one, but we ought not to forget that even play is not divorced from cultural influences. The amount of time one has in any society for leisurely experimentation or individual pursuits is very much conditioned by other social priorities. And the question of the integration of the specific results of any play activity into the scheme of men's lives is, of course, the question of the acceptance of change.

Although it is undoubtedly difficult to realize the personal aspects of an individual's life through a full understanding of his material world, that material world is all that we have to work with and we have to wring out of it as much as we can. In the absence of historical records, the only approach to some confrontation with the individual is through his art and his technology. However, we are not unfortunate in that, for a man's art and his technology are the tangible products of an intimate interaction among so many of his senses that they are perhaps the best tools we have for understanding him.

Style in Technology

In the foregoing, we have alluded several times to technological "determinism" and to "styles" of technology, both of which are important in our view of technology-culture interactions, although we strongly suspect that the cultural exigencies of technological systems far outweigh those determinants that seem to be wedded to specific materials or energy sources. The subject of style in technology is worthy of closer study, and in closing this discussion we would like to make some tentative remarks about its potential usefulness as a way of thinking about or of describing technologies.

A style—in the sense that we use it here—be it in art, technology, or life (as in "life-style"), consists of a series of traits (such as form, function, elaboration) that are culture specific. By this we mean that there are enough discernible traits in an object or an institution to make it unmistakably recognizable as belonging to a certain culture or culture area. Thus a pot has a distinctive style and is representative of the culture that produced it because it is a complex human product reflecting elements of function, value, organization of labor, perhaps ritual significance, aesthetic content, and a specific selection of materials and technical processes. Along with all those other cultural traits, the technological system that went into its manufacture is clearly a cultural trait. When we can describe that technology as culture specific we are speaking of technological style.

It is often quite difficult to extract from archaeological objects many of the traits we have enumerated—and there are surely still others to be considered—but their technologies can generally be described quite accurately as we have argued above. We wonder whether those technologies cannot also tell us more about the individual cultures that produced them, more than simply what materials were used, how they were processed, and possibly how the work was organized. Is there a more general cultural pattern that can be elicited from a technology that reiterates or reinforces other cultural patterns observable in other ways? It certainly seems to us that this line of inquiry is worth pursuing, particularly at first through ethnographic studies in which, observing technologies wholly integrated into a cultural setting, we could test the viability of a concept of technological style. We might begin to see if there are, in fact, ways of "reading" technologies that could be applied in archaeological contexts where most of the rest of the culture complex is missing. It is precisely from such ethnographic work that we have already begun to see how certain technologies may be expressive of general cultural characteristics.

In a recent lecture at M.I.T.,[36] Monni Adams suggested, on the basis of her intensive study of East Sumba society,[37] that there is a whole complex of linguistic, agricultural, ceremonial, and social-structural behavior peculiar to this Indonesian people that is also manifest in one of their most important technologies—the production of ikat textiles. When Sumbanese institutions are studied in sufficiently diverse aspects, it becomes clear that certain symbolic structures running through the entire culture are summarized in the production of ikat—that ikat itself, in the very formal aspects of the techniques it employs, mirrors other formalisms in the culture as a whole. For example, the act of tying and binding, essential to the formation of the warp yarn patterns in ikat and peculiar to the ikat technique, is repeated over and over again not only in other technologies on Sumba but in ceremonial and ritual observance—for instance, the binding together of the hands of bride and groom during the marriage ceremony or the physical connection made between religious images carried in processions and their worshippers through the medium of long textile streamers that bind image and individual together. Furthermore, "in myth, ritual and social rules on Sumba, the stages of textile work are consistently linked to the progressive development of individual human life. These stages provide an overarching metaphor for the phases of the Sumbanese life cycle." [38] In this single example, we see how a particular technology, albeit one to which cultural energies are largely directed, appears to manifest a style that serves almost as a microcosm of the entire culture.

Cyril Smith has argued that often the most interesting experimentation leading to technological developments comes in the arts, rather than in the more mundane, functional areas of a technology.[39] One might begin to test his argument by looking closely at some of the more interesting and reasonably well studied technologies archaeologists confront most frequently—the ceramic and building technologies of ancient societies. It might prove extremely useful, for example, to contrast the Chinese development of high-temperature pottery glazes and all that this entailed with the totally different but equally important ceramic technology of ancient Athens to arrive at some system for characterizing the distinctiveness of those two technological styles. Or perhaps the Roman predilection for brick and concrete as building materials might be contrasted with the building technology of another imperial state, Assyria, to seek further evidence for the expression of Roman values and the Roman concept of empire.[40]

There are also other technologies, particularly evident in complex societies, that appear to have culture-specific, role-reinforcing functions in and of themselves and which similarly manifest their own styles. We refer to those technologies, often metallurgical (an interesting observation in itself), used by ruling elites as status symbols that serve to maintain their authority and elitism. We might enumerate technologies such as Etruscan and Greek gold working, Japanese sword making, Chinese bronze casting, Peruvian gilding, and Sumerian goldsmithing, though one could obviously extend the list greatly. The interesting features of all these technologies are that they all involve rare or luxurious materials; they appear to produce objects only for a ruling group and are probably controlled by that group as are the sources of the materials themselves; their products suggest a formal system of quality control; and, above all, they show a degree of technical virtuosity that is staggering to the observer, as impressive to us now as it must have been to the audience for whom such objects were made or to the audience meant to see them. Their impressiveness was clearly an important aspect of their social function.

By looking at such elite technologies in groups we may find that there are useful universals that can be seen in the way that certain technologies are manipulated by elites, useful generalizations that can help us to characterize more fully an elite and how it maintains its position. However, once we have isolated the "universals" of rare materials, resource control, luxury goods, exaggerated technical virtuosity, perhaps also ritual function (burial gifts, religious objects, and so on)—in other words, all the status-bearing characteristics of an elite—are there still culture-specific stylistic traits that are observable and that can be informative about the larger society in which the elite emerged?

We might cite here the Japanese development of special bronze alloys and inlaying techniques expressly for the elaboration of *tsuba* (sword guards) as but one example among many of that culture's extraordinary investment in exploiting to the fullest the aesthetic and working properties of materials. The fact that both the Japanese sword and its maker are esteemed as national treasures is testimony to the value placed upon impeccable craftsmanship and the identification of man with material. Similarly, we sense that the extraordinary development of sectional molds for bronze casting in Chou China [41]—where they were used for the basic forming of the ritual vessels as well as for the elaboration of the various appendages to these vessels—to the exclusion of other feasible techniques such as lost-wax casting or welding is to be seen as a culture-specific trait, and that the Peruvian predilection for depletion gilding [42] to the exclusion of other possible techniques is a characteristic of Late Intermediate north Andean culture. The last two examples are much more difficult to analyze and substantiate because of their archaeological nature, but in neither case can the specific technological styles be seen in isolation. They must be taken as one more element, albeit an important one, in piecing together the mosaic of the culture under consideration.

We have tried simply to focus on what we consider a potentially important aspect of our approach to technology. Since technologies are a basic element in culture, once they are properly understood they can be used as a rich resource in the study of culture. Taken together with other cultural traits, the styles in which technologies are expressed can be helpful in evaluating cultural idiosyncracies as well as the more universal elements of man's technological adaptations to his physical and social environment. Certainly in looking for style in technology we may have a chance to understand a primary mode of cultural behavior in a manner that has not often been attempted.

Notes

1. David Kaplan, "The Superorganic: Science or Metaphysics?," *American Anthropologist*, vol. 67 (1965), 958-76 (quotation on p. 959).
2. See, for example, Paul Bohannan's article, "Rethinking Culture: A

Project for Current Anthropologists," and the reactions to it evoked in *Current Anthropology,* vol. 14, no. 4 (1973), 357-72.

3. Kaplan, p. 959.

4. Julian Steward, *Theory of Culture Change* (Urbana, Ill., 1963).

5. Robert S. Merrill, "The Study of Technology," in *International Encyclopedia of the Social Sciences* (New York, 1968), vol. 15, 576-89 (quotation on pp. 577, 582).

6. Leslie A. White, *The Evolution of Culture* (New York, 1959).

7. Merrill, "Study of Technology," p. 582.

8. Roger Keesing, "Paradigms Lost: The New Ethnography and the New Linguistics," *Southwestern Journal of Anthropology,* vol. 28, no. 4 (1972), 299-332.

9. Jacques Maquet, "Introduction to Aesthetic Anthropology," *Current Topics in Anthropology,* vol. 1, module 4 (1971), 1-38; Daniel Biebuyck, *Lega Culture* (New York, 1955); George Mills, "Art and the Anthropological Lens," in *The Traditional Artist in African Societies,* Warren L. d'Azevedo, ed. (Bloomington, Ind., 1973).

10. Since this paper was presented, a group of articles written around the theme of technological style has been published in *Material Culture: Styles, Organization, and Dynamics of Technology,* Heather Lechtman and Robert S. Merrill, eds., (St. Paul, Minn., 1977). The lead article by Heather Lechtman, "Style in Technology—Some Early Thoughts," discusses the phenomenon of technological style and attempts to evaluate its usefulness in the interpretation of archaeological data.

11. Mark P. Leone, "Issues in Anthropological Archaeology," in Leone, ed., *Contemporary Archaeology: A Guide to Theory and Contributions* (Carbondale, Ill., 1972), pp. 14-27.

12. Lynn White, Jr., *Medieval Technology and Social Change* (Oxford, 1962), p. v.

13. Michael W. Spence, "The Obsidian Industry of Teotihuacán," *American Antiquity,* vol. 32, no. 4 (1967), 507-14; Spence, "The Development of the Classic Period Teotihuacán Obsidian Production System," paper read at the 38th annual meeting of the Society for American Archaeology, 1973; William A. McDonald and George R. Rapp, Jr., eds., *The Minnesota Messenia Expedition: Reconstructing a Bronze Age Regional Environment* (Minneapolis, Minn., 1972).

14. Heather Lechtman, "A Metallurgical Site Survey in the Peruvian Andes," *Journal of Field Archaeology,* vol. 3, no. 1 (1976), 1-42; Heather Lechtman and Michael E. Moseley, "The Scoria at Chan Chan: Non-metallurgical Deposits," *Ñawpa Pacha,* vol. 10-12 (1975), 135-85; Radomir Pleiner, *Stare Evropské Kovářství* (Alteuropaishes Schmiedehandwerk), (Praha, 1962); Pleiner, "Rediscovering the Techniques of Early European Blacksmiths," *Archaeology,* vol. 16 (1963), 234-42; Pleiner, "Die Eisenverhüttung in der 'Germania Magna' zur römischen Kaiserzeit," in *45. Bericht der Römisch-Germanischen Kommission, 1964,* 1965, pp. 11-86; Pleiner, "Die Technologie des Schmiedes in der grossmährischen Kultur," *Slovenská archeólogia,* vol. 15 (1967), 77ff; Pleiner, "Středověká Výroba Smoly v Krásné Dolině u Rakovníka" (Die Technologie der Mittelalterlichen Teerbrennerei in Krásná Dolina bei Rakovník, Böhmen), *Památky Archeologické,* vol. 61 (1970), 472-518; Arthur Steinberg and Frank Koucky,

"Preliminary Metallurgical Research on the Ancient Cypriot Copper Industry," in *Excavations in the Idalion (Dhali) Region, Cyprus: Preliminary Results, 1971-72,* G. E. Wright et al., eds., supplement to *Bulletin of American Schools of Oriental Research* (1973); Beno Rothenberg, *Were These King Solomon's Mines?* (New York, 1972); R. F. Tylecote, "Iron Smelting on the Nigerian Early Iron Age Site at Taruga, Abuja Emirate," *Bulletin of the Historical Metallurgy Group,* vol. 2, no. 2 (1968), 81-82; Tylecote, "Ironworking at Meroë, Sudan," *Bulletin of the Historical Metallurgy Group,* vol. 4, no. 2 (1970), 67-72; Theodore Wertime, "A Metallurgical Expedition through the Persian Desert," *Science,* vol. 159 (1968), 927-35.

15. Brooke Hindle, *Technology in Early America* (Chapel Hill, N.C., 1966), p. 24.

16. K. R. Maxwell-Hyslop and H. W. M. Hodges, "Three Iron Swords from Luristan," *Iraq,* vol. 28 (1966), 164-76; Albert France-Lanord, "Le fer en Iran au premier millénaire avant Jésus-Christ," *Revue d'histoire des mines et de la métallurgie,* vol. 1 (1969), 75-127; Cyril S. Smith, "The Techniques of the Luristan Smith," in *Science and Archaeology,* R. Brill, ed. (Cambridge, Mass., 1971), 32-54.

17. D. Schmandt-Besserat, "The Use of Clay before Pottery in the Zagros," *Expedition,* vol. 16 (1974), 11-17.

18. Radomir Pleiner, *Iron Working in Ancient Athens* (National Technical Museum, Praha, 1969).

19. Flannery has made a similar suggestion, though with less emphasis on the technological aspects, in Kent V. Flannery, "Origins and Ecological Effects of Early Domestication in Iran and the Near East," in *The Domestication and Exploitation of Plants and Animals,* P. Ucko and G. Dimbleby, eds. (Chicago, 1969).

20. Douglas S. Byers, ed., *The Prehistory of the Tehuacan Valley* (Austin, Tex., 1967), vols. I, II.

21. Kent V. Flannery, ed., *Preliminary Archaeological Investigations in the Valley of Oaxaca, 1966-69* (Ann Arbor, Mich., 1970).

22. Brooke Hindle, "The Transfer of Power and Metallurgical Technologies to the United States, 1800-1880," paper given at ICOHTEC Conference, Pont-à-Mousson, France, 1970.

23. Frederick R. Matson, "The Glazed Pottery of Afghanistan—Variations on a Theme," paper given at Archaeological Institute of America Annual Meeting, Philadelphia, Pa., 1972.

24. Joseph Needham, *The Development of the Iron and Steel Technology in China,* (London, 1958).

25. Olga Linares de Sapir, "Diola Pottery of the Fogny and the Kasa," *Expedition,* vol. 11 (1969), 2-11.

26. Dean E. Arnold, "Ethnomineralogy of Ticul, Yucatan, Potters: Etics and Emics," *American Antiquity,* vol. 36 (1971), 20-40.

27. Especially K. Nicklin, "Stability and Innovation in Pottery Manufacture," *World Archaeology,* vol. 3, no. 1 (1971), 13-48; and M. J. Rowlands, "The Archaeological Interpretation of Prehistoric Metalworking," *World Archaeology,* vol. 3, no. 2 (1971), 210-24.

28. Cyril Stanley Smith, "Metallurgical Footnotes to the History of Art," *Proceedings of the American Philosophical Society,* vol. 116 (1972), 97-135.

29. Frederick R. Matson, "The Potters of Chalkis," in *Classics and the Classical Tradition,* Eugene N. Borza and Robert W. Carruba, eds., essays presented to Robert E. Dengler on the occasion of his eightieth birthday, Pennsylvania State University (University Park, Pa., 1973), 117-42; Matson, ed., *Ceramics and Man* (Chicago, 1965); Franz Boas, *Primitive Art* (New York, 1955).

30. James Deetz, "The Inference of Residence and Descent Rules from Archaeological Data," in *New Perspectives in Archaeology,* Sally R. Binford and Lewis R. Binford, eds., (Chicago, 1968).

31. Hindle, *Technology in Early America.*

32. See William Bascom, "A Yoruba Master Carver: Duga of Mẹkọ," in *The Traditional Artist in African Societies.*

33. Arthur Steinberg, "Foreword," in *Art and Technology, a Symposium on Classical Bronzes,* S. Doeringer, D. G. Mitten, and A. Steinberg, eds. (Cambridge, Mass., 1970).

34. Cyril Stanley Smith, "Matter versus Materials: A Historical Review," *Science,* vol. 162 (1968), 637-44.

35. Smith, "Matter versus Materials," p. 638.

36. In April, 1973.

37. Monni J. Adams, "Work Patterns and Symbolic Structures in a Village Culture, East Sumba, Indonesia," *Southeast Asia Quarterly,* vol. 1 (1971), 321-34; Adams, "Structural Aspects of a Village Art," *American Anthropologist,* vol. 75 (1973), 265-79.

38. Adams, "Work Patterns and Symbolic Structures," p. 322. Some of these points are elaborated upon in a more recent article by Adams, "Style in Southeast Asian Materials Processing: Some Implications for Ritual and Art," in *Material Culture.*

39. Cyril Stanley Smith, "Art, Technology and Science: Notes on Their Historical Interaction," *Technology and Culture,* vol. 11 (1970), 493-549; Smith, "Metallurgical Footnotes to the History of Art."

40. William L. MacDonald, *The Architecture of the Roman Empire* (New Haven, 1965); Heather Lechtman, "The Technology of Roman Building," paper presented at the Cambridge Archaeology Seminar on the Hellenistic and Roman Periods, 1972.

41. Noel Barnard, *Bronze Casting and Bronze Alloys in Ancient China,* Monumenta Serica Monograph, XIV (Tokyo, 1961); Barnard, "Chou China: A Review of the Third Volume of Cheng Te-K'un's *Archaeology in China,*" in Monumenta Serica Monograph, XXIV (Tokyo, 1965), 307-459; R. J. Gettens, *The Freer Chinese Bronzes, Vol. II. Technical Studies* (Washington, D.C., 1969).

42. Heather Lechtman, "The Gilding of Metals in Pre-Columbian Peru," in *Application of Science in Examination of Works of Art,* William J. Young, ed. (Boston, 1973).

Part II The Philosophy of Technology

Carl Mitcham

Philosophy and the History of Technology

Philosophy relates to history as a confessor to the penitent, and, like a confessor, it ought to have a supple and searching ear for the penitent's secrets. . . .
—Sören Kierkegaard, *The Concept of Irony*

I. The History of Technology as a Philosophical Problem

Like economics, political science, anthropology, and other humanistic disciplines, philosophy carries on a running dialogue with the past. In this dialogue are the foundations of its continuing life. Clearly, there is a sense in which history is the most fundamental of the humanities. The relation between the philosophy of technology and the history of technology is thus a subspecies of the relation between philosophy and history.

But the relation between philosophy and history is problematical. At one extreme, philosophy does no more than approach history to gather grist for its analytic mills; at another, it looks upon history (with Hegel) as itself the supreme embodiment of a progressively developing reason. Or, intermediately, philosophy approaches history with a desire to discover, in the apparently ceaseless flux of events, ideas, and arguments, perennial or ahistorical problems and truths; to use the words of one eminent historian of philosophy, it seeks out the unity of its various philosophical experiments.

Different philosophies of technology can be found to exhibit each of these basic attitudes toward the history of technology—each time bringing to light different sets of needs and expectations. First, representing what may be called the analytic approach, the philosophic studies of technology by Mario Bunge, James K. Feibleman, I. C. Jarvie, and Henryk Skolimowski center on conceptual or methodological issues. These philosophers are particularly interested in historical case studies of individual technologies and their special technical problems as sources for testing or illustrating their conceptual analyses.

Second, the Hegelian or historicist approach is perhaps best illustrated in the philosophy of technology of Karl Marx, the man who proposed to turn Hegel "rightside up"—to replace the phenomenology of spirit (and its idealist dialectic) with what can be called a phenomenology of technology (and its corresponding materialist dialectic). Hegel's vision of the history of ideas as the progressive actualization of spirit binds philosophy to history, at least to the writing of philosophical history. Similarly, Karl Marx's optimistic vision of technical development as the progressive realization of human nature generates a tendency not just to read the history of technology, but to write it. The Marxian philosophy of technology, with its essentially anthropological or man-centered character, is characteristically embedded in a historical presentation of the triumph of *homo faber*. (See, e.g., the histories of Samuel Lilley and Alexsandr Zvorikine.) Of course, this is no less true of those capitalist ideologies of technology which, deriving more immediately from Francis Bacon, dominate American popular thought. (See, e.g., the histories of James Kip Finch, Courtney R. Hall, and John W. Oliver.)

But Marx, too, can be turned "rightside up." The development of technology can be viewed as the progressive dehumanization rather than the humanization of man. It is noteworthy that the pessimistic (not to say antitechnological) theories of, say, Lewis Mumford and Jacques Ellul also tie the philosophy of technology into its history, but in presentations which interpret the history of technology as the progressive obscuring or warping of human nature. In a more metaphysical fashion, Martin Heidegger's philosophy of technology approaches the history of technology with similar comprehensive expectations: the desire to find in the history of technology an ongoing anthropological or ontological revelation, a clarification about the nature of man and being.

Finally, representing what may (with no prejudicial intentions) be called an Aristotelian approach, there are the historico-philosophical studies of Freidrich Dessauer and H. van Riessen. Like Aristotle, who surveys the history of his predecessors' opinions in order to elucidate the basic problems of ethics and metaphysics and their alternative solutions and then often returns to them in developing his own arguments, both Dessauer and van Riessen philosophize about technology against the background of the history of technology, especially against the background of the history of other ideas about technology (a subject which will be referred to again shortly). Dessauer, for instance, in *Streit um die Technik* (1956) surveys both the classical philosophical theories about the nature of *techne* and human making and the ideas of late-nineteenth- and early-twentieth-century German scientists, engineers, economists, historians, philosophers, and theologians in the course of developing his own philosophical theories. Van Riessen's *Filosofie en Techniek* (1949) is, if anything, even more comprehensive in this respect, surveying not only explicitly developed ideas about technology but also implicit attitudes to be found in thinkers as diverse as Comte and Nietzsche. In English, the works of Andrew G. van Melsen and Hans Jonas exemplify this same historico-philosophical approach to the philosophy of technology.

These three categories are not strictly homological. In the second the relationship between the philosophy and history of technology is closely tied to a specific philosophical content, whereas the relationship found in types one and three is open to a variety of contents. Moreover, the three categories are not mutually exclusive. Mumford and Ellul contain historico-philosophical as well as conceptual analysis; the conceptual analyses of Bunge, Feibleman, Jarvie, and Skolimowski have definite historico-philosophical elements; and the historico-philosophical approach itself cannot do without conceptual analyses. But accepting for a moment this broad typology (with its inherent limitations), what can be said about these different philosophical approaches to the history of technology? Is the structure of the history of technology itself such as to invite one attitude to predominate over others? Which is most fruitful for the elaboration of a philosophy of technology—and in what way? Let me approach these questions through a series of outsider's observations about the character and structure of the history of technology.

First, it is worth noting that philosophy of technology and history of technology have similar sociological roots. It was existentialist philosophers such as José Ortega y Gasset, Karl Jaspers, and Heidegger along with technologists (Dessauer and others associated with the Verein Deutscher Ingenieure) who, out of a desire to understand what was being recognized as the dominating influence on our times, gave birth to the philosophy of technology as a distinct discipline. Likewise, it was social historians (such as Mumford) along with technologists (e.g., R. J. Forbes, Cyril Stanley Smith, and other members of German and American engineering societies) who, out of a similar desire to bridge the gap between industrial science and the humanities, gave birth to the history of technology.

But sociological similarities should not be allowed to mask important differences. The history of technology, much more than the philosophy of technology, may well be a preeminently modern study. At the beginning of the modern period Francis Bacon (1561-1626) suggested that man could only truly understand what he could construct or make—that is, that nature had to be taken apart and put back together by experiments in order to be known. Such knowledge, Bacon argued, equals power. One hundred years later, Giambattista Vico (1668-1744) argued, even more directly, that to know a thing is to be able to produce it. But whereas Bacon directed attention to the arts and crafts, Vico made history the object of his new science. History is a human product even more than the arts. The history of technology obviously unites these two strains in such a fashion that one can only wonder why it did not happen sooner.

To say the same thing in a different way, history and technology are both somehow characteristic of modernity. The modern period has been called the "age of technology"; it has also been called the "age of historical consciousness." These views are both grounded in modern man's affirmation of himself. Classical thought could not have taken such an interest in either technology or history, although not (as is often asserted) because it looked down upon them. It would be more accurate to say that because it refused to affirm the unique importance of man, the classical mind naturally directed its attention elsewhere. Modern consciousness arose out of Renaissance humanism and the affirmation of the singular value of man above everything else in nature. Technology is designed to give this modern man his rightful power over nature; history is the study of this man's life independent of nature. The history of technology unites these two aspects of modernity in a paradigmatic fashion.

The history of technology is not, however, some one thing. Like philosophies of technology, histories of technology fall into three different types. First, there is what is called technological history—the history of hardware, how artifacts are made and used. Second, there is what has been called the social history of technology—that is, a study of the influence of technology on man and his social institutions. Third, there is what, for want of a better phrase, I will refer to as history of ideas about technology—that is, the study of how different periods and individuals have conceived of and evaluated the human making activity.

Now if I correctly understand a contemporary debate in the historiography of technology, until recently historical studies have been largely confined to technological history. This, for instance, is the subject of the monumental Singer five-volume *A History of Technology* (1954-58). Some others, especially those associated with the Society for the History of Technology, have expanded the subject in the direction of the second type, the social history of technology. This, for instance, is the thrust of the Kranzberg-Pursell two-volume *Technology in Western Civilization* (1967). These are the two major areas of research in the history of technology. Although some interest has been shown in the history of ideas about technology, work on this has been even more limited.

It is my suggestion that these three types of history of technology correspond remarkably well with the needs and expectations of the three types of philosophy of technology outlined above. The needs of analytic philosophies of the methodology of technology are largely met by technological history. Marxian (plus capitalist-American) philosophies, along with their pessimistic counterparts, are committed to social histories of technology. And, as already mentioned, the philosophies of technology of thinkers such as Dessauer and Jonas are closely involved with the history of ideas about technology.

The first relationship, as I see it, raises no unique problems. Definitional and epistemological analyses of technological invention and production necessarily make use of historical studies in order to test the generality of their concepts. Any philosophical problems which arise are permanent features of relating theory and empirical reality (e.g., the problem of induction). Moreover, technological history itself is much simpler in structure than either general history or other specialized histories like, say, the history of science or art. Technological history is closer to chronicle, the straightforward temporal enumeration of discoveries and inventions, and freer of narrative interpretation. Insofar as narrative interpretation and causal explanations are constructed, while technological history is easily construed as linear and progressive in a way that the history of art, for instance, never could be, the norms for such a history do not need to be imported from outside; the historical actors, that is the technologists themselves, seem to share a univocal ideal of technical efficiency which can be used to judge technological development. At least this is true in the modern period. The only real question concerns how self-conscious these same ideals were in premodern periods, and whether in fact they are shared in anything like the same form by both ancients and moderns.

The social history of technology and its relationship to historicist-oriented philosophies of technology pose greater difficulties. First, with regard to the social history itself, it is much more complex than technological history. Not only are more complex causal factors involved in relating technology and society, but the interpretation which turns the chronicle of technological history into the narrative of social history inevitably brings into play human values and questions about the nature of man. Ideally, perhaps, social history should be able simply to describe the various social influences of technology at different historical periods—relating them, at most, to alternative human possibilities. Ideally, that is, the social history of technology should be a historical sociology of technology, but almost invariably it has failed to remain within these limits, probably for two reasons. One is the strong influence of the idea of technological progress in the strict sense, which easily tips over into notions about human progress in general. The other is the fact that this transposition of the notion of progress from the strictly technological to the general human realm is positively encouraged by modern theories about the nature of man and his ontological primacy. If one affirms the theoretical primacy of man and sees technology as the means for realizing this primacy in practice, then when one finds progress in technological history, it is just natural to interpret all social influences of technology as necessary and beneficial to human progress as well.

What this means is that to a greater or lesser degree virtually all social histories of technology contain built-in philosophies of technology—that is, theories about the nature and meaning of technology, especially about the anthropological meaning of technology. On the other hand, because of the clear progressive development of technology itself, all theories about the anthropological meaning of technology almost have to contain a theory of history. Both points suggest that the structure of the history of technology itself is such as to encourage particular philosophical attitudes.

Most commonly the particular philosophy invoked is an optimistic, not to say triumphalist view of both the history of technology and the nature of man. How this relationship works, however, can be illustrated by looking at the so-called pessimistic theories about the anthropological meaning of technology. Consider, for example, the work of Mumford, Ellul, or Heidegger. Each of these thinkers has a nontechnological theory of man, but in each case arguing for it appears to require explaining the progressive development of technology as an obscuring or falling away from his true nature or essence. Some kind of dehumanization has to have been involved. At the very least, as with the existentialist Jaspers, the advent of modern technology must be seen as a crucial turning point in human history. Thus in every case a theory of the relationship between technology and the nature of man appears to require a philosophical evaluation of technological history.

But the social history of technology is not only a history of the social consequences of technology; a number of historians have argued that social history should also include, if not even be primarily concerned with, the history of how society itself affects technology. Different societies, in this theory, because they have different material bases and different spiritual ideals, develop different technologies in pursuit of their self-realization. In fact, this was also the philosophical position of two of the earliest contrbutors to the philosophy of technology, the existentialists Ernst Jünger and Ortega y Gasset. For Jünger, modern technology is "the mobilization of the world through the Gestalt of the worker" or the means by which the ideal type of the worker realizes itself. For Ortega, technology in general is "the system of activities through which man endeavors to realize the extranatural program that is himself." Ortega, particularly, stresses the historical differences between the technology created, for instance, by Tibetan society in pursuit of the Buddhist bodhisattva ideal and the English technology of the bourgeois ideal.

The historiographical debate about whether the social history of technology should focus primarily on the effects of technology upon society or the effects of society upon technology raises a number of philosophical issues. At least three of these are the following: (1) the problem of periodization in the history of technology—that is, when and upon what basis are different historical periods to be distinguished in the history of technology? Are social effects or social ideas the key? Or is power? Or the mode of production? (2) What is the relationship between technology and science? On the hypothesis that society is primarily influenced by technology, technology is typically thought to develop independently, usually being conceived as a practical application of science. However, if society influences technology as well, through either its economic base or its cultural matrix, then technology must be conceived as more independent of science. (3) What is the relationship between technology and the nature of man—or, more specifically, what is the relationship between technology and human action? This last, clearly, is the most critical of questions, yet one which has been largely ignored by historians and philosophers alike. Most studies of technology simply begin with stipulative definitions which, when examined, fail to distinguish between action in general and technological action in particular or to articulate the distinctions between various types of technology. The result is that in most cases man is necessarily if unwittingly defined as a technological animal—with the understanding of man and his technology predetermined and obscured.

Because answers to these questions properly influence our historical perspectives, each needs to be undertaken independent of historicist pressures—that is, with full awareness of various alternative solutions, their historical bases, and their philosophical implications, but without conceptually predetermined conclusions. What is needed are metaphysical investigations along the lines of Ortega's early "Excursion to the Substructure of Technology" in *Meditación de la técnica* (1933). Actually, it was Dessauer who, with *Philosophie der Technik* (1927), can be said to have initiated such studies. However, to the present time most metaphysical studies have remained either historically naive (e.g., Dessauer, Ortega) or lacking in proper philosophical rigor (e.g., Mumford). What is called for is a philosophical as opposed to a historical wedding between philosophy and history.

This needed marriage points toward the third type of history, a history of ideas about technology. Besides economic and cultural determinants, one other way society can influence technology is through ideas, as is indicated in the second issue raised by the historiographical debate. Also, clearly any periodization ought at least to be correlated with a history of ideas. As already mentioned, however, this is the least developed of the histories. My suggestion is simply that from a philosophical perspective it is the type which would be most beneficial for confronting the problems outlined above. Let me be more specific by sketching an example of how this might work.

II. *Techne* and Technology

Virtually all historians (except Mumford, who prefers the term *technics*) use the word *technology* to refer to both the ancient and modern, primitive and advanced, making activity or knowledge of how to make and use artifacts, or the artifacts themselves. For instance, Singer defines technology as "how things are commonly done or made . . . [and] what things are done or made." While rightly objecting to this definition as so wide as to include even legislation or the making of laws, Kranzberg and Pursell define technology as "man's effort to cope with his physical environment . . . and his attempts to subdue or control that environment by means of his imagination and ingenuity in the use of available resources." Indeed, they specifically reject any limiting of the term to "those things which characterize the technology of our own time, such as machinery and prime movers." Technology "is nothing more than the area of interaction between ourselves, as individuals, and our environment, whether material or spiritual, natural or man-made"; it is "the most fundamental aspect of man's condition."

But Kranzberg and Pursell contradict themselves in these two formulations. In the first instance they limit technology to action on the physical environment; in the second the environment is allowed both material and spiritual dimensions. Nor do they escape, even in the strict formulation, their own quite valid objections to Singer. If technology is "nothing more" than they suggest, it is hard to imagine anything that would be left out. Certainly not legislation which, being concerned with "who gets what, when, how" (Lasswell), is in some sense an "effort to cope with . . . physical environment." On their own terms technology becomes veritably coextensive with human action and fails to exclude the fine arts, spiritual disciplines such as yoga and meditation, or even language (although in practice they clearly do exclude such subjects). But more to the present issue, such a conceptual framework forces historians to equate Greek τέχνη (and Latin *ars*) with the English *technology* (and all of these with the German *Technik*, French *technique*, and others). I submit, however, that a historical study of the word *technology* (for etymology is a good place to begin the history of ideas) would show this to be questionable. At the very least it should help clarify our conceptions of the essence of technology as much as technological or social histories.

The Greek τέχνη, commonly translated as "art," "craft," or "skill," has behind it the Indo-European stem *tekhn-*, meaning probably "woodwork" or "carpentry," and is akin to the Greek *téktōn* and Sanskrit *tákṣan,* meaning a "carpenter" or "builder," and the Sanskrit *tákṣati,* "he forms," "constructs," or "builds." One could compare also the Hittite *takkss-*, "to join" or "build," and the Latin *texere,* "to weave," hence figuratively "to construct," and *tegere,* "to cover," hence "put a roof onto." In nonphilosophical literature *techne* is used to refer to cleverness and cunning in getting, making, or doing, as well as specific trades, crafts, and skills of every kind.

In philosophical works, however, *techne* comes to be conceived not only as an activity of some particular sort or character, but as a kind of knowledge. In Plato, who is the first to deal at length with this notion, *techne* and *episteme,* art and systematic or scientific knowledge, are closely associated. (Note, too, that in nonphilosophical usage *episteme* itself commonly means "acquaintance with," "skill," or "disciplined experience," as in the *episteme* of archery or war.) In the *Gorgias,* for instance, Socrates argues that every *techne* is involved with *logoi* (words, speech) bearing upon the specific subject matter of the art (450b). Moreover, Socrates goes on to distinguish between two types of *techne,* one which consists mainly of physical work and requires minimal use of language (such as painting or sculpture) and another which is more intimately bound up with speech and requires little physical exertion (such as arithmetic, logistic, or astronomy) (450c ff.).

On the other hand, those human actions which are devoid of art, which are nontechnical, *atechnos,* are activities such as cooking and persuading—each being a mere knack or routine way of operating, a *tribé,* based simply on experience, *empeiria* (501a). (For the association of *atechnos* with *tribé* see also *Phaedrus* 260e.) Such activities are not art because they have no awareness of the nature, *physis,* or cause, *aition,* of the things they make or do; they are *alogos* (cf. 465a). To say that such actions are nonlogical is to say that they are not based on a consciousness of the true nature of the things with which they deal; they are simply means. In modern parlance, they are "pure technique." In the *Ion,* the poet who exercises his craft of making, *poesis,* by virtue of divine inspiration is also said to be devoid of *techne* or art (cf. 533d); if the poet possessed an art he would be able to explain his creations to others and to teach (532c). Evidently, then, *techne* in the early Plato refers to all human actions which involve some speech or reasoning—all actions, that is, which are neither spontaneous nor the result of some unconscious drive or intuitive perception. If such usage seems at first to make technology coextensive with human action in a way reminiscent of Kranzberg-Pursell, it also stresses the "logical" character of *techne*—not, of course, in the modern sense of mathematicized deduction, but in the Greek sense of being involved with language and hence with consciousness or knowledge of the inner nature of things. One might even venture that, in marked contrast to Kranzberg-Pursell, *techne* lays emphasis on the nonhuman or nonutilitarian aspects of that action.

This general understanding of *techne* is, however, extended in the direction of modern notions of technology by a classification of knowledge developed in the later dialogue *Philebus*. Here Plato takes knowledge and divides it into two classes: that involved with education and upbringing, and that involved with making or producing (55c). Of the second, technical knowledge, there are again two types: those (such as music, medicine, and agriculture) which proceed by conjecture and intuition based simply on practice and experience, and those (such as carpentry) which consciously involve the use of numbering, measuring, weighing (55e-56c). The latter possesses greater exactness or precision, *akribeia* (a word which also implies deeper insight), and it is this which is *techne* in the primary sense. Thus *techne* is clearly distinguished from all human action and knowledge of a political sort (education and, by extension, governing) so as to be associated more closely with the activities of making or producing which operate upon the nonhuman material world. Of these activities those are most truly *techne* which involve the greatest degree of quantitative precision.

Up to this point in his classification scheme Plato has used *techne* and *episteme*, art and systematic knowledge, almost interchangeably. Now he proceeds to speak of a "philosophic" arithmetic, which differs from the arithmetic of the carpenter in that it deals not with numbered things but with numbers alone (56d). But in referring to this latter, still more precise or penetrating *episteme* he no longer employs the word *techne*. Thus *techne* is also conceived to be distinct from what we would call pure theory, or any knowledge which does not bear upon the material world in some practical manner. (Cf. the distinction between pure and applied arts or sciences indicated at *Statesman* 258e.) While tying *techne* into consciousness, and indicating the primary type of *techne* as that which can use mathematics to express itself, *techne* is nevertheless distinguished from pure consciousness or consciousness of a nonmaterial reality. (This ultimate reality is, of course, only grasped in a provisional or inadequate way through mathematics. The deepest cognition, *gnosis*, of being is only to be had through dialectic, *dialegein*. *Philebus* 58a ff.)

Plato's discussion clearly points toward a conception often associated, at least intuitively, with modern technology—that of rationalized production, or production made maximally efficient through mathematical analysis. And yet the Greek term τεχνολογία has yet to appear. The first appearance of *technologia* (or one of its cognates) is to be found in Artistotle, although not in any of his major discussions of what today would commonly be considered *technai*. For Aristotle, as for Plato, *techne* is a special knowledge of the world which informs human action accordingly. As a type of awareness of the world, it lies between unconscious experience and knowledge of first principles; *techne* is part of that continuum which moves from sense impressions and memories through experience to systematic knowledge, *episteme* (*Metaphysics* I, 1; 980b25 ff.). "From experience, that is the universal come to rest in its entirety in the soul—the one alongside the many, the unity that is a single identity within them all—originate art *(techne)* and science *(episteme):* art in the realm of coming to be, science in the realm of being" (*Posterior Analytics* II, 19; 100a6-9). Yet while continuing to stress the epistemic character of *techne*, Aristotle thinks of it not solely as a kind of knowledge, but partially returns to the common-sense Greek notion of *techne* as a type of activity. *Techne* is not strictly activity, but it is a capacity for action, founded in a special kind of knowledge.

According to Aristotle's formal definition, *techne* is ἕξις μετὰ λόγου ἀληθοῦς ποιητική (*Nicomachean Ethics* VI, 4; 1140a1). Translated literally, this defines *techne* as a habit (or stable disposition to act in a specific manner) with a true *logos* concerned with (or ordered toward) making (the human production of material objects). To paraphrase, *techne* is an ability to make which is dependent upon correct awareness of or reasoning about the thing to be made. The absence of *techne* in an ability to make involves either the absence of any *logos* (consciousness) or the presence of a false *logos* (consciousness) (*Nicomachean Ethics* VI, 4; 1040a20-23). Once again the nonutilitarian character of technical knowledge comes to the fore; insofar as it is true, this *logos* is based on a mental grasping or cognition, *gnosis,* of causes, *aition* (*Metaphysics* I, 1; 981b6-7).

Mention of the connection between *logos* and *aition* recalls Aristotle's distinction of the four causes—a distinction which, incidentally, is consistently illustrated by reference to technical products. According to this discussion (cf. *Physics* II, 3) the "why" of a thing is only answered by grasping the "that out of which" it comes (material cause), its *eidos* or archety.pe (formal cause), the "what makes of what is made and what causes change of what is changed" (efficient cause), and the *telos* or "that for the sake of which" a thing is made (final cause). What is important in such a discussion is that Aristotle does not limit the technical, as we might be tempted to do, to efficient causation. The making of artifacts involves all four causes.

Techne, then, is *episteme* in a general sense in that it involves true consciousness of the world and hence can be taught or communicated (*Metaphysics* I, 1; 981b8-10); but it is to be distinguished from *episteme* in the strict sense in that it bears upon changing rather than unchanging things (cf. *Nicomachean Ethics* VI, 6; 1141b31-36). Aristotle agrees with the later Plato in stressing the "logical" character of *techne* while divorcing it from knowledge of human affairs, on the one hand, and pure theory, on the other. What is absent from Aristotle's understanding of *techne* is any suggestion that part of its "logic" need be the use of quantitative or mathematical concepts; *logos* is not restricted even in its highest aspects to mathematical reasoning. Compare, for example, Aristotle's reference at *Politics* I, 11 (1258b35-40) to those activities which are most truly *techne* as those in which there is the least element of chance, with *Politics* VIII, 6 (1341a17), where he explicitly refers to flute playing as requiring great *techne.* Unlike Plato, Aristotle is able to think of medicine as *techne* in the primary sense (cf. *Metaphysics* I, 1; 981a13 ff.).

Plato and Aristotle agree, then, in stressing the "logical" character of *techne,* even when they disagree somewhat on their understandings of the character of the *logos* involved. Yet neither feels drawn to join these two words—to speak of a *logos* of *techne.* *Techne* simply uses *logos.* Here Plato's distinction between numbering, measuring, and weighing in carpentry and a philosophic numbering, measuring, and weighing is suggestive. For Plato, carpentry merely uses a more general or universal arithmetical *episteme*. Although arithmetic is a *logos* to be used by carpentry, it is not a *logos* of carpentry in the sense of being derived from or limited to this particular *techne*, nor is it all of the *logos* of carpentry. There are elements of consciousness in carpentry which cannot be expressed through this *logos*, which are not subject to being expressed in the language of arithmetic. Furthermore, there are elements of any *techne* which, because of its involvement with the material world, cannot be expressed at all.

To put it simply, that which is able to be grasped or known by *techne* through *logos* is the form or idea, *eidos*, the whatness of the thing to be made. What is not so graspable is the process, the how of the actual making, *poesis*. Here Plato's example from the *Cratylus* (389a-390b), of the carpenter who repairs a broken shuttle, is instructive. In making a repair he looks not to the broken shuttle, but to the form, to "that which is fitted by nature to act as a shuttle." It is this which he must re-embody, *apodidomi* (literally "give back to," "restore") in the "that out of which he makes" (i.e., the material) "not according to his own will, but according to its nature." Again, in the *Timaeus* (29a) Plato says that the *demiourgos* made the world by looking to an eternal and unchanging form or pattern, *paradeigmatos*, which is apprehended by reason, *logos*. As to the how or process of making, the becoming as opposed to being, this can only be grasped through *pistis*, belief or trust, the mental disposition which in the *Republic* (511d and 534a) Plato associates with the sensation of material things. Clearly, then, it is the *eidos* which is grasped by the *logos* —and sometimes by a mathematical *logos*—which is operative in *techne*. But the matter, that out of which a thing is made, and the consequent process of making do not fall within the logical structure of the art.

Aristotle's analysis of carpentry makes the same point. In an artifact, he argues, no material part comes from the carpenter, nor is any part of carpentry (the art) in that which is produced. Instead, the shape, *morphe*, and form, *eidos*, are engendered in matter by motion. It is the soul, in which is the form of knowledge, which moves the hands (or some other part of the carpenter) with a definite motion, one varying with the varying character of the object produced, the hands which in turn move the tools, and the tools the matter (*On the Generation of Animals* I, 22; 730b10-20). (On the hand as an instrument or tool, *organon,* see *Parts of Animals* IV, 10; 687a2-24.) Elsewhere, even more pointedly (if more abstractly) Aristotle argues that it is part of *techne* "to know the form and the matter," but the matter, *hyle,* only "up to a point" (*Physics* II, 2; 194a23). Or again, "Matter is unknowable *(agnosis)* in itself" (*Metaphysics* VII, 10; 1036a9). Only as informed, or related to form, is matter able to be grasped by the mind. And yet, relative to every work of *techne* there is a matter and a form, and it is "the *matter* (my emphasis) which governs the making *(poesis)* and generation of any work of art" (*Metaphysics* VII, 9; 1034a10-11). "*Techne* imitates nature *(physis)*" (*Physics* II, 2; 194a21; *Meterology* IV, 3; 381b6; *On the Cosmos* 5; 396b12) by uniting form and matter in a particular something (cf. also *On the Generation of Animals* II, 4; 740b25-29). The form is the idea in the mind of the artist (*Metaphysics* VII, 7; 1032a35), but its union with matter is, as it were, at the mercy of matter and its specific receptivity. Form cannot be forced into or imposed upon matter; an artisan must let the matter guide the way it will receive form. The ultimate decision in action rests not with reason but with sensation, *aisthesis* (*Nicomachean Ethics* II, 9; 1109b23; cf. *Nicomachean Ethics* II, 2; 1104a1-9. Indeed, on one occasion Aristotle goes so far as to describe the coming together of form and matter, the becoming of an entity, as dependent on a "desire" or "reaching out" on the part of matter for form (*Physics* I, 9; 192a18).

What is at issue, as St. Thomas Aquinas notes in his commentary on this last passage, is the fact that matter, at least any specific piece of matter, is not just pure privation of form, but a real something in its own right. Although with respect to the object to be made it can be spoken of as formless, in fact it is itself an object which "seeks form or further form according to its proper nature" (*Commentary on Aristotle's Physics* I, lecture 15, paragraph 8). As Thomas says elsewhere (*Summa Theologiae* I, question 85, article 7), "Act and form are received into matter according to the capacity of the matter." Without a deep sensitiveness to the particular characteristics of this ordering toward form, this "desire" of matter, the result will almost surely be a weak unity, one tending to either rapid physical decomposition or aesthetic disorientation (which is but decomposition of another sort), or both. The classical artisan was interested in bringing about as perfect a union as possible, while recognizing that he could never completely duplicate the substantial union of form and matter to be found in nature (cf. *Physics* II, 1; 193a12-17). *Techne* in the classical understanding—and this cannot be emphasized enough when comparing ancient and modern making activities—is thus fundamentally oriented toward particulars instead of toward the efficient production of many things of the same kind in order to make money. Mass production would be unthinkable to the classical mind, and not just for technical reasons. In Book I of the *Republic* Thrasymachus puts forth a conception of *techne* as that power which pursues its own interests or the interests of its possessor—as a means by which the stronger dominates the weaker. Thrasymachus's *techne,* like modern technology, is oriented toward the extrinsic end of making money (not to say efficiency). In response Socrates rejects such a view and argues that *techne* as *techne* "does not consider its own advantage . . . but the advantage of that of which it is the art." "There is no kind of knowledge *(episteme)* that considers or commands the advantage of the stronger, but rather of what is weaker and ruled by it" (342c-d). *Techne* and *episteme* are both fundamentally oriented toward some otherness and its good, its "desires," and its "proper nature." When this otherness is material (as it is in the case of carpentry) and when matter is understood as inherently particular (as it is by Aristotle), then *techne* will be radically limited in its use of *logos.* Because it is matter which gives a particular its particularity (cf. *Metaphysics* VII, 8; 1033b20-1034a7), particular individuals themselves cannot be known in their particularity through the logical universal. *Logos* breaks down before particulars.

This inability of a technical *logos* to comprehend particulars in their particularity can be further elucidated by comparison with the inability of law, *nomos* (for the connection of *nomos* with *logos,* see *Politics* III, 15; 1286a15-17), to take account of every political circumstance. "Well laid down laws should themselves determine all the things they can," affirms Aristotle, "and leave as few as possible to the decision of the judges" (*Rhetoric* I, 1; 1354a32-34). Yet law is always a universal or general statement, and about some things it is not possible to make a universal statement which will be correct (*Nicomachean Ethics* V, 10; 1137b13-15); law is unable to speak with precision, *akribeia,* because of the difficulty any general principle has in embracing all particulars (*Politics* III, 11; 1282b5). Or again, "Some things can and others cannot be comprehended under law. . . . For matters of detail about which men deliberate cannot be included in legislation" (*Politics* III, 16, 1287b19-23). (For Plato's concurrence on this issue see *Statesman* 295a ff. and *Laws* VI 769d.) Indeed, so much is this the case that (at *Nicomachean Ethics* VI, 8) Aristotle feels compelled to distinguish between legislation, *nomothetikos,* and politics, *politike,* as a daily activity of deciding what is right in particular cases. The proper operation of both is grounded in φρόνησις or *prudentia,* which, in a definition which exactly parallels one given earlier for *techne,* is described as a habit with a true *logos,* περὶ τὰ ἀνθρώπινα ἀγαθὰ πρακτικήν—that is, concerned with humanly good action, *praktikos* (*Nicomachean Ethics* VI, 5; 1140b21). (Cf. the previous distinction between making, *poesis,* and the general process of acting or doing, *praxis,* at *Nicomachean Ethics* VI, 4; 1140a1-6; plus the use of *poiein* in *Categories* 4 and 9 and *Topics* I, 9 to refer to action in general, including both making and doing.) Politics as an active involvement in public affairs, by contrast, is concerned with action and deliberation about particulars (*Nicomachean Ethics* VI, 8; 1141b27-30). Men are forced to deliberate, to consider and take counsel with others (see *Politics* III, 15; 1286a26-30 and *Politics* III, 11) with regard to processes that happen in a definite way for the most part but not always, so that the results are indeterminate, contingent. In short, deliberation is about means, processes, not about ends or ideals (*Nicomachean Ethics* III, 3). Political wisdom thus focuses on *nomos,* law; politics on *dike,* that which is just or right in particular cases. Justice may be in accordance with *nomos;* still the specifications of law do not exhaust *dike* (*Nicomachean Ethics* V, 1). On the other hand, although *nomos* cannot fully determine *dike,* normally only one formed by or grounded in *nomos* will be able to do justice. (Compare the relation between virtue, *arete,* and *logos* described in *Nicomachean Ethics* VI, 1.) Judges are educated by the law in order to perfect or complete it (*Politics* III, 16; 1287a25-28 and 1287b25-26). They are thus the functional equivalent of artisans (cf. *Nicomachean Ethics* X, 9). Therefore, "as in relation to the other arts *(technai),* so in relation to the political" art, and its product, the political "organization, it is impossible that everything should be written down with precision" (*Politics* II, 8; 1269a10). Besides, "in any art of any kind it is absurd to govern procedure by written rules" (*Politics* III, 15; 1286a10). Once again, in making and in doing insofar as it involves making, the logical element breaks down before particulars.

As Jacques Maritain has argued in commenting upon a later development in this traditional understanding of *techne:*

> In the . . . arts the general end . . . is beauty. But . . . as an individual and original realization of beauty, the work which the artist is about to make is for him an end in itself: not the general end of his art, but the particular end which rules his present activity and in relation to which all the means must be ruled. Now, in order to *judge* suitably concerning this individual end, that is to say, in order to conceive the work-to-be-made, reason alone is not enough, a *good disposition of the appetite is necessary.* . . . The artist has to love, he has to love *what he is making.* . . .

And,

> because in the . . . arts the work-to-be-made is . . . an end in itself, and because this end is something absolutely individual, something entirely unique, each occasion presents to the artist a new and unique way of striving after the end, and therefore of ruling the matter. . . . [Thus] it is by using prudential rules not fixed beforehand but determined according to the contingency of singular cases, it is in an always new and unforeseeable manner that the artist applies the rules of his art. . . .

Recognition of the issue involved is not limited to the scholastic tradition. To cite one philosopher with different affinities, James K. Feibleman, in discussing the place of skill in technological activity, makes the following supplementary observations:

> The human individual is also a material object and if in harmony with his tools is capable of depths of understanding of them *as* material

objects when he has used them long enough. Such love for particular kinds of material objects comes only through a prolonged familiarity with their use and is not confined to their form but extends more deeply into their material.

However,

> It takes a man a long time and a great deal of concentration to become deeply acquainted with any material object. The mystical knowledge of matter has long been practiced but seldom recognized. Abstract knowledge is easy to acquire and to identify as such, but concrete knowledge is a different thing. Concrete knowledge uses quite different channels. It is absorbed by means of the sense organs and muscles. It comes through exteroceptors, such as the eye and the ear, and also through proprioceptors in the muscles.

In other words, there is at the heart of technical activity if not of *techne* itself an irreducible, nonlogical component; there is an aspect of *techne* which necessarily cannot be brought into consciousness except through the immediacy of a singular, direct encounter, an encounter which takes place through sensorimotor activity and is properly grounded in one of the various forms of love, *storge, philia, eros, agape.* Only love can encompass or grasp the singular.

Here, then, is the most fundamental difference between Greek *techne* and modern technology. *Techne* involves *logos,* but only in grasping form, not in directing the actual process of production, the activity *qua* activity. There is no *logos* of this activity. But is this not precisely what modern technology proposes to furnish, a *logos* of the activity, a rationalization of the process of production, independent of if not actually divorced from any particular conception of *eidos* or form? Is this not precisely why it can so vigorously claim to be neutral, to be dependent in use on what man wants to do with it, on purely extrinsic ends?

All this can be thrown into relief once more by considering the teachability of *techne,* something both Plato and Aristotle wish to affirm. Although *techne* is involved with language, and is hence teachable, we must be careful about reading into this notion our own conceptions of teachability. It is not teachable in the way modern engineering schools teach technology. What is teachable are the forms of beauty, not the processes of production. As for the practice of *techne,* for Aristotle *logos* is not enough. Like virtue, *techne* is learned primarily through practical imitation; "men become builders by building" (*Nicomachean Ethics* II, 1; 1103a35). Indeed, it is this fact which explains the absence of any general treatises on *techne* in the Aristotelian corpus—an absence, that is, except for one instance. It is in this instance, in Aristotle's treatise on the *techne* of rhetoric, that the words *techne* and *logos* are first joined together into a single term.

In writing on rhetoric, Aristotle four times ventures to unite *techne* and *logos* (*Rhetoric* I, 1; 1354b17, 1354b27, 1355a19, and I, 2; 1356a11). The precise meaning of each occurrence is debatable. That the term does not mean the *techne* of *logos,* the art of words, a synonym for *techne retorike,* is indicated by the parallel use of *techne tou logou.* In each case Aristotle could mean only "words about *techne*" or "speech concerning art," although even this weak sense would be significant. Yet in each instance there is intimation that *logos* of *techne* means something stronger, that Aristotle is trying to refer to a *logos* of the process of the *techne* of persuasion. This is suggested, for example, by the way Aristotle, unlike Plato, argues for the divorce of rhetoric from considerations of truth; rhetoric is a *techne* of the "means" of persuasion (1355b10). The *Rhetoric* is a treatise as much on "how to" as "what"—even "how to" never mind "what." Apparently when dealing with the art of persuasion, which operates with the medium of words or language—a very rarefied, not to say "artificial" material—there can be systematic discourse not only about forms or ends, but also about means or processes. And while Aristotle here argues that this is equally true of other arts such as medicine, one cannot help but suspect that this itself is a use of rhetoric. Is it not enough to note that Aristotle wrote no other such technological treatises? Be that as it may, it is in fact the case that words alone, divorced from reason, can acquire power simply as a means. Indeed, it is because of this divorce that Plato criticizes rhetoric (cf. *Gorgias and Phaedrus*) in a way which can easily be applied to modern technology. So it is not without raising pregnant questions that the Greek term "technology" comes to mean the study of grammar or rhetoric, and that we find "technologist" used to refer to the grammarian or rhetorician.

The full history of this usage need not be attempted here. Suffice it to say that occurrences involving these or closely related meanings are to be found in the works of Dionysius of Halicarnassus and Pseudo-Longinus (1st century B.C.), Hermogenes (2nd century A.D.), Porphyry (3rd century), Sextus Empiricus (3rd century), Iamblichus (3rd century), Gregory of Nazianzus (4th century), Photius (9th century), Zonaras (12th century), and Eustratius (12th century). The Greek term is used in a different sense once by Cicero (*Letters to Atticus* IV, 16), but its classical and medieval Latin adoption is obscure. The fact that dictionaries of medieval Latin do not go beyond the twelfth century—a need being filled by preparation of the *Index Thomisticus*—makes research in this area particularly difficult. Interestingly enough, however, the term does have an obsolete English usage which corresponds to the ancient Greek and must have been derived either directly or from some late Latin source. John Twell, in the preface to his *Grammatica Reformata, or, A General Examination of the Art of Grammar* (1683), writes, "There were not any further Essays made in Technology, for above Fourscore years; but all men acquiesced in the Common Grammar."

One of the earliest uses of *technology* in a more general sense occurs at the close of Sir George Buck's "breife report of the sciences, arts, and faculties" of schools "within and about the most famous cittie of London," when he speaks of an "apt close of this general technologie" in his *The Third University of England* (1615). Tobias Venner, in his *The Baths of Bathe* (1628), gives the term a similar meaning when he says he "cannot but lay open Baths Technologie." However, it is not until 1706, in John Kersey's edition of Edward Phillips's dictionary, *The New World of English Words,* that *technology* is clearly given its modern usage to refer to "a Description of Arts, especially the Mechanical." The fact that this same definition does not occur in the 1658 edition of Phillips's dictionary strongly implies that it was in the last half of the seventeenth century that the term acquired its present English denotation. It is from this usage which refers to the mechanical arts that extensions can readily be made first to the industrial arts and then to the practical arts collectively. As Jacob Bigelow says in the preface to his *Elements of Technology* (1831), he has adopted a word "found in some of the older dictionaries" in order to refer to "the principles, processes, and nomenclatures of the more conspicuous arts, particularly those which involve applications of science."

But what is to be made of this etymological development? What, if anything, is implied? Right away let me admit the incompleteness of the present study. It needs to be supplemented with etymologies of the German and French cognates of *technology* (the Buck and Venner spellings alone suggest French influence) as well as etymologies of *technique* and its cognates in German and the Romance languages. More than that, there is the influence of the concept of technology given by Christian Wolff in his *Preliminary Discourse on Philosophy in General* (1728), that which developed out of René Réaumur's posthumous *Descriptions of the Arts and Crafts* (beginning in 1761) and the great French *Encyclopédie* (1751-72), and that of the German educator Johann Beckmann in his *Anleitung zur Technologie* (1777). For Wolff *Technologie* is "the science of the arts and of the works of art" or the use of physics to give "the reason of things which occur through art." In the French Enlightenment technology is informed by the general rationalist ideals of criticism and power. With Beckmann technology is a functional description of the processes of production; this "general technology," as he calls it, provides a basis for the sound economic and political regulation of trade. The Puritan theological notion of *technometrica* or *technologica* as the science of defining all arts, with the consequent conception of the individual arts as "eupraxia methodically delineated by universal rules" also had its influence. Nevertheless, while recognizing the failure of the present work to take these and other influences into account, let me venture a speculative interpretation of the transformation from *techne* to technology.

At the foundation of the difference between ancient *techne* and modern technology (and ultimately I would argue that the adjectives "ancient" and "modern" are redundant here) is a conception of matter, an ontology or metaphysics of matter. Making *qua* activity or process, viewed solely as production rather than as production of some one thing, is dependent on this ontology. It is the theory about the nature of what one is working with that is a primary determinate of *how* one works, the structure of the working itself, although not necessarily *what* one works to do. The ancient or classical ontology involves looking upon matter as a living reality ordered toward the taking on of form—in accord with whatever form it already possesses and the potentialities contained therein. There is a hierarchy of form to be articulated in thought and to be responsible to in action. Plato envisions the cosmos itself as an organic unity, a living creature; at one point he specifically characterizes it as divine, a god (*Timaeus* 30c-31b and 34a-b). In the system of his Neoplatonic followers, all multiplicity—including its principle, matter—emanates from and is involved in a continual retroversion to the one source of all being, God (see Proclus, *The Elements of Theology,* Prop. 57 and Cor, with Prop. 72, Cor. Cf. also Plato's own myth of cosmic reversal at *Statesman* 268d-74d). This is not to deny the fact that matter or its functional equivalent (i.e., Plato's receptacle, *Timaeus* 49a) is often conceived by Neoplatonists as resistant to or opposing form—in the extreme case, as the principle of evil (Plotinus, *Enneads* II, 4, 16). Yet even Plotinus admits (*Enneads* IV, 8, 6) that everything, including matter, participates in the good itself "in the measure that each is capable of doing so." And St. Augustine, arguing that Christianity is the perfection of Platonism, states unequivocally that "matter participates in something belonging to the ideal world, otherwise it would not be matter" (*De vera religione* xi, 21). Thus matter is caught up in a cosmic process, and in that sense living.

For Aristotle and Aristotelians, in their less dramatic manner, something similar is involved. No matter, not even that strictly logical construction prime matter, is a purely neutral or lifeless stuff able to be imposed upon at will; it seeks or is related to form—in any particular case, in some particular way. This is why Aristotle can quite legitimately speak of a "desire" on the part of matter. It is also why, traditionally, moral discipline could not be divorced from the making activity; it is moral rather than intellectual discipline which cultivates man's receptivity to the needs and desires of another, which develops his ability to respect another being (whether human or not) for what it is in itself. This, incidentally, is also why alchemy is a sacred rather than profane activity: it is a "work" taking place at once in the subject and in his materials. The transmutation of some base metal into gold is but the exterior correlate of an interior spiritualization or divinization of soul. Only under circumstances where matter is cut off from this cosmic process—in a case, that is, such as language—is the idea of a technology thinkable for the ancients.

In the latter half of the seventeenth century, however, Western man's ontology of matter underwent a radical transformation. Perhaps it is not too much to say that under the influence of Galileo (1564-1642), Descartes (1596-1650), Newton (1642-1727), and their followers the material world began commonly to be regarded in much the same way as Aristotle looked upon words. Instead of a potentiality unknowable in itself yet ordered toward something higher, matter began to be conceived of as separated from any cosmic process. This trend is easily exemplified by the Cartesian theory of matter as pure, lifeless extension, in itself ordered toward nothing else, something to be done with as one pleases. Put more succinctly, matter ceased to be thought of as in any sense living—as having, as it were, any spiritual aspirations of its own. We must refer once more to alchemy as illustrative of the ancient worldview; for the alchemist, matter was conceived as an aspect of God. It was not so much opposed or indifferent to spirit as it was its necessary complement. In the modern scientific theory, however, matter does come to be conceived as wholly inert, totally devoid of spirit. (Note how, from this perspective, Spinoza's conception of extension and thought as two aspects of one divine substance was an attempted return to the ancient worldview.) Finally, it was on the basis of the modern hiatus that men began to imagine the possibility of a *logos* of *techne,* so that it made sense to use a term originally applied to the study of the manipulation of words to name the study of the manipulation of nature. Modern technology may, as Heidegger maintains, be the last stage of metaphysics, but not of ancient metaphysics.

I am not, of course, suggesting that this transformation of the term *technology* took place consciously; I have looked for evidence that this might have been the case but have not found any—except for things like Galileo's description of the natural world as a book whose language needs to be correctly understood. (Indeed, a history of the "book of nature" metaphor might well contribute insight along this line.) However, unintentional transferences are often more significant than strictly intentional ones. A full defense of my idea about the etymology of *technology* following upon conceptions of the ontology of matter would again, of course, entail plunging deeply into a discussion of the philosophy of nature—something which is clearly beyond the possibilities of this presentation. Nevertheless, perhaps this brief study can serve to substantiate my questioning of the facile historical identification of *techne* and technology in a way which is beneficial to larger issues.

III. Conclusion

What, more exactly, does my questioning mean? What I have argued by means of an etymo-philosophical study of the term *technology* is that, despite certain material continuities in the history of technology, there might well be formal discontinuities of greater significance. The history of technology is not nearly so linear and progressive as technological history has sometimes led us to believe. To borrow a suggestion from Thomas Kuhn's *The Structure of Scientific Revolutions* (2nd ed., Chicago: University of Chicago Press, 1962), perhaps the development of technology, like the development of science, should be viewed as proceeding within the framework of what he calls "paradigms." The making of artifacts—what things are made, how they are made and used —is not the result of some straightforward accumulation of technical knowledge or power; it is conditioned not only by social needs and values (as the social historian of technology would argue) but also, and perhaps more significantly, by philosophical ideas. Indeed, it may well be that technology is more akin to art, in the history of which, while there are change and multiplicity, there is no simple progress. There are periods of achievement and periods of failure, advance and decline, some periods in which ideals are realized and others in which possibilities are missed or lost. And there are different historical periods, marked off by differing ideals and practicalities along the physical and historical horizons. Approaching the world of artifacts from such a pluralistic perspective, through the conditioning of different social and conceptual circumstances, would in turn, I think, free us from the kinds of historicist pressures mentioned in the first part of this paper.

To summarize the argument, let me offer the following two theses for consideration: (1) The history of technology in its simplest form, i.e., technological history—especially in conjunction with the common presuppositions of our technological culture about the nature of man—is such that it invites a triumphalist philosophy of technology. So much is this the case that even when one wishes to object to this triumphalism—given, say, a different theory of the nature of man—one invariably feels called upon to present an opposing view of history. (2) But the philosophy of technology must itself overcome both biases, at least as biases. One good way to do this is through a study of the history of ideas about technology. And insofar as this is the case, philosophy would welcome from the historian more of this particular kind of history in the belief, as well, that it would throw new light on other problems in the history and philosophy of technology.

Let me close, however, by acknowledging that what I have said about the history of technology is not strikingly original. It simply restates, in different terms, a line of thought which has become part of discussions concerning the historiography of technology. Perhaps it is of some interest to find that, when looking at the history of technology, philosophy can sometimes reach conclusions similar to those of historians themselves. Let us welcome those occasions when the Muses meet in harmony.

Bibliographic Note

I.

For representative analyses by Bunge, Feibleman, Jarvie, and Skolimowski see essays by each author in C. Mitcham and R. Mackey, eds., *Philosophy and Technology: Readings in the Philosophical Problems of Technology* (New York: Free Press, 1972).

For the basic Marxist philosophy of technology see *Capital,* Vol. I, chapter 15, "Machines and Modern Industry." In English one could also consult the *Notes on Machines* (Leicester: Sublation, Students' Union, Leicester University, 1966), a brief translation by B. Brewster from Marx's *Grundrisse der Kritik der politischen Ökonomie* (1857-58). Portions of this same material can also be found in Karl Marx, *The Grundrisse,* edited and translated by David McLellan (New York: Harper and Row, 1971), chapter 18, "The Labour Process and Alienation in Machinery and Science." Representative Marxist ideologies of technology can be found in Samuel Lilley, *Men, Machines, and History* (New York: International Publishers, 1966) and the following articles by Alexsandr Zvorikine: "Science as Direct Productive Force," *Impact of Science on Society* 13:1 (1963), pp. 49-60; "Technical Progress and Society," in G. S. Metraux and F. Crouzet, eds., *The Evolution of Science* (New York: New American Library, 1963); "The Laws of Technological Development," in C. F. Stover, ed., *The Technological Order* (Detroit: Wayne State University Press, 1963); and, with E. I. Rabinovich, "Technology and Society," *Journal of World History* 12:1-2 (1970), pp. 103-26. Zvorikine is the general editor of the official Soviet history of technology.

For representative capitalist-American ideologies of technology see James Kip Finch, *Engineering and Western Civilization* (New York: McGraw-Hill, 1951); Courtney R. Hall, *History of American Industrial Science* (New York: Library Publishers, 1954); and John W. Oliver, *History of American Technology* (New York: Ronald, 1956).

Mumford's two major works on the history and philosophy of technology are, of course, *Technics and Civilization* (New York: Harcourt Brace, 1934) and *The Myth of the Machine,* 2 vols. (New York: Harcourt Brace Jovanovich, 1967-70). For Ellul's theories see his *The Technological Society* (New York: Knopf, 1964). For Heidegger, "Die Frage nach der Technik," in *Vorträge und Aufsätze* (Pfullingen: Neske, 1954).

Freidrich Dessauer has written two major works on the philosophy of technology: *Philosophie der Technik* (Bonn: F. Cohen, 1927) and *Streit um die Technik* (Frankfurt: J. Knecht, 1956). H. Van Riessen's work is *Filosofie en techniek* (Kampen: J. H. Kok, 1949). For Andrew G. van Melsen, see *Science and Technology* (Pittsburgh: Duquesne University Press, 1961). Hans Jonas has written the following important studies: "The Practical Uses of Theory," in his *The Phenomenon of Life* (New York: Harper and Row, 1966); "Philosophical Reflections on Experimenting with Human Subjects," *Daedalus* 98:2 (Spring, 1969), pp. 219-47; "The Scientific and Technological Revolutions: Their History and Meaning," *Philosophy Today* 15:2 (Summer,

1971), pp. 76-101; and "Technology and Responsibility: Reflections on the New Tasks of Ethics," *Social Research* 40:1 (Spring, 1973), pp. 31-54.

For basic existentialist reflection on technology see the two works by Ortega and Jünger translated in Mitcham and Mackey, *Philosophy and Technology;* as well as Karl Jaspers, *The Origin and Goal of History* (New Haven: Yale University Press, 1953), Part II, chapter 1.

For an account of the Verein Deutscher Ingenieure discussion of technology see Dessauer, *Streit um die Technik,* chapter 6. For mention of the involvement of American engineering societies see "At the Beginning," *Technology and Culture* 1:1 (Winter, 1959/60).

For my account of the historiographical debate I am indebted to "Symposium: The Historiography of American Technology," *Technology and Culture* 11:1 (Jan., 1970), especially George H. Daniels's contribution, "The Big Questions in the History of American Technology." The two major histories referred to are: Charles Singer, E. J. Holmyard, A. R. Hall, and Trevor I. Williams, eds., *A History of Technology,* 5 vols. (New York: Oxford University Press, 1955-58); and Melvin Kranzberg and Carroll W. Pursell, Jr., eds., *Technology in Western Civilization,* 2 vols. (New York: Oxford University Press, 1967). See also Maurice Daumas, ed., *A History of Technology and Invention: Progress through the Ages,* vols. 1 and 2, translated by E. B. Hennessy (New York: Crown Publishers, 1969); two further volumes are in preparation.

II.

Charles Singer's definition of technology (from *A History of Technology,* vol. 1, p. vii) is quoted, criticized, and then reformulated in Kranzberg and Pursell, "The Importance of Technology in Human Affairs," in *Technology in Western Civilization,* vol. 1, pp. 4-11. In the Singer history see also Gordon Childe's "Early Forms of Society," where (vol. 1, p. 38) he restates the definition of technology as "the study of those activities, directed to the satisfaction of human needs, which produce alterations in the material world." Immediately he adds, "In the present work the meaning of the term is extended to include the results of those activities." With Childe's commentary on the concept of human need as historically, socially, and even individually determined, his addition to the definition becomes vacuous. Daumas, in *A History of Technology and Invention,* is slightly more careful. The history of technology, he argues (vol. 1, p. 7), is the "description of techniques and their development." Techniques are then limited to "only those human activities whose object it is to collect, adapt, and transform raw materials in order to improve the conditions of human existence." This eliminates the techniques of accounting, banking, the conduct of military operations, and so forth. Furthermore, unlike Singer, who would explicitly include language as a technology (Singer, vol. 1, p. v.), Daumas only deals with "the methods of transmitting, recording, and writing it—paper, the proliferation of written texts, and so on." Still, by his qualifying to "improve the conditions of human existence," not to mention the problematic character of the ideal of improvement, does Daumas really intend to exclude the invention of nerve gas and nuclear weapons from his history?

The basic etymological information on *techne* and technology is compiled from standard reference works in the field: Liddel and Scott, *Greek-English Lexicon, Thesaurus Graecae Linguae,* the *Oxford English Dictionary,* and Eric Partridge, *Origins,* among others, although wherever possible original sources have been investigated. So far the only works specifically on this subject are to be found in German. See especially Wilfried Seibicke, *Versuch einer Geschichte der Wortfamilie um τέχνη in Deutschland vom 16. Jahrhundret bis etwa 1830* (Dusseldorf: VDI-Verlag, 1968); Johannes Erich Heyde, "Zur Geschichte des Wortes 'Technik,'" *Humanismus und Technik* 9:1 (1963), pp. 25-43; and Wolfgang Schadewaldt, "Die Begriffe 'Natur' und 'Technik' bei den Griechen," in *Natur, Technik, Kunst* (Gottingen: Musterschmidt, 1960). Seibicke's chapter III, "*Technologia*—Von der philosophischen 'Kunstlehre' zur 'Handwerkswissenschaft'," is especially relevant to parts of my argument toward the end of this paper, but because I am preparing an English translation of this work for publication, I have refrained from summarizing or commenting on its conclusions. A translation of Schadewaldt's essay is in preparation as well.

For other commentary on Plato's and Aristotle's understandings of *techne,* cf. Simon Moser, "Toward a Metaphysics of Technology," *Philosophy Today* 15:2 (Summer, 1971), pp. 129-56, with the supplement available from the editor-translator. Also, for a counterbalance to my strictly textual and conceptual analysis, one should compare the provocative work of the Plato scholar Robert S. Brumbaugh who, in *Ancient Greek Gadgets and Machines* (New York: Thomas Crowell, 1966), on the basis of archaelogical evidence, seeks to correct the notion that Greeks were devoid of an interest in and involvement with technology. "The truth is," he writes, "that invention and gadget-design gradually blossomed to a nearly contemporary fruition by the second century A.D." (p. 131).

The quotations from Maritain are to be found in *Art and Scholasti-*

cism and the Frontiers of Poetry (New York: Scribner's, 1962), pp. 46-48. Note that in the scholastic adaptation of Aristotle's definition of art which Maritain employs—"art is the right rule about things to be made"—"rule" is a translation of the Latin *ratio,* which in turn translates the Greek *logos.* Thus when, in the last sentence quoted, Maritain speaks of "prudential rules" he is evidently using the term in a different sense.

Art et scolastique was first published in 1920. Three decades later, in *Creative Intuition in Art and Poetry* (New York: Pantheon, 1953), Maritain returned to this theme. After rewording the scholastic definition of art as "the straight intellectual determination of works to be made" (p. 48), he goes on to discuss once again the character of rules in artistic activity. In this further discussion (pp. 56-58), Maritain makes three points relevant to the present argument. First, such rules "are subjected to a law of perpetual renewal." The eternal laws of art are "not to be found at the level of the particular rules of making." Second,

> the work to be made, in the case of the fine arts, is an end in itself, and an end totally singular, absolutely unique. Then, every time and for every single work, there is for the artist a new and unique way to strive after the end, and to impose on matter the form of the mind. As a result, the rules of making—which, as concerns art in general, are fixed and determined, as opposed to the rules used by prudence—come in the fine arts to share in the infinite suppleness and adaptability of the rules used by prudence, because they deal every time with the utter singularity of a new case, which is, in actual fact, unprecedented. It is, then, with prudential rules not fixed beforehand but determined according to the contingency of singular cases, it is with the virtues proper to prudence—perspicacity, circumspection, precaution, industry, boldness, shrewdness, and guile—that the craftsmanship of the artist succeeds in engendering in beauty.

Third,

> because the work to be made is an end in itself and a certain singular and original, totally unique participation in beauty, reason alone is not enough for the artist to form and conceive this work within himself in an infallible creative judgement. . . . To produce in beauty the artist must be in love with beauty. Such undeviating love is a supra-artistic rule—a precondition, not sufficient as to the ways of making, yet necessary as to the vital animation of art—which is presupposed by all the rules of art.

Here, in slight contrast to his earlier remarks, Maritain stresses love, not of the object made, but of the ideal of beauty. One may wonder whether, in the case of the practical arts, a corresponding love of the ideal of the useful (as a practical manifestation of the good) is not to be found. Evidently each of these elements mentioned by Maritain (and Feibleman) needs to be clarified by a phenomenological analysis of their interrelationships. The second work by Maritain, in its discussion of the conscious and unconscious aspects of making and of the various ways in which art focuses its attention upon particulars, points in such a direction. See, e.g., chapter I, "Poetry, Man, and Things," and chapter III, "The Preconscious Life of the Intellect."

For the quotation from Feibleman, see his "Technology as Skills," *Technology and Culture* 7:3 (Summer, 1966), pp. 327-28. Along this line compare Abraham Maslow's distinction between nomothetic and idiographic knowing—the one concerned with laws and classes, the other with unique individuals, especially persons—in his *The Psychology of Science: A Reconnaissance* (New York: Harper and Row, 1966), pp. 8-11. Cf. also John Julian Ryan, *The Humanization of Man* (New York: Newman Press, 1972), especially p. 22, and two articles by Cyril Stanley Smith: "Matter versus Materials: A Historical View," *Science* 162 (Nov. 8, 1968), pp. 637-44, and "Metallurgical Footnotes to the History of Art," *Proceedings of the American Philosophical Society* 116:2 (Apr. 17, 1972), pp. 97-135.

The point being made in these various references from Maritain, Feibleman, Ryan, and Smith can on occasion be found expressed by artists themselves. For instance, although in many of his letters and poems Michelangelo likens his work to the heroic imposition of form upon matter—an image of the artist which has been dominant in the West since the Renaissance—in one of his sonnets he speaks of merely releasing the form imprisoned in a block of marble. (See, e.g., Joseph Tusiani, ed. and trans., *The Complete Poems of Michaelangelo* [New York: Noonday Press, 1960], poems 83 and 84, pp. 76-77.) In this case his work was no more than instrumental, at the service of matter. This dialectic between active and passive responses to matter needs further elucidation, yet for present purposes it is enough to correct an imbalanced picture simply by noting that even the most "heroic" artists sometimes speak of "following the materials." The notion is more prevalent, however, in the crafts. Particularly in the discussion of oriental crafts one can often find this theme quite consciously formulated. See, e.g., Bernard Leach, *A Potter's Book* (New York: Transatlantic, 1965), as well as other works by the same author; and D. T. Suzuki, *Zen and Japanese Culture* (Princeton: Princeton University Press, 1959), especially his remarks on the insufficiency of technique in the art of swordsmanship, pp. 14, 113, 173, and 212. One relevant quotation is the story of a woodcarver in *Chuang Tzu* XIX, 10. Asked by the king how he was able to create for him such a perfect bell stand the artisan replied:

> When I began to think about the work you commanded
> I guarded my spirit, did not expend it
> On trifles, that were not to the point.
> I fasted in order to set
> My heart at rest.
> After three days fasting,
> I had forgotten gain and success.
> After five days
> I had forgotten praise or criticism.
> After seven days
> I had forgotten my body
> With all its limbs.
>
> By this time all thought of your Highness
> And of the court had faded away.
> All that might distract me from the work
> Had vanished.
> I was collected in the single thought
> Of the bell stand.
>
> Then I went to the forest
> To see the trees in their own natural state.
> When the right tree appeared before my eyes,
> The bell stand also appeared in it, clearly, beyond doubt.
> All I had to do was to put forth my hand
> And begin.
>
> If I had not met this particular tree
> There would have been
> No bell stand at all.
>
> What happened?
> My own collected thought
> Encountered the hidden potential in the wood;
> From this live encounter came the work. . . .

From the "imitations" by Thomas Merton in *The Way of Chuang Tzu* (New York: New Directions, 1965), pp. 110-11. Cf. also *Chuang Tzu* III, 2 and XIII, 10.

On the relevant aspects of Puritan theology see Keith L. Sprunger, "Technometria: A Prologue to Puritan Theology," *Journal of the History of Ideas* 29:1 (Jan.-March, 1968), pp. 115-22.

With regard to my brief mention of alchemy—that hermetic science in which the Aristotelian imagination is transmogrified—there are two works which are especially helpful to reflection on technology against this background: Mircea Eliade, *The Forge and the Crucible: The Origins and Structures of Alchemy* (New York: Harper and Row, 1971); and Titus Burckhardt, *Alchemy: Science of the Cosmos, Science of the Soul* (Baltimore: Penguin, 1971). It is interesting to note, moreover, that in an alchemical work attributed to St. Thomas Aquinas matter is raised to the status of a second principle and, along with form, is given divine significance. See *Aurora Consurgens* (New York: Pantheon, 1966), the "Commentary" by Marie-Louise von Franz, p. 385. As this Jungian interpreter observes in a footnote (p. 341), the importance of matter in the alchemical tradition is attested to by the figure of Hyle, a mother goddess found in Asclepius Latinus, as well as by the teachings of Hermogenes and Numenius. Perhaps it would not be too bold to suggest that the reason is because alchemy is more a practical than a theoretical science, one concerned with the actual process of generation—whether it be of gold or of the psychic reality for which this metal is a common symbol. If one may be even bolder and venture a comparison, in the theory of knowledge the scholastic principle *quodquod recipitur, recipitur per modum recipientis* (whatever is received must be received in accordance with the manner of the recipient) can be left as an abstract principle in a way which allows considerations of form to dominate an epistemological discussion; in educational practice, however, this principle will require an attention to the particular student which has not always been appreciated even by scholastics themselves. In practice, matter becomes of virtually equal importance. More succinctly, from the perspective of thought there is but one principle, form (matter is only a relationship to form); but from the perspective of practice there are two, form and matter (although the exigences of practice may at times make form seem to be no more than a relationship to matter). This interpretation is borne out by Giordano Bruno's description of matter as "the divine and excellent progenitor, generator and mother of natural things" (see *Concerning the Cause, Principle, and One,* Fourth Dialogue, in Sidney Greenberg, *The Infinite in Giordano Bruno* [New York: King's Crown Press, 1950], p. 156). However, the nature of Renaissance alchemy is ambivalent, to say the least. The relation of alchemy to the transition from ancient to modern conceptions of matter should not be investigated without reference to Frances A. Yates, *Giordano Bruno and the Hermetic Tradition* (New York; Random, 1964). According to Yates, "The basic difference between the attitude of the magician to the world and the attitude of the scientist to the world is that the former wants to draw the world into himself, whilst the scientist does just the opposite . . ." (p. 454). Donald Brinkmann in *Mensch und Technik: Grundzüge einer Philosophie der Technik* (Bern: A Franke, 1946) and William Leiss in *The Domination of Nature* (New York: Braziller,

1972) both consider the relation between alchemy and modern technology from a somewhat different angle. For the accepted interpretation by technological history see R. J. Forbes, *Studies in Ancient Technology,* vol. 1 (Leiden: E. J. Brill, 1955), pp. 121-44.

Galileo's famous metaphor is to be found in *The Assayer* (1623): "Philosophy is written in this grand book, the universe, which stands continually open to our gaze. But the book cannot be understood unless one first learns to comprehend the language and read the letters in which it is composed. It is written in the language of mathematics, and its characters are triangles, circles, and other geometric figures without which it is humanly impossible to understand a single word of it; without these, one wanders about in a dark labyrinth." Translated by Stillman Drake, in *Discoveries and Opinions of Galileo* (Garden City, N.Y.: Doubleday, 1957), pp. 237-38. Thomas Browne in *Religio Medici* (1643) makes a similar reference to "nature, that universal and publick manuscript, that lies expansed unto the eyes of all" and concludes by denying the fundamental Greek distinction between nature and artifice: "All things are artificial; for nature is the art of God" (First Part, xvi). Lest it too readily be assumed that this follows simply from the Christian doctrine of creation *ex nihilo,* it can be noted that St. Augustine explicitly rejects the metaphor. In a sermon from the *Ennarationes in psalmos* (psalm 103, part I) Augustine, with obvious reference to Romans 1:20, describes nature as a visible image of the invisible God, but then contrasts the word of Scripture with the silence of nature and argues for a need to read the natural descriptions in Scripture and nature as well in an allegorical manner. Applying his own principle, in *Epistle* 55 (viii, 13) he writes that "we make use of parables, formulated with reverent devotion, to illustrate our religion, drawing freely in our speech on the whole creation, the winds, the sea, the earth, birds, fishes, flocks, trees, men; just as, in the administration of the sacraments, we use with Christian liberty, but sparingly, water, wheat, wine, oil." Quoted from *Fathers of the Church,* vol. 12 (Washington, D.C.: Catholic University of America Press, 1950), p. 271.

Finally, for some collaborative confirmation of my speculative thesis regarding ancient versus modern ontologies of matter, one could compare R. G. Collingwood, *The Idea of Nature* (New York: Oxford University Press, 1945), pp. 111-12:

> For the early Greeks quite simply, and with some qualification for all Greeks whatever, nature was a vast living organism, consisting of a material body spread out in space and permeated by movements in time; the whole body was endowed with life, so that all its movements were

> vital movements; and all these movements were purposive, directed by intellect. This living and thinking body was homogeneous throughout in the sense that it was all alive, all endowed with soul and with reason; it was non-homogeneous in the sense that different parts of it were made of different substances each having its own specialized qualitative nature and mode of acting. The problems which so profoundly exercise modern thought, the problem of the relation between dead matter and living matter, and the problem of the relation between matter and mind did not exist. . . . There was no material world devoid of mind, and no mental world devoid of materiality; matter was simply that of which everything was made, in itself formless and indeterminate, and mind was simply the activity by which everything apprehended the final cause of its own changes.
>
> For the seventeenth century all this was changed. Science had discovered a material world in a quite special sense: a world of dead matter, infinite in extent and permeated by movement throughout, but utterly devoid of ultimate qualitative differences and moved by uniform and purely quantitative forces. The word "matter" had acquired a new sense: it was no longer the formless stuff of which everything is made by the imposition upon it of form, it was the quantitatively organized totality of moving things.

From what has been argued in the main body of my text, it can be seen that I would strongly disagree with certain elements of Collingwood's idea of the ancient ontology of matter. But, as Whitehead has remarked in one of his own studies on this theme, "The history of the doctrine of matter has yet to be written" (*The Concept of Nature,* [Cambridge: Cambridge University Press, 1920], p. 16).

Supplementary Note (1978)

With regard to Bibliographic Note I, paragraph 2: Marx's notes on machines can now be consulted more easily in Martin Nicolaus, ed. and trans., *The Grundrisse: Foundations of the Critique of Political Economy* (New York: Random House, 1973), especially pp. 692-704.

Ellul's analysis of technology has been revised and updated in *Le Système technicien* (Paris: Calmann-Levy, 1977).

The last three papers by Jonas, cited on pp. 190-91 have been included, along with five others on this same theme, in Hans Jonas, *Philosophical Essays: From Ancient Creed to Technological Man* (Englewood Cliffs, N. J.: Prentice-Hall, 1974).

For further historiographical discussion along the lines indicated here, see the symposium "The Historiography of Technology," *Tech-*

nology and Culture 15:1 (Jan., 1974), which contains the following papers: Robert R. Multhauf, "Some Observations on the State of the History of Technology"; Eugene S. Ferguson, "Toward a Discipline of the History of Technology"; Edwin T. Layton, Jr., "Technology as Knowledge"; and Derek de Solla Price, "On the Historiographic Revolution in the History of Technology: Comments on the Papers by Multhauf, Ferguson, and Layton." One earlier reference along the same line is Reinhard Rürup, "Historians and Modern Technology: Reflections and the Development of Current Problems of the History of Technology," *Technology and Culture* 15:2 (April, 1972), pp. 161-93.

With regard to Bibliographic Note II, paragraph 2: Schadewaldt's essay can be found translated as "The Concepts of 'Nature' and 'Technique' According to the Greeks" in *Research in Philosophy and Technology,* vol. 2 (1979).

The German essay that contains Simon Moser's original discussions of *techne* in Plato and Aristotle (from his *Metaphysik einst und jetzt* [Berlin: DeGruyter, 1958]) has been slightly revised and reprinted as "Kritik der traditionellen Technik-philosophie," in Hans Lenk and Simon Moser, eds., *Techne, Technik, Technologie: Philosophische Perspektiven* (Pullach: Verlag Dokumentation, 1973), pp. 11-81.

For an interesting etymological footnote on "technique-technology" in English see D. L. Munby's "Note on 'Technology,'" in his *God and the Rich Society* (London: Oxford University Press, 1961), pp. 153-56.

Another extremely suggestive etymo-philosophical discussion of this topic is Richard McKeon's "Logos: Technology, Philology, and History," in *Proceedings of the XVth World Congress of Philosophy: Varna, Bulgaria, September 17-22, 1973* (Sofia: Sofia Press Production Center, 1974), vol. 3, pp. 481-84.

For more on technology as skill, see Stanley R. Carpenter, "Modes of Knowing and Technological Action," *Philosophy Today* 18:2 (Summer, 1974), pp. 162-68; and the chapter "Craftsmanship and Knowledge" in Nathan Rotenstreich, *Theory and Practice* (The Hague: Martinus Nijhoff, 1977).

Further support for my attempt to distinguish *techne* from technology can be found in George Sturt's *The Wheelwright's Shop* (Cambridge: Cambridge University Press, 1923; reprinted Cambridge, Mass.: Harvard University Press, 1963) from which, had I known about it, I would have quoted in conjunction with my references to Michelangelo, Bernard Leach, D. T. Suzuki, and Chuang Tzu. Christopher G. Williams's *Craftsmen of Necessity* (New York: Random House, 1974) also contains good material along this same line. Recent discussions of alternative technology and historical studies such as Arnold Pacey's *The Maze of Ingenuity: Ideas and Idealism in the Development of Technology* (Cambridge, Mass.: M.I.T. Press, 1974) also tend to confirm my thesis.

In a remarkable book entitled *Zen and the Art of Motorcycle Maintenance: An Inquiry into Values* (New York: William Morrow, 1974), Robert M. Pirsig points in the same direction when he describes a good mechanic's sensitivity to materials:

> The craftsman isn't ever following a single line of instruction. He's making decisions as he goes along. For that reason he'll be absorbed and attentive to what he's doing even though he doesn't deliberately contrive this. His motions and the machine are in a kind of harmony. He isn't following any set of written instructions because the nature of the material at hand determines his thoughts and motions, which simultaneously change the nature of the material at hand. The material and his thoughts are changing together in a progression of changes until his mind's at rest at the same time the material's right. (p. 167)

Later he speaks more explicitly of "what's called 'mechanic's feel,'" something

> which is very obvious to those who know what it is, but hard to describe to those who don't; and when you see someone working on a machine who doesn't have it, you tend to suffer with the machine.
>
> The mechanic's feel comes from a deep inner kinesthetic feeling for the elasticity of materials. Some materials, like ceramics, have very little, so that when you thread a porcelain fitting you're very careful not to apply great pressures. Other materials, like steel, have tremendous elasticity, more than rubber, but in a range in which, unless you're working with large mechanical forces, the elasticity isn't apparent.
>
> With nuts and bolts you're in the range of large mechanical forces and you should understand that within these ranges metals are elastic. When you take up a nut there's a point called "finger-tight" where there's contact but no takeup of elasticity. Then there's "snug," in which the easy surface elasticity is taken up. Then there's a range called "tight," in which all the elasticity is taken up. The force required to reach these three points is different for each size of nut and bolt, and different for lubricated bolts and for locknuts. The forces are different for steel and cast iron and brass and aluminum and plastics and ceramics. But a person with mechanic's feel knows when something's tight and stops. A person without it goes right on past and strips the threads or breaks the assembly. (pp. 323-24)

Similar descriptions of the intuitive element and material factors can be found in John Jerome, *Truck: On Rebuilding a Worn-Out Pickup, and Other Post-Technological Adventures* (Boston: Houghton Mifflin, 1977).

For a discussion that supplements my references to alchemy and to traditional crafts see Frithjof Schuon, *The Transcendent Unity of Religions,* rev. ed. (New York: Harper and Row, 1975), chap. 4, "Concerning Form in Art," especially p. 66: "The treatment of the material must be in conformity with the nature of that material in the same way that the material itself must be in conformity with the use of the object." Diogenes Allen, *Finding Our Father* (Atlanta, Ga.: John Knox, 1974), also contains an interesting and relevant discussion of Christian love as necessarily bearing on particulars. A strictly epistemological discussion of "Knowledge of the Singular: Aquinas, Suarez, and Recent Interpretations" is given by Francisco L. Peccorini, *Thomist* 38:3 (July, 1974), 606-55.

The elaboration of the *techne* (but not the technology) of psychology in works such as Milton Mayeroff, *On Caring* (New York: Harper and Row, 1971), also bears out my thesis about the nonlogical character of *techne.*

Jean-Claude Beaune

12

Technology from an Encyclopedic Point of View

My aim is to isolate and circumscribe the field of technology and then, if possible, to evaluate it as a unit within the global field of knowledge. As this subject seems to me too large and complex for this short space, I shall simply outline my work and indicate some important elements to treat elsewhere at greater length.

The reader may wonder why this research apparently depends more on compilation of references than on original thinking. Two principles I have laid down for myself account for this method:

1. This research can only be objective and honest when effected by anonymous thinking free of any complacent subjectivity. (As such, it deliberately breaks with a certain European philosophical tradition.) Such thinking is a relative or, shall we say, irreal language—a provisional solution and, perhaps, an easy way out of the problem of objectivity.
2. Form and substance overlap in my proposed definition of technology, especially in the linguistic function I assign to it insofar as it is a transphenomenal field.

The concept of technology to be developed here represents the formal code of an interpretation of technics coextensive with possible objective conditions. However, this interpretation also implies an organized whole, autonomous and significant in itself, free of any external attributes—and, in particular, free of any shadow of humanism. In this sense we can speak of the autonomy of the technological world as an operative notion, a working hypothesis or, in Max Weber's words, an "ideal type."[1] In my opinion, this notion of autonomy is more operative, objective, and profound than the traditional idea of the political or moral neutrality of technology.

This structuralist point of view implies a historical approach, but history rewritten and idealized. To avoid contamination by the ideology of continuity and progress, this history is based on the axiom of recurrence. For example, the recent emergence of a unified technological language—cybernetics—gives us an archaeological understanding of past forms of technical language, even if by comparison they seem archaic, unconscious, and faltering.

Confronted with the notion of technology, we find ourselves in a labyrinth, and the problem is to find Ariadne's thread. We shall consider the idea of the encyclopedia for this clue, but this only intensifies the difficulties. This Ariadne's thread is, in fact, a second labyrinth which intersects with the labyrinth of technology, through the historical and structural multiplexity of the encyclopedic phenomenon. Can we escape this fantastic situation which even Borgès would not disown? How can we penetrate to the center of meaning where a legislative act will change technology from outlaw to positive law?

This problem of relating technology to encyclopedism, expressed in terms of fields, consists in varying two sets of meanings, each as a function of the other, and in considering their interactions, unions, and disjunctions. To change images, history intervenes indirectly, as an objective landscape so to speak, not carrying with itself any metaphysical postulates. Philosophy stands out like the actor's silhouette in a shadow play, as a dark projection of technology, not only its mask or substitute but also the ultimate boundary conditions of impossibility around technology's definition of itself. Naturally I am speaking of the subjectivist, humanistic, and theological philosophy we are well acquainted with in Europe. As for other philosophy, occasional flickers of light permit a glimpse of it, but not accurate characterization. At the center of meaning of our labyrinth is science. However, this is a fragmented and mosaicized science: in 1959 J. T. Thykouner counted 1,150 existing sciences.[2] Furthermore, this science is questioned by the society of man which retains, even when silent, the bitterness of frustrated hopes.

Then why not let technology be? What is the point of justifying, through forgotten antecedents, the authority technology possesses naturally? Well, the attempt seems to me worthwhile. For one thing, it is justified by technology's ambiguous pedagogic standing among the traditional university subjects. In Europe, at least, technology remains a subordinate subject in the university, somewhat despised both by traditional intellectuals and by managers who show little enthusiasm for hiring students trained in this discipline.

Technology, in the French context at least, tries to conceal but in fact reveals a twofold incapacity of the traditional pedagogic system: first, its incapacity to define clearly the relationship between industry and the university; second, its inability to define a strict organization of the different bodies of knowledge. This inability is concealed by vague concepts of complementarity and interdisciplinarity, the result of a withdrawal from outmoded, impregnable, but metaphysically overestimated positions such as pure science, humanistic and reflective philosophy, art for art's sake, and so on.

My goal is therefore (1) to add positive criteria to technology as a subject taught in universities and as a systematic and partly autonomous organization of a body of knowledge; then (2) to constitute an endogenous model for it, a model that permits us to understand its relationship with external fields of knowledge as expressions which reflect back on technology itself and make possible an investigation of its intrinsic meanings. This idea entails a development at several related levels.

It appears to me that the crux of the matter is to be found in a juridical evaluation of the status of technology. All its problems and meanings are crystallized there in the strictest and most abstract way. This essential legalism has significant repercussions on other more empirical levels:

1. A linguistic level—the relationships between the different scientific syntactic and semantic fields and the parascientific fields of technics defined according to their hypothesized autonomy;
2. An epistemological level—the relationships between scientific research and the technological world, provided with at least a partial autonomy;
3. A sociological and "deontological" level—that of the fall-out

of science onto society and of the general diffusion of knowledge, considered as a strategy of communication.

Let me now indicate the logical order of this paper:

1. Definition of a privileged touchstone to consider the problem: Diderot's Encyclopedia,[3] a political act which removes all indignity from craftsmanship and productive work, and the first act institutionalizing technics. From this virtually alchemical affinity between the two fields of encyclopedia and technology, we move to a more systematic and stricter conception of encyclopedia, more capable of resisting historical contingency. Lastly, we catalogue the different meanings of encyclopedia and technology in their different historical modes.

2. Introduction of science to study its reciprocal relationships with technology. Through a projection of the problem onto science, we study borderline cases of this relationship. Next we define the role of technology in relation to science and, by extrapolation, in relation to human knowledge as a whole and organized system. Then we evaluate the relationships between the abstract establishment of technology as autonomous and its encyclopedic function, which is to fix and unify knowledge.

3. By way of counter-proof, we define the present state of technology and apply this definition retrospectively to different historical expressions of technology, especially *technological "language,"* this definition being a code and a basis for a new decoding of well-known meanings.

I. The Encyclopedia of Diderot and His Contributors

Is Diderot's Encyclopedia a summary of human knowledge prompted, as Groethuysen maintains, by a will to possess the world? [4] Or it is, as Butor says, a "gigantic but entirely useful mystification?" [5] The *Description des Arts et Métiers* of the Royal Academy of Science would seem to provide just as good an example for the historian of technology.[6] However, apart from the obvious affiliation between these two works, it is the spirit of Diderot's Encyclopedia which interests us here, or rather its possibly questionable function of systematizing and popularizing technics.

Throughout the Encyclopedia, one critical problem is that of technical knowledge and power being withheld by a group of men responsible for their use in work and production. The Pythagorean conception of secret knowledge as the privilege of a few initiated specialists disappears, shattered by a great hope. With the Encyclopedia, the ignominious etymological meaning of "work" as torturous fatigue (*tripaliare* = travail) is obliterated. The working class now has a means to define itself in a positive way, even if this definition will only be given later. To begin at the beginning, the Encyclopedia is a number of printed plates for craftsmen and in addition an opportunity for these craftsmen to write some of the articles.

Technics speaks, probably for the first time, and can be heard by all. Naturally, this language is not elaborate, but it is the language of reason and, moreover, all the other nontechnical languages of this work bear its mark and influence. Two sentences by Diderot raise the curtain on a rational and already partially autonomous technology. "Let a man from the academies come and visit the workshops, who would gather up the phenomena of the arts and expose them to us in a book which would urge artists to read, philosophers to think in a useful way, and powerful people to make sound use of their authority and rewards." [7] "In what system of physics or metaphysics can one find more intelligence, sagacity and importance than in a machine spinning golden threads or a machine making socks, and in the craft of lace makers, clothiers or silk makers?" [8]

In this work there is a remarkable convergence of old and new. The old encyclopedic ambitions of Blount, Chambers and his Cyclopedia, Descartes, and Leibnitz, as well as all the breviaries, dictionaries, compendia, and libraries of that time—all these meet the new mechanical creations such as Vaucauson's automata for "demonstration," John Kay's shuttle of 1733, and Wyatt and Paul's weaving machine of 1738. This convergence is expressed in a prophetic poetry reminiscent of the utopian texts of those days, all those imaginary recreations of a social world made rational by technical devices.

The Encyclopedia will cause a later reclassification of values, activities, and elements of knowledge. It undertakes to substitute order for disorder, planning for chance, and systematic thought for unfounded speculation. In fact, it resembles a complex but harmonious machine. Despite the Encyclopedia's many deficiencies and oversimplifications, the image which survives takes on a mythical value. Like myths which are boundless though closed, it recalls intellectual imperatives that are still vital: abandoning subjectivity, accepting artificiality in order to reject dogmatism, de-dramatizing the human world, and perhaps even a social mask-swapping comparable to the great communication game Rousseau had already mentioned in his *Lettre à d'Alembert sur les spectacles* (Letter to d'Alembert on the theater), which strikingly characterized politics as organizing a reciprocal social space.[9] Both a retrospective and a prophetic work, the Encyclopedia—defined by Diderot as "an organism renewing its cells"[10]—deserves a better fate than being considered an anachronistic and exotic literary document.

Imagine a comprehensive science in which all terms used would be totally defined and through which the practical function of the mind could be reintegrated into science. Technology would play the part of a science, or even perhaps of an element unifying all the sciences, and would be a source of models for general knowledge. Leibnitz and the spirit of his work are implied, of course, by these words. The encyclopedic perspective we started with takes on a new meaning. We move from noting the historical convergence between encyclopedia and technics to the quest for a universal form for scientific knowledge—a "tree of knowledge" or classification of sciences. This formalistic and systematic encyclopedism aims to substitute order for barbarism, to fill the large gaps in knowledge which accompany extreme specialization, and to substitute structure for fragmentation. Then technology could appear as a guide, a "governor" in this systematizing activity. It could take up again the work left unfinished by philosophy (in the sense of Auguste Comte), since philosophy at present too often reduces its epistemological vocation to working out the chronology of knowledge, a history *in subjectò* of the methods of science, or making an herbarium of scientific things.

Is technological encyclopedism a very remote and archaic approach to solving the problem? Not at all, for the encyclopedic spirit is not dead. A concrete and outstanding fact of civilization today is the need for encyclopedism, a need felt by both scientists and laymen to rely on a *summa*. This wish or this sense of inadequate knowledge partly explains the confusion of isolated specialists, of disillusioned students, and of culturally alienated workers and technicians. The encyclopedic ideal of universal science has altered its meanings, but it is still not dead.

We need anchoring places, crossroads sciences, sectors where knowledge can be fixed, at least provisionally. Technology as a study of general forms of communication would play a synthesizing role in the total ensemble of knowledge. Perhaps also, as we shall see later, it could replace classical logic in formalizing knowledge at the foundation of the sciences and arts.

Here are roughly parallel lists of various meanings of the two ideas *encyclopedia* and *technology,* showing their more than circumstantial similarity. The idea of encyclopedia connotes:

1. To Diderot: a rational glossary and operational catalogue of recipes or tricks of the trade
2. To Leibnitz: an encyclopedia of general science and universal characterization
3. To Ampère, Cournot, Comte, and Spencer: classification of sciences and specialization of knowledge
4. To Linnaeus, Jussieu, Cuvier, et al.: a method of systematizing and classifying things
5. A "Rabelaisian Summa," tables of anecdotes and recipes, genealogies
6. Interdisciplinarity, communication of knowledge, codes for deciphering, and structural constants

To this empirical and noncomprehensive list, we can compare a list of suggestive meanings of technology. We can even superimpose one list on the other:

1. Reflective and rational technics
2. Catalogue of artisan's recipes
3. Applied science, quantitative science, and its commercial applications
4. An autonomous and axiomatized science
5. Universe of communication with its own rules and methods
6. Social and economic worlds, either local or global, such as trade guilds or the technostructure

Anyone can establish both historical and structural correspondences in this double list, taking into account the more or less conscious nature of the relationships. However, it seems to me that definite results cannot be obtained without introducing science.

II. Relationships between Science and Technology

This is not the place to consider the eternal problem of the relationship between science and technics, or between applied science and technology. Their relationships remain particular and contingent relationships. To sum it up in an image, the scientifico-technical world is a double sun around which research workers revolve. Even if it is possible to be closer to one sun than to the other, attraction to the second one still persists however close one may be to the first. We need to leave the multiplicity of facts and consider a problem of scientific law. There is a juridical lacuna as far as technology is concerned. Establishing institutions and problems requires a unanimously acceptable, clear, and precise definition of technology. It is no doubt possible, as some authors have done, to grant technology the status of science, but this implies a reduction if not an assimilation.

It is the opposite process which should interest us: the unifying function of technology for science as a whole and perhaps for knowledge as a whole. This function is not contradictory to the idea of autonomy presented above. The idea of technology from the point of view of an internal crisis of scientific law will serve as a guide to the field of technology in some of its historical expressions and its organizational structures. It is worth noting that Alsted in his Encyclopedia (1630) defined the word *technologia* as "doctrina praecognoscenda de affectionibus, ordine et divisione disciplinarum" [11] Alsted was mentioned by Leibnitz in *De Arte combinatoria* as one of his forerunners.[12]

Among the legislators of science, one of the most remarkable is Ampère. Technology is defined by this author as a science of the first order inside the branch of physical sciences and the kingdom of cosmological sciences.[13] Technology is distinguished from the natural sciences which consider their inorganic subject "apart from its uses." [14] Then the field of technology defines four sciences of an inferior rank. This classification of sciences, explicitly derived from Jussieu's classification of plants, is based on a "key" the writer develops for us in his preface. Ampère's attention is first focused on "general physics" and the specific elements which distinguish it from other sciences. The point is to encompass a field. From the very first page, three related sciences appear, which are distinct from general physics: (1) natural sciences, which study inorganic properties of bodies; (2) technology, which considers the uses of these bodies; (3) physical geography, which deals with variations in these bodies according to time, place, and climate. We may say that a science is not only defined by its subject, but also by the point of view and the relative flexibility of this subject in time and space.

Therefore, Ampère stands at the level of scientific language in which subjects acquire the value of signs authorized by a general consensus from their context. His approach is not theory or intellectual operations on subjects, but lawful organization of scientific language. These rules must be justified by changing from linguistic consensus to definite law. In the last part of his preface, Ampère defines his task as foundation and legislation without neglecting language. Cataloguing the sciences therefore consists of assigning names and designating each science by an appropriate term. Undoubtedly sciences and words interact in a reciprocal way which affects the legal quality of this catalogue, but this very reciprocity establishes a relation between language and law, between custom and science. Two main consequences of this example are to be kept in mind: first, the first appearance of technology clearly distinguished from general physics already determines its autonomy as a science; and second, this is only an idealized autonomy, since it has meaning only in its systematic context of a theoretical and legalizing encyclopedia; but this very legalism justifies technology by clarifying its relation to other sciences. Although Ampère's program is very formal, it suggests a richer interpretation of technology than does traditional reductionism.

Let us consider two characteristic borderline cases. The first concerns Diderot's assessment of the interplay of technology and mathematics. This implies a more or less clear consciousness of the dogmatism of traditional classifications of knowledge. Inspired by an insight that is both profound and naive, Diderot attacks pure or "transcendent" mathematics as a "kind of general metaphysics in which the bodies are deprived of their individual qualities." In the "Interprétation de la nature," Diderot does not hesitate to predict, "We are on the point of a great revolution in science. I would even dare to assert that before a hundred years has passed, not three geometricians will be counted in Europe. This science will stop short. We shall not go beyond it." [15] Thus, even in the Age of Reason and at the climax of *mathesis*, Diderot foresaw that science would break with it. Abandoning its imperialistic role in science, *mathesis* was to give way to *physis* and *techné* in particular.

An even better example is the second borderline case, Reuleaux's attempt to axiomatize technology. Lewis Mumford characterizes it as "a work so remarkable it discouraged its successors." [16] Indeed the structure, if not the contents, of Reuleaux's work will not be excelled. It is a systematic, axiomatic, universal, and logically formal treatise on machinery. To paraphrase Leibnitz, it is both an "*ars probandi*" and an "*ars inveniendi*." Analysis and construction are inseparable, for practical activity is contained in technical knowledge. The point is to find an elementary structure of the machine simple enough to be general and exhaustive enough to provide designs for special constructions. Reuleaux wishes to substitute unity for multiplicity, models for images, structure for analogy. His fundamental concept is that of coupled elements. Two objects always interact through structures and laws that allow only one movement for each one in relation to the other. Combining these couples of elements produces a mechanism when one of the members of the kinematic chain is fixed; and one or more of these kinematic chain mechanisms comprise a machine. From this elementary structure may be deduced the fundamental elements of real machines: (1) material organs of tension and transmission such as chains and straps or their hydraulic counterparts or (2) linguistically fixed concepts or symbols able to retain their meaning in different languages. With this analysis, more than thirty apparently dissimilar rotary machines can be deduced from the ordinary steam engine.

Mathematical elements appear in analysis and construction to crystallize their language. But mathematics does not predetermine the machine, and the machine cannot be reduced to a linguistic field different from the one its existence defines. Simple terms such as couple, strap, or fluid are technological; mathematics appears in the organization of these terms. Reuleaux asserts: "The science of machinery must be based on deduction. To this end, it is necessary to give the teaching system of this science a form which allows it to rest on a few fundamental truths which it possesses in its own right." In a strictly scientific way, the study of machinery is dependent on theoretical mechanics of a mathematical type. However, "It is no longer the same, in my opinion, if we consider a practical point of view; in that case indeed, since the study of machinery draws from a large number of sources and since it is enriched with many theories of its own, it actually constitutes a special field of activity capable of many subdivisions. This being so, it is obviously not only possible but necessary to distinguish the scientific problems it can bring about from the problems of general mechanics."[17] Nature has a share in the machine itself. One of the main originalities of Reuleaux is to assert that the cosmic freedom of natural phenomena is transformed, in machines, into a system of laws in which external forces no longer apply. In addition, natural latent forces prevent perturbations. The teleology of the machine is thus opposed to natural chance by (1) excluding the possibility of any perturbation through the mechanical use of nature and (2) assigning a unique and necessary movement to the whole.

In the field of autonomous technology, Reuleaux proposes a classification of mechanical sciences. "Practical mechanics" can be divided into four "secondary sciences" of machines: (1) the general science of machines, determining suitable ways to execute prescribed functions; (2) the special science of machines, studying perceptible forces and movements; (3) the science of machine building, studying the resistance of bodies; and (4) the science of machine motions, studying the possibility of producing predetermined results.

This classification is not arbitrary, but deduced from the qualities of the subject (or what Reuleaux calls "the properties of the machine"), which are the axes onto which the elementary structure of the machine is projected.[18] Here we are very close to the assertion of technological autonomy. If we remember that, for Reuleaux, to think and to build were one and the same, never has Diderot's encyclopedism been so close to that of Leibnitz.

Now we come upon two important extensions of the problem: the image of the machine, and the functions of technology. The very image of the machine determines the organization of the field of technology in different cultural contexts. In the eighteenth century, the technical complex as a combination of elementary apparatuses is still limited to the workshop. The machine is interpreted as an extension of gesture, as the sum of instruments to extend and to arm the body of the craftsman. Consequently, the machine has not acquired full power over matter and nature. Nature still dictates its conditions. Among these conditions, some very numerous and essential ones "cannot be submitted to that part of the calculations which extend to the most subtle differences in quantities." [19] The machine thus poses the crucial question of the relationship between the science of quantity and the qualitative realities of nature. Mathematics can be accused of depriving bodies of their "individual qualities." Diderot, we can imagine, finds himself confronted with a vitalist obstacle he cannot overcome. The machine is assimilated to a natural creation whose origin, genesis, and cycle of development are animal-like. The technical world remains completely dependent on a hylozoist naturalism.

Similarly, if Reuleaux's exhaustive, precise, and systematic program is not outdated, the science he defined has been found defective only because of the evolution of its subject, the machine itself. Wiener rediscovers this notion of determining the field of the technology by the machine as its central model.[20]

Second, what functions does technology bear in relation to knowledge as a whole? Let us consider four aspects of this question.

1. According to Wiener, "For us, a machine is a device for converting incoming messages into outgoing messages." [21] Rediscovery of the field of technology is contemporary with the problem of communication. Stress is laid on the intersection of scientific linguistic ensembles which technics brings together. The "second Industrial Revolution" substitutes a combinatory process for a linear one, and interdisciplinarity for drastic specialization. Arborescent encyclopedism gives way to an encyclopedism of reciprocity and feedback. Today, scientific discoveries made by isolated specialists are scarce; invention is the result of intersections of people working toward a goal. Intellectual nomadism and many-sided instruction are dispersing the vague ideology of

general culture. To prevent interdisciplinarity from returning to that kind of vague ideology, two conditions are helpful:

a. The real and imaginary journeys of the scientists can only lead to mutual understanding if there are caches for depositing a code authorizing an exchange. Since its first if not only objective is communication, technology defines the privileged places where scientists from different subject contexts can meet. This, it seems to me, is the main point of this symposium.

b. By technological decree, the complex instrument and the science laboratory are the places where scientific "truth" is crystallized, in terms of operationality, efficiency, and unanimity.

2. This "new" encyclopedic function of technology is expressed in cybernetics, which offers an original way to unify the various fields of science. The language of cybernetics can be considered as a composite one since it includes preexisting scientific and parascientific language such as mathematics, statistics, physiology, medicine, military art, and so forth. Wiener deplores this disparity.

The word *description,* as applied to automata by Von Neumann, designates both the outward appearance of the object and its internal functioning, and hence the rules of its construction.[22] Description also plays the part of rays of light through which the eye penetrates an object to disclose its mechanism, thus constructing a model of the machine. It may be said that language and its object interpenetrate, and that words themselves are like cybernetic machines. As Saumjan says, "Cybernetics is only interested in machines from the point of view of the highly abstract system they represent, and not from the point of view of their concrete form." [23]

Thus a machine may be the symbol of another machine. When the functioning of a machine is made clear, that understanding becomes a machine in the cybernetic sense—a machine which in turn will find its symbol in another machine, and so on. Machines serve as pretexts for experience and models for theory; theory and action are interdependent. In cybernetic language, words may lose their usual symbolism to take on another, just as machines are both the machines we can see and models of other machines. This enables us to say that in cybernetic language a word is an operative model—echoing Von Neumann's self-reproducing automaton which embodies both the Golem and the robot Utopia. Words or machines constitute independent circuits and function like automata. It cannot be expected that cybernetics should unify the different sciences it resorts to, yet it offers a new use of scientific language where words and signs achieve autonomy through their heuristic function toward the traditional terms to which they are bound in the linguistic context.

3. Can we define the general impact of technology upon the intellectual method? In other words, would a precise definition of this subject and of its methods enable us to reinterpret certain texts? Perhaps a look at technics, unhampered by genetic preconceptions (that is, "Who created the machine?"), would elucidate the operative function of machines. To the mind, a machine is, sequentially or all at once, a concrete thing, a mathematical theory, an image, an ideal type, a model for thinking, an interpretative model, or a constructed image. We have more or less forgotten the etymological meaning of the word επιστήμη taken from ἐφιστήμι, a way of placing oneself relative to the object. Scientific knowledge itself is dependent on these real or symbolical problems of postures; it is fundamentally technicist. The problem of astronomy put by Plato, "to find what circular and regular movements must be supposed to save the appearance of the wandering stars," defines a constructivist way of thinking which is logically not very far from Rutherford's atomic model.

Technical thinking grants the mind the very precious ability to reduce the object of dimensions imagination can control. The selection taking place when an object is being built, which reveals its essential characteristics by ignoring variations and irregularities, is comparable to the elaboration of scientific fact. Finally, technics preserves the object by producing a copy, which resolves the difficulties of experimenting on living material. When the model becomes a simulator in the cybernetic sense, its pedagogic function is paralleled by a heuristic one which expresses itself, for one thing, as the possibility of testing the rationality of a science and of establishing connections between specialized sciences. "The brain-computer is a test, it tests the internal consistency of psychology as a unitary science", says Masanao Toda.[24] Wiener believes that living material can be imitated in its structure, functioning, and behavior. The structural and functional models concern biologists, then, and the models of behavior lead to psychology.

Modeling theory which has an obvious technological origin is more a method than a universal theory. Is it perhaps only a transitory step in knowledge? It is no longer appropriate to undermine the role played by mathematics in building an object. A machine becomes a model all the more as it is the subject of a mathematical treatment. In the last analysis, this mathematical treatment makes it possible to replace analogical correspondences between concrete objects with relationships between general laws. But the machine is the first step in the course of analysis; and perhaps also the last state of the synthesis. The act of building transcends itself into a product. The undetermined object whose model is its construction has been determined by this model, "modeled," so to speak. The fruitfulness of the model appears when it becomes poorer and by transferring its operative qualities to the object enriches the object, organizes, builds, and endows it with intelligible limits.

4. Let us go back to the juridical point of view. The working hypothesis of the autonomy of the field of technology is strengthened by encyclopedism. The clue to the problem is the quarrel of scientism, and the fact that any metaphysical conception of the unity of science forces one to look for this unity outside science.

This is the case with Husserl's phenomenology. "The unification of all conceivable sciences . . . can be achieved on the new and nondogmatic basis of phenomenological method."[25] But, to Husserl, the unity of science depends in fact on philosophy, the only absolute and universal science. Therefore, science proper is separate from philosophy, which can never be transformed into science but remains fundamentally different, while remaining fully a science. It is easy to realize, from all this, the setbacks and ambiguities induced by this perspective. Husserl's questionable approach leads directly to the more or less open suppression of science. How does the idea apply to our context? The field of technology opens out of the exact and natural sciences but develops separately.

The so-called human sciences will give technology a new impetus. It is worth mentioning the programmatic definition by

Espinas of general technology and its species. This is a sociological study and Espinas appears to echo Comte's thinking. He declares: "[General technology] is symmetrical in the field of action to logic in the field of knowledge, for in the same way it examines and classifies the various sciences, whose conditions and laws it determines and whose development and history it records. Sciences are social phenomena like the arts. . . . The ensemble of opinions we call Reason is a collective thing." [26]

However debatable, these ideas offer a very interesting critical approach derived from the human and social sciences. The technology-logic symmetry is very instructive. The parallelism of ways in which technology and logic are unaware of each other is striking.

Logic is not directly mentioned in various famous classifications. It does not appear in Ampère's catalogue. Comte merely relates the most abstract part of mathematics to its hidden source in "natural logic." As for Cournot, although he saves an explicit place for logic inside the noological sciences in the theoretical series, it is nonetheless trapped between biopsychology and sociology, even though, as with Comte, it also lies at the root of mathematics.[27] For our subject, logic's and technology's ignorance of each other is even more striking if we do away with the familiar distinction between knowledge and action, and if we grant a fundamental unity of human behavior, whether intellectual or practical. In the most fundamental sense, this quotation from Leroi Gourhan can be taken for granted: "Man makes both concrete objects and symbols from the same basic mental equipment. . . . Language and tools express the same property in man." [28]

The rift opened by science is a cultural and historical phenomenon; it does not alter the basic data. We must consider science in a less dogmatic and metaphysical way than the classical rationalists did, according to (1) its linguistic structure and its levels of language, and (2) its cultural and historical expressions within society.

For a linguistic definition of science, a distinction between syntax and semantics, and a reductionist point of view, the reader should refer to Carnap's "Logical Foundations of the Unity of Science," and to his encyclopedic intention.[29]

From the point of view of unified science, logic represents a borderline case, an ultra-science partly duplicating mathematics and partly retaining an autonomy of function which reflects the unifying goal. Its role of organizing and peacemaking is also a polemic one, for logic challenges each science and all of them taken as a whole. It leads them to the vanishing point where a subject disappears into the statement of itself. Therefore we can speak of symmetry between the function of logic (or linguistics) and that of technology. Always united to the applied sciences and always breaking away from science, technology does not provide a redundant rational account of what it must be—that is, to be aware. Rather, establishing itself by its own laws, it connects with science and mathematics in a new way. It changes from *confirmation* to *re-creation,* from applied rationalism to constructive rationalism which builds itself up by language games of communication and mutual compatibility.

The intuition of a structural unity of science prompts us to favor an externality of science toward itself, something above and beyond logic and technology as the ultimate legitimization of scientific language.

Traditional philosophy is inherently defective toward logic and technology. It mistakes empirical logic for ontology and hides technology under the mask of a science, vague and all powerful but seen as pure reason produced by subjective and solitary thought.

III. Technology Today and in History

Now we must clarify this idea of self-sufficiency in the technical world and apply this model to particular historical examples of techics following the recurrent method. In this method, historical perspective cannot be based on any preconception of the continuity or discontinuity of technical progress, or of the existence of gaps or obstacles, but only on the axiom that the present illuminates the past. The recurrent method—developed for the history of science—asserts that each scientist is very close to the scientists preceding him. As humanity's best memory, science achieves its encyclopedic role. But this memory must be located somewhere. Technology defines the objective conditions of this location, particularly by adding to science communication between scientists and the relationship between science and society. Technology, however, acts from outside. The best proof of this externality is the varying and flexible nature of the relation of science to technology.

Let us try to define the essential characteristics of existing technology in the industrially developed countries and the most striking connections between science and technology. We arbitrarily choose the following categories: (1) cybernetics, (2) scientific internationalism, and (3) management and technostructure.

In this limited paper, it is not necessary to justify these choices. The point is really to show that these three parameters characterizing modern technology can provide a guide for understanding some important events in the history of technics. We shall look along these lines: (1) the varied quality of the link between science and technology, and (2) the coalescence of the technical environment, a system of signs and functions which characterizes a culture. Pedagogy, internal social stratifications, and political organizations would be among the best sectors to study this phenomenon and to analyze its encyclopedic function as the stronghold of knowledge in its connections with society as a whole.

One can foresee a development of technics which would cross the development of science and sciences only occasionally and briefly. Although the two curves would have some points of intersection and zones of superposition, they would be more or less far apart—the farther apart, the more significant.

A traditional and autonomous milieu for crafts, provided with religious and mythological connotations, coherence, rules, and its own language, emerges from the scientific long night of the Middle Ages. Lefebvre-Desnouettes lists inventions or reinventions, century after century:

ninth century: saddle harness;

tenth century: modern harness, shoulder horse-collar, shoeing with nails;

twelfth century: watermill and windmill, mechanical sawmill, forge, bellows with plates and valves, ogival transept, stained-glass windows, fireplaces, the paving of roads;

thirteenth century: telescope, plow with wheels and moldboard, stern-post rudder;
fourteenth century: lock, gunpowder, weight-driven clock, plane;
fifteenth century: printing press, vocal polyphony.[30]

The central theme of this list is the production of means of communication and social cohesion. Therefore, the most advanced technology of the present time is not a phenomenon *ex nihilo* but affords new answers to a very old corpus of problems.

Knowledge of traditional craftsmanship was and still is preserved in exclusive societies, connected by a common language and a real, initiatory, and pedagogical journey such as the famous Tour de France. In this journey through culture, learning was acquired by varied experience, while the ancients taught formal patterns and the code needed to organize the various and labile meanings of everyday experience. Trade guilds imitate from a distance today's City of Science. Past dilemmas and ambiguities reveal the working of today's technical culture, multicellular and yet unified by a shared code of signs and tricks of the trade. Solemnly condemned by the Sorbonne on September 21, 1645, and legally murdered by the Le Chapelier Act of July, 1791, trade guilds bequeathed to a culture which rejected them a sense of achieved product and masterpiece. This qualitative appreciation of the product contrasts with the normalization and standardization associated with scientific ideas of measure, measure of time, and division of labor. The guilds also bequeathed the memory of an esoteric apprenticeship and esoteric teaching. We are aware that these problems arise perennially. A science which pretends to be pure and universal lapses back into esoterism. In my personal experience, one may still find, in some centers of craftsmanship, traces of acts and signs bearing witness to a technical culture surviving autonomous and authentic outside science, even if these sporadic remains constitute what Lévi-Strauss would call "pseudo-archaic phenomena of regressive evolution."[31]

This idea could be confirmed by assessing the technological cohesion of some populations we call primitive, but which sometimes still prove to be technological wholes with a cultural function for their society. For example, one might analyse the internal coherence and the social function of the caste of blacksmiths in the African grassland countries of the Bambara or Dogon people.

When science took over technics, the latter already possessed compact fields of language and sociological expressions which were the result of long development. Common craftsmanship testified to a vast amount of ingenuity before scientific imagination managed to reduce its scope. The view that technics necessarily depends on science must therefore be limited to a well-defined scientific category and a comparatively short period. One may show that it is possible, without too many contradictions, to reverse the traditional direction of the connection between science and technics. Furthermore, the autonomous sociological and linguistic expressions of technology bring science to reflect on its own significance, consistency, and encyclopedic function, in the light of its history and language.

"Recurrent history" suggests lines of descent, but it reveals more subtle historical mutations than the traditionally acknowledged "scientific revolutions" treasured at the museum of epistemology. The usual opinion that science conquered technics in the seventeenth century, or the sixteenth at the earliest, is relative. All the main elements of the scientific revolution, in the modern sense, already existed in Leonardo de Vinci's work: transforming observation into experience (examining the stability of buildings, showing the general cause of meanders, drafting his flying machine from detailed and synthesized observations on the flying of birds); mathematical formulation of problems (resuming the work of Alberti and Francesco di Giorgio on the solidity of beams); and rejecting earlier experience (analysis of experience with steam). It must also be noted that his research work is almost completely restricted to the preoccupations of engineers of the time, and that his solutions are not especially original compared to those of his contemporaries.

Moreover, when one knows that it was a tradition of his time to use mathematics only for mechanical arts, especially for the work of carpenters, architects, and surveyors, it appears more plausible to consider Leonardo's work as an attempt to systematize the technical problems of his time, that is to say, an encyclopedic attempt. The wish to do so must be related to other wishes of this kind. Perhaps the most important aspect of the Renaissance is that it did not choose clearly between catalogue and formalization. We see now why Leonardo does not really achieve an intellectual revolution. Unlike Leibnitz who transmutes the Ars Magna by Lulle or by Kircher into combinatorial analysis, Leonardo cannot change from alchemist into mathematician. Because of this, Leonardo remains dependent upon a "Rabelaisian Summa" concept of the encyclopedia as a catalogue of concrete skills. However, his interest in technical problems was great. In a letter in the Codex Atlanticus which was probably sent to Ludovico Sforza, Leonardo lists his talents. Sculpture and painting are only mentioned in tenth place, after the building of mobile and removable bridges and various machines.[32]

By means of breaks, errors, and catastrophes, Leonardo tries to grasp technological rationality. This last idea gains importance when compared to our technology, whose subject appears to it in a negative way through its lacunae, because of imperfect or even impossible communication. Without farfetched comparisons, we understand that, as Valery said, "this myth deserves to be replaced in the fable of [recurrent] history." [33]

Williams Carlos Williams answers Ezra Pound in *Paterson* [34] by giving him a summary of culture in the form of technical data to build an artesian well. However, the natural quality of the sand in various layers of the earth holds far more interest than the size or power of the machine drill. The old image of the well at the bottom of which treasure and understanding are found could prompt us to reconsider this paper and explore the technical and intrinsic double life of its subjects. Perhaps there we might regain hope for philosophy.

Afterword

I should like to note briefly the sources of the issues discussed in this paper and, in so doing, pay a tribute to the French scientists and teachers who have provided me with appropriate material and incentives for its development.

The major themes of this paper are borrowed from the works

of Professor Dagognet, for whom I am substituting here, and to whom I wish to do homage. Those themes stress the following: a systematic organization of scientific linguistic data; a code of apprenticeship and communication; the "artless makers of classifications" in science, those anonymous but efficient workers who catalogued nature and established rational lexicons; the emergence of that essential technical principle in science, "Only weighing guarantees sound chemical reasoning," to quote merely his works on medication and prosthesis; [35] and finally the very technological idea that any reproduction is production and any organizing activity in scientific language defines the change from a pictorial image to an operative image (to paraphrase Wiener). A technological justification of science and a search for an encyclopedic framework for it provide the basis for this paper.

Professor Lévi-Strauss confirmed my opinion that it was necessary to define a systematic and precise encyclopedic perspective, to reject the subject of philosophy for methodological reasons, and also to discard all the theological temptations which would predetermine the language of research.

Professor Serres guided me to Leibnitz's works and to *De Arte combinatoria,* in particular. Let us note this exemplary effort to unify knowledge by Leibnitz, who preferred the craftsman to the artist, knew the work of weavers, goldsmiths, and even alchemists, and asserted in a profound prediction, "The human mind shines in games more than in anything else." [36]

Notes

1. Max Weber, *Essai sur la théorie de la science,* trans. Julien Freund (Paris: Plon, 1965). "The ideal-type is a table of thoughts, it is not a historical reality and certainly not the authentic reality. . . . it has no other significance but that of a concept-limit purely ideal . . . it is a utopia" (p. 185).

2. Lucien Gérardin, *La bionique* (Paris: Hachette, 1968). "In 1959 J. T. Thykouner published a catalogue where the names of all the sciences existing at the time are listed in alphabetical order: he had noted 1,150 sciences" (p. 8).

3. *Encyclopédie ou dictionnaire raisonné des Sciences, des Arts et des Métiers par une société des gens de lettres. Mis en ordre et publié par D. Diderot, de l'Académie Royale des Sciences et Belles-Lettres de Prusse; et, quant à la partie mathématique, par M. d'Alembert de l'Académie Royale des Sciences de Paris, de celle de Prusse et de la Société royale de*

Londres, 17 vols. (Paris: Briasson, David, Le Breton, Durand, 1751); *Recueil de Planches sur les sciences, les arts libéraux et les arts mécaniques avec leurs explications,* 11 vols. (Paris: Le Breton, 1762-72).

4. Groethuysen's comment on Diderot is found in the article "Encyclopédie de Diderot," by Yvon Belaval, in *Encyclopaedia Universalis* (Paris: La Société Encyclopaedia Universalis, 1968–), vol. 6, p. 184. Groethuysen is one of the best commentators on "le Siècle des Lumières" (cf. his works on J. J. Rousseau and on the philosophy of the French Revolution).

5. Ibid. Butor is a contemporary French philosopher. See his *Essai sur les modernes* (Paris: Gallimard, 1967), and *La Modification* (Paris: Editions de Minuit, 1957).

6. Academie des Sciences, *Description des Arts et Métiers* (Paris, 1761-88; 2nd ed., Neuchâtel: E. Bertrand, 1768).

7. Diderot, *Oeuvres complètes, édition chronologique,* introduction by Roger Lewinter (Paris: Le Club Français du livre, 1969), vol. II, p. 349.

8. Ibid., p. 348.

9. J. J. Rousseau, *Lettre à d'Alembert sur les spectacles* (1758) (Paris: Librairie Hatier, 1935). See in this work the theory of feast as an arrangement of social space with reciprocal exchanges.

10. *Oeuvres complètes.*

11. *Johannes Henrici Alstedii Encyclopaediae* (Herborn, 1630; 1st ed., 1609). "Quid est Technologia? Est doctrina praecognoscenda, de affectionibus, ordine et divisione Disciplinarum" (p. 27). Cf. also: "Liber secundus exhibens technologiam id est Doctrinam de proprietatibus, ordine et numero disciplinarum" (title of Book II).

12. G. G. Leibnitz, *De arte combinatoria* (1666), in *Opera philosophica* (Berlin: J. E. Erdmann, 1840), pp. 7-44. The text was written by Leibnitz in his youth and is full of nearly magical ideas ("ars magna" of R. Lulle), which take on with this writer the form of a *mathesis universalis.* In it you can already find the outline for a Universal Language that he will maintain until his death.

13. André-Marie Ampère, *Essai sur la philosophie des Sciences ou exposition analytique d'une classification naturelle de toutes les connaissances humaines* (Paris: Bachelier, 1834); also in *Impression anastaltique. Culture et civilisation* (Brussels: G. Lebon, 1966). Ampère was no doubt the most exacting and systematic classifier of nineteenth-century knowledge, considered from the viewpoint of the zoological and botanical classifiers. The tree of life and the tree of knowledge (cf. Descartes) have the same intellectual origins.

14. Ampère, *Essai sur la philosophie.*

15. Diderot, "De l'interprétation de la Nature," in *Textes choisis,* ed. J. Varloot (Paris, 1952), vol. II. See also "La pensée II" and "La lettre à Voltaire du 18-2-1758."

16. Lewis Mumford, *Technique et civilisation,* trans. J. Moutonnier (Paris: Ed. du Seuil, 1950). In "Bibliographie commentée," p. 409, Mumford adds, "the most important of the systematic morphologies of machines."

17. Frantz Reuleaux, *Cinématique. Principes d'une théorie générale des machines,* trans. A. Debize (Paris: Librairie F. Savy, 1877), p. 30. Reuleaux's work appears today to be the only authentic try to bring a systematic and axiomatic quality into the mechanical and technical realm.

18. Reuleaux, "Classification des sciences mécaniques," in *Cinématiques,* pp. 38-44. The classification of the technical species, for Reuleaux, discloses the ambiguity between theory and practice. Reuleaux dreams of reducing technology to pure science, but also hopes to maintain its irreducible material originality.

19. Diderot, "De l'interprétation de la nature. Pensée II." Many of Diderot's texts confirm the same viewpoint: e.g., "La lettre sur les aveugles," the article "Enclopédie," the article "Art," etc. Like many other philosophers and scientists of the eighteenth century, Diderot is against the limitations of the theoretical mechanism, such as Descartes's, and wants to restore the language of craftsmanship, more technical and less scientific.

20. Norbert Weiner, "God and Golem, Inc.: A Comment on Certain Points Where Cybernetics Impinges on Religion." Unpublished manuscript, Cambridge, Mass., Massachusetts of Technology. Pp. 29-32.

21. ". . . as the engineer would say in his jargon, a machine is a multiple-input, multiple-output transducer" (ibid., p. 32).

22. J. Von Neumann, "The General and Logical Theory of Automata," in *The World of Mathematics,* ed. James R. Newman (New York: Simon and Schuster, 1956), vol. 4, pp. 2070-98, esp. pp. 2094-95.

23. Sebastian K. Saumjan, "La cybernétique et la langue," *Diogène,* no. 51 (1965), pp. 137-52. This is an important article concerning the relationship between language and cybernetics. It could be further enriched by a debate among the contemporary generativists: Saumjan, Bar-Hillel, and Chomsky.

24. Masanao Toda, "Brain Computer Approach to the Theory of Choice," in *Logic, Methodology and Philosophy of Science. Proceedings of the 1960 International Conference* (Stanford, Calif.: Stanford University Press, 1962). The artificial brain was not intended to imitate exactly the human brain but rather to test the scientific quality of a knowledge simulated by psychoindividual functions.

25. See E. Husserl, *La philosophie comme une science rigoureuse,* introduction, translation, and commentary by Q. Lauer (Paris: Presses Universitaires de France, 1955); *Logische Untersuchungen;* and *Die Krisis der europaïschen Wissenschaften.* Husserl insists on the importance of turning to philosophy and in particular to transcendental phenomenology to avoid scientism by placing outside the realm of science the "principe d'enracinement" of the latter.

26. Alfred Espinas, *Les origines de la technologie* (Paris: F. Alcan, 1897). This is a somewhat forgotten work, very interesting nonetheless because of its historical and social import. It also marks one of the outcomes of the nineteenth-century French positivism.

27. A. Comte, "Discours sur l'ensemble du positivisme," *in Systeme de politique positive,* 4 vols., 5th ed. (Paris: Vrin, 1975); A. Cournot, *Matérialisme, vitalisme, rationalisme* (Paris: Hachette, 1875), pp. 313 ff; and *Traité de l'enchaînement des idées fondamentales dans les sciences et dans l'histoire* (Paris: Hachette, 1861), pp. 192, 222, 224, and final table. Cournot is both a pedagogue and a philosopher of the nineteenth century, concerned with learning and legal and encyclopedic knowledge.

28. André Leroi-Gourhan, *Le geste et la parole,* 2 vols. Vol. 1, *Technique et langage.* Vol. 2, *La mémoire et les rythmes* (Paris: A. Michel,

1964, 1965). Quote is from Vol. 2, p. 24. These are fundamental texts for all genetic analysis of techniques related to biology.

29. R. Carnap, "Logical Foundations of the Unity of Science," in *International Encyclopedia of Unified Sciences* (1938), I, 1, p. 5. Beyond Carnap's reductionism, his most important contribution is his attempt to determine an encyclopedic classification and knowledge, which revives old attempts by some empiricist philosophers of the eighteenth century, Condillac in particular.

30. Lefebvre-Desnouettes, "La Nuit du Moyen-Age et son inventaire," *Bulletin du Mercure de France,* April 1, 1972. See also, by the same author, *L'attelage, le cheval de selle à travers les âges,* 2 vols. (Paris: A. Picard, 1931); and a text annotated at great length by Lynn White, Jr., *Medieval Technology and Social Change* (New York: Oxford University Press, 1962).

31. The notion of regressive evolution was developed by C. Lévi-Strauss to describe certain Amazonian populations (in particular the Nambikwara and the Tupi-Kawahib). We find it exemplified in his famous work, *Tristes Tropiques* (Paris: Plon, 1955).

32. Leonardo da Vinci's letter to Sforza, *Codex Atlanticus.* See the studies of B. Gille on da Vinci and di Giorgio in *Les Ingénieurs de la Renaissance* (Paris: Hermann, 1964). See also Centre national de la Recherche scientifique, *Léonard de Vinci et l'expérience scientifique au 16è siècle* (Paris: Presses Universitaires de France, 1953), in particular the articles by G. Sarton, P. Francastel, R. Dugas, and B. Gille.

33. It's possible to consult Valery's very interesting opinions on technique in his works, *Eupalinos ou l'architecte* (Paris, 1921; Paris: Gallimard, 1944); *Introduction à la méthode de Léonard de Vinci* (Paris, 1894; Paris: Gallimard, 1957); and *Variété V: L'homme et la coquille,* vol. 5, *Discours aux chirurgiens* (Paris: Gallimard, 1944).

34. W. C. Williams, *Paterson* (New York: New Directions, 1963), p. 139.

35. The principal scientific works of Francois Dagognet are *La raison et les remèdes* (Paris: Presses Universitaires de France, 1964); *Le catalogue de la vie* (Paris: Presses Universitaires de France, 1970); and *Tableaux et langages de alchimie* (Paris: Ed. du Seuil, 1969). Dagognet analyses the methods of classification and organization of scientific knowledge. His most recent works are concerned with the analysis of the aesthetic characteristics of one type of reflection and its era.

36. Leibnitz, "The human mind shines in the realm of games more than in any other discipline," *Lettre à un ami en France,* in *Opera philosophique,* p. 700. Leibnitz may as rightly be considered the father of the scientific theory of probabilities and games as Pascal or Bernouilli.

Peter Caws

13

Praxis and *Techne*

In spite of the fact that many people think it was invented at about the time of the Industrial Revolution, technology has a much longer history than either science or philosophy, and the philosophy of technology can be traced back to the earliest philosophers. Aristotle, in the *Parts of Animals,* recounts that when some visitors surprised Heraclitus in the kitchen he invited them to come in with the remark "for here too are gods." [1] Where modern philosophers would naturally appeal to scientific examples, Plato habitually chooses technological ones; his dialogues are full of references to agriculture, medicine, shipbuilding and navigation, and the training of animals—technologies whose origins go back to neolithic culture. We may speak of them as technologies and not mere techniques because there was a lore that went with the techniques; they were passed on—and no doubt also originated—not only through trial and error and imitation but also through discussion and instruction. We have a tendency to think of civilization before the invention of writing as silent, but it was not. Nor was it irrational. "As for me," says Socrates in the *Gorgias,* "I do not give the name *techne* to something lacking in reason." [2]

Technology, after all, is not merely the theory of the practical arts; it is the practical arts themselves, regarded as an activity of reason—the *logos in* the *techne,* rather than the *logos of* the *techne.*[3] Nevertheless, technology has continued for the most part to be relegated to the kitchen, and even philosophers who are beginning to pay attention to it do not always manage to avoid the patronizing tone of those who, to their surprise, have found hidden talent below stairs. The assumption, all too readily made and accepted, that technology is to be defined as the practical application of scientific theory is symptomatic of this. The opposition of theory and practice, and the scorn of the latter, also has ancient roots, but we have perhaps not understood them. It was certainly not, as is sometimes supposed, a simple question of slavery and nobility; rather it was perceived that there are two different and independent manners of relating to the world. Thus Aristotle in the *Nicomachean Ethics* says, "For a carpenter and a geometer investigate the right angle in different ways; the former does so in so far as the right angle is useful for his work, while the latter inquires what it is or what sort of thing it is; for he is a spectator of the truth." [4]

There is no evidence that the Greeks despised carpenters. There is plenty of evidence that they admired geometers and indeed mathematicians in general, but the reason for this can easily be traced back to religious motives. Pure contemplation in Aristotle is a function of the divine; God is the archspectator, the *theoros* of himself and the world, not merely—like ordinary *theoroi*—of limited events like the games or the consultation of oracles. The essential distinction is between the perishing and the unchanging, between the transient and the eternal. Mathematics, the paradigm of theory, gives access to another world, where the soul is released from the body, the geometrical right angle from the wooden one, a world which in Plato almost certainly derives indirectly from an older philosophical tradition in India. It is significant for our purposes that when Plato's God turns his attention to our world he does so in the guise of a craftsman. (It may also be significant that some connotations both of *techne* and of its more commonplace relation *praxis* were extremely down to earth. One of the meanings of *praxis* given in the standard lexicons is "sexual relations," and there existed a minor branch of learning known as *erotike techne,* exemplified in a work of Paxamus called *Dodecatechnon* because it dealt with "obscene positions to the number of twelve." [5])

I mention all this not just in order to observe the usual philosophical pieties toward the Greeks, but in order to stress that the concepts invoked by the philosophy of technology are embedded in a linguistic tradition that we ignore at the risk of talking nonsense. Our problems are not nearly as new as we think. A further development of the linguistic tradition, and one to which we ought to be sensitive, has taken place in the last hundred years or so. If the theoretician often tends to look down upon the merely practical, there has grown up since Marx an opposite tendency to look down, in a moral sense at least, on the merely theoretical. The fact that the widespread use of the term *praxis* in an exclusively political sense rests upon an etymological mistake (since it was the plural form *pragmata* that chiefly carried this connotation in Greek) does not blunt the force of the argument that a balance may need to be struck between the luxury of theoretical detachment and the utility of practical involvement. This is not to say that practical involvement can ever be a criterion for the truth or adequacy of theory in its own domain. But, the society of men has other values besides truth, values which in the end truth may rightly be expected to serve without compromising its status as truth: it is rather a question of choosing how much of one's life to devote to the pursuit of truth, and which truths to pursue.

Mao Tse-tung, whose works it would be quite wrong, I believe, to dismiss as merely demagogical (they are quite as much pedagogical, presenting surprisingly orthodox philosophical views in a simplified form accessible to the young or ignorant), offers in his essay *On Practice* a typically Marxist inversion of the theory/praxis relationship, reminiscent of Marx's own treatment of the money/commodity relationship in *Capital.* Money, Marx maintains, began its career quite reasonably as a form of mediation between commodities (C–M–C), but in the capitalist economy it becomes an end in itself, while commodities are reduced to mediating between its manifestations (M–C–M).[6] Similarly, for Mao the proper role of theory is as a form of mediation between praxes (which might be represented analogously as P–T–P), whereas the over-intellectual and over-theoretical habits of the West elevate theory to the dominant place and put (experimental) praxis at its service (T–P–T).[7]

My own conviction, which is of long standing, is that the attempt to establish conceptual priority between theory and praxis is futile; the relation between them is dialectical in the strict sense, in that both historically and conceptually they alternate in the development of knowledge and of its applications. But if we are to understand this dialectic, its elements need to be sharply distinguished, not run together with one another. Therefore, I should now like to explore the senses in which theory and praxis can be thought of as parallel and autonomous. An incidental but striking tribute to this parallelism is provided by an illustrative analogy used by Spinoza in his essay *On the Improvement of the Understanding:*

> In the same way as men in the beginning were able with great labour and imperfection to make the most simple things from the instruments already supplied by nature, and when these were completed with their aid, made harder and more complex things with more facility and perfection, and thus gradually proceeding from the most simple works to instruments, and from instruments to other harder pieces of work, they at last succeeded in constructing and perfecting so many and such difficult instruments with very little labour, so also the understanding by its native strength makes for itself its intellectual instruments wherewith it acquires further strength for other intellectual works, and with these makes others again and the power of investigating still further, and so gradually proceeds until it attains the summit of wisdom.[8]

This citation serves my purpose in another way because it stresses the slow and complex character of the evolution of human knowledge and competence, which I believe to be a constant characteristic, although it has been concealed in recent history by quantitative increases in the capacity and facility of use of external storage systems and effectors, the speed and capacity of communication and its channels, the absolute numbers of human beings, and the proportionate numbers of them engaged in production and research. It is necessary to remember that if, as some anthropologists and linguists maintain, the complexity of the human mind is relatively constant over all known cultures and hence over a very long period of cultural development, men have since early prehistory confronted the world on roughly equal terms. The difference between us and prehistoric men is that we find in our world a great many things—buildings, clothes, books, and instruments—left in it by our predecessors, and that our immediate ancestors—parents, teachers, and the like—take pains to introduce us to the use and sometimes to the meaning of these things, as well as to the use and meaning of various activities, like speaking a language, that they learned from their immediate ancestors. We also find in the world, of course, many other people already versed in all this, with beliefs, prejudices, and the rest.

Our basic relation to the world, I repeat, is constant. In order to make this clear, it is necessary to complicate matters slightly. On the practical side as well as on the theoretical side, we need to distinguish between a mode of immediate interaction of man as knower or agent with the world, and another mode of activity that is independent of such immediate interaction and which in *both* cases I would call intellectual. On the practical side, I associate *praxis* as immediate with *technology* as mediated by the intellect. In parallel on the theoretical side, the notion of the *empirical* (from *empeiria,* meaning, roughly, an experimental acquaintance with things) provides the immediate basis for the intellectually mediated activity of *theory.* There is no point in trying to make all this too perfect and symmetrical. *Empeiria,* for example, is not wholly passive; a case might be made for appealing to *episteme* rather than *theoria,* and so on. My point is only to stress that coming up against the world physically, on the one hand, and looking at and talking about it, on the other, represent two complementary and to some degree separable kinds of involvement with it, and that each leads to its own variety of mental activity, which emerges again on a higher level in a more complex form, in the first case technological and in the second case scientific.

The parallel can be, and has been, drawn out to considerable lengths. An explicit version of it is to be found in the late-nineteenth-century work of Alfred Espinas, *Les Origines de la technologie,* in which the sequence sensation–perception–knowledge–science is matched with the sequence reflex–habit–custom–art or technique.[9] Philosophy has concentrated almost exclusively on the theoretical side, of course, as the relative states of development of the philosophy of science and the philosophy of technology clearly show. The philosophy of science would never have reached its present level, however, without the basis laid down by the analysis of empirical knowledge carried out by epistemologists since the seventeenth century.

It is my belief that a developed praxiology is just as essential to the philosophy of technology as epistemology has proved to be to the philosophy of science. Praxiology, however, is in its infancy. Apart from the works of Marxists, who interpret the concept in an arbitrarily narrow way, and of pragmatists, who mistakenly suppose that the problem is really after all an epistemological one, hardly anything has been written except the disappointingly anecdotal treatment by Kotarbinski.[10] That is why the philosophy of technology can scarcely as yet be said to exist; what passes for it usually amounts to no more than inserting technology, taken straightforwardly as the application of scientific theory, the proliferation of machines, and the like, as a boundary condition into some other branch of philosophy, such as value theory or political philosophy. Some philosophical disciplines, like logic and automata theory, have been greatly stimulated by technology, but to regard them as its philosophy is, in my opinion at least, to underestimate the philosophical interest of technology.

Let me then explore the parallel between the practical/technological and the empirical/theoretical at its most humble level, the level at which the epistemologist would be talking about simple perceptions or basic observation statements. A problem arises for the praxiologist which is analogous to that of deciding among phenomenology, sensationalism, and physicalism, and I shall suppose it resolved much as standard empiricism resolved it on the epistemological side. That is, I shall assume a capacity to recognize *things* and shall not insist on beginning with pure sensory elements such as resistance to touch and the like. This, however, presumes a prior learning. Just as physicalism in epistemology involves the learning of a descriptive language, an insertion into a culture of names, so physicalism in praxiology involves the learning of habits of manipulation, an insertion into a culture of objects. It requires a certain effort to see past the transparency of our habitual praxes, but if we reflect for a moment on the vast repertoire of elementary cultural praxes we have all acquired—things as banal as buttoning buttons, tying shoelaces, using knives and forks, opening and closing drawers or doors or boxes, brushing teeth, shaving, and the like—and on the way in which we acquired it, the parallel with language learning should be obvious enough. Special technological praxes, like special scientific terminologies, are acquired subsequently in the context of special training.

An elementary praxis, like an elementary observation, picks out a bit of the world and operates on it. The great difference, of course, is that while the observation leaves the world as it is, the praxis alters it. (At quantum levels, therefore, the distinction gets blurred, but that does not mean that it is not a perfectly good distinction.) A simple but often neglected corollary is that the world must be in a quite specific state before praxis becomes operative. People will not be found buttoning buttons unless they are confronted with unbuttoned buttons and with buttonholes to button them into; they will not be found doing it if buttons have not yet been invented, or if they are wearing clothes that do up with zippers, or if their buttons are already buttoned. Also, even if all the conditions are right, they will not be found buttoning buttons unless they want them buttoned. This inevitable incursion of *value* into questions of praxis—and *a fortiori* of technology—is familiar enough. It does not, however, change the character of the praxis; it only decides whether or not it will be practiced.

Every praxis, in other words, has a domain, a domain of ordinary recognizable macroscopic objects that can be altered, arranged, connected up, stored and the like. Things become interesting from the point of view of our present discussion when the objects in the domain of the praxis are not natural objects or simple cultural objects but things like light switches, air-conditioner buttons, gearshifts, or triggers. Such praxes obviously depend on a prior technology, where by "a technology" is meant a planned, purposive, relatively complex, probably collaborative, structured sequence of praxes. The important thing to notice is that the man who turns on the light or shifts the gear need know nothing whatever about all this; all he needs to know is the practical effect of his action. For all he cares, the device might have come into being naturally. It is a remarkable truth, when one comes to think of it, that an artificial eye, had we constructed one out of carefully replicated tissue, would work exactly like an organic eye; so an organic watch, were one to grow accidentally or miraculously in some improbable metal-rich environment, would work exactly like an artificial watch (artificial in the literal sense of being an artifact; the complexities of the ordinary-language behavior of this term need not distract us). There could be no technology, in other words, if there were no laws of nature. But this, of course, is a quite different claim from the claim that there could be no technology if there were no science. The difference that science makes is a difference of efficiency, both in selecting what technologies to try and in deciding how to go about trying them. If you know how the relevant laws of nature operate, you go straight to the desired result and do not need to spend millions of years trying this and that. The point is, again, quantitative.

For the consumer of technology, in fact, the relation between a given praxis and its outcome may be wholly magical. People who know nothing about the internal combustion engine, when starting automobiles on cold mornings, seem to me to be in a position strictly comparable to that of dancers dancing for rain: sometimes the ritual, which consists of a certain learned sequence of pulling out chokes, turning keys, and depressing accelerators, pleases the god, and sometimes it doesn't. Apart from the complexity and predictability of his environment—the range of different options, and the facility with which the right choice leads to the right results or the wrong choice to the wrong one—contemporary man is pretty much in the position of his remotest ancestors. He still has to decide what to do. The difference that technology makes is to give him a greater freedom of choice and greater responsibility. The risk of unexpected consequences—getting cancer from smoking or killing lakes with industrial effluents—is more dramatic and yet, once explained, more tractable than similar risks in past times. Such consequences as denuding whole territories through cultivation and irrigation simply could not have been predicted on the basis of prehistoric knowledge.

It is, I think, pure irresponsibility to claim that technology has made an *essential* difference in the condition of man as knower and agent. Even the difference it has made in his circumstances is a quantitative rather than a qualitative one, and I do not believe that the dialectical law of the transition of quantity into quality applies. The moral and political problems that result from technology are not, I repeat, problems in the philosophy of technology. The trouble is that in claiming that man has been changed by technology we encourage men to think themselves at its mercy. In fact they are not. They may be at the mercy of other men who misuse technology, but the remedy for that lies outside technology, and a Luddite solution is not, in the end, satisfactory.

In conclusion I wish to turn back from this set of external problems to the internal, conceptual ones that constitute, in my mind, the essence of the philosophy of technology. In particular, I want to consider the transition from praxis to technology proper as analogous to the transition from empirical acquaintance with the world to theory.

A technology was said earlier to be a planned, purposive, relatively complex, probably collaborative, structured sequence of praxes. The technologist is the man who knows which praxis to carry out when, and under what objective conditions; he knows what resources of material and energy it will require and what its outcome within the technological context will be. He has learned all this from other technologists or practitioners. The question is, what sort of activity is it? It is my contention that technologies in this sense are just as much an evidence of human intellect as scientific theories are, and that our tendency to think of the latter as superior is just cultural prejudice arising out of the dominance of the verbal among the leisured classes. (We no longer, as a rule, have the religious excuse I earlier attributed to the Greeks.) It seems to me clear that technology, in this sense, represents a form of insight into the workings of the world, a form of practical understanding, that takes as much talent and application as the most rigorous theoretical calculation.

The most dramatic cases come from other cultures. I will cite only one, that of the Truk navigators of the Pacific. These islanders, working from traditional recipes which consist of a few lines drawn on a piece of bark, are capable of making landfalls within a half mile or so after voyages out of sight of land for hundreds of miles, in variable weather and at all seasons. They learn during a long training how to take in and process staggering quantities of information—not only winds and currents, the feel of the boat and of its rigging, and the appearance of the sky, but at any point the sequence of these things over the previous course of the voyage in relation to their value at that point.[11] (It is no argument against this achievement to say that migratory birds, sea turtles, and salmon perform feats just as staggering. In the latter cases it is a species-specific activity, not a culture-specific one. Also, the birds and turtles may be more intelligent than we think.) In our own culture, it is clear that the whole development of the plastic and performing arts is an example of this kind of thing. The point is that what is involved is a form of representation inside people's heads of the behavior of things in the world, a representation that is not descriptive but rather, we might say, operational. It is not just a question of manual dexterity, since there are well-attested cases—J. J. Thomson was one—of people who can see how the thing goes without being able to do it. The story is that Thomson's laboratory assistants refused to let him touch the apparatus, because whenever he did so it broke, but that they depended on him to show them what to do with it. It is a question of making our way about in a world whose physical and causal properties we know not only or even mainly by catching them in formulae, but also by the daily practice of a form of learning embodied in a structure of behavior that has been accumulated and transmitted over many millennia.

Along lines like these I think we might hope eventually to come to some philosophical understanding of technology. One contrast I have tried to stress, and one that is essential if the whole subject is not to fall into confusion, is between the relatively unintelligent praxis of the button-pusher and the true technology of the engineer himself. The button-pusher has nothing interesting to do with technology, and his use or abuse of it, even if it has disastrous consequences, does not touch its essence. A moral question is thereby posed for the engineer: the question for whom he is working. There are, after all, risks involved in offering buttons to be pushed. It was remarked earlier that people will not button buttons unless they want them buttoned, but they may be tempted to push buttons without having considered very carefully what they want, or whether there is anything they really want. The trouble is that the results of button pushing are out of proportion to the effort, and this is likely to produce also a disproportionality between decision and purpose, whereas in elementary praxes such as button buttoning these things are generally in a kind of natural balance.

Also, people may want the wrong thing. These are problems that technology poses, although they are not technological problems, and the philosophy of technology cannot by itself deal with them adequately. There is an overlapping of the philosophy of technology with the theory of value, as with the philosophy of purposive action in general; technology, indeed, might be represented as the systematic working out of the hypothetical imperative, putting into the consequent of the hypothetical anything we can lay our hands on—scientific theory if we have it, but other forms of knowledge and competence as well. So technology is not value free, but there is still a sense in which it is value neutral. The antecedent of the hypothetical remains to be filled in as we collectively prefer; nothing in technology itself compels us one way or the other.

Notes

1. Aristotle, *Parts of Animals.*
2. Plato, *Gorgias.*
3. The fact that, as Carl Mitcham has pointed out, Aristotle uses *technologia* to mean "grammar"—the *techne* of the *logos*—does not invalidate this argument, which rests on current usage in English. Etymological analyses are helpful because they show how terms are articulated and sometimes how they have changed, not because classical usage supports our own.
4. Aristotle, *Nicomachean Ethics,* 1098a28.
5. Friedrich-Karl Forberg, *Manuel d'érotologie classique,* trans. Alcide Bonneau (Paris, 1959), vol. 1, p. 6. Forberg contrasts the treatise *Dodecatechnon* with the courtesan (Cyrene) known as "Dodecamechanos," because the former talks about the twelve positions while the latter knew how to practice them—a further reinforcement of the distinction between *techne* and *praxis* and an illustration as well of the classic forerunner of the notion of the machine. Whether (as some commentators maintain, see V. de Magalhaes-Vilhena, *Essor scientifique et technique et obstacles sociaux a la fin de l'antiquité* [Paris, n.d.]) this use of *mechanos* meant that the courtesans practiced the art of love in a "mechanical" way, or whether, which seems more likely, it reflects a conception of the human body as a kind of living machine, Forberg does not say.
6. Marx, *Capital.*
7. Mao Tse-tung, *On Practice.*
8. Benedict de Spinoza, *De Intellectus emendatione,* trans. A. Boyle, published with the *Etica* (London, 1910), p. 236.
9. Alfred Espinas, *Les Origines de la technologie* (Paris, 1897), p. 10.
10. Tadeusz Kotarbinski, *Praxiology,* trans. O. Wojtasiewicz (Oxford, 1965).
11. See Thomas Gladwin, "Culture and Logical Process," in *Explorations in Cultural Anthropology,* Ward H. Goodenough, ed. (New York: 1964), pp. 167 ff.

David Wojick

14

The Structure of Technological Revolutions

Studies of technological revolutions have tended to concentrate either on the act of invention or on the societal impact of new technology. Considerably less attention has been given to the processes by which new technology is initially adopted by the engineering community. This neglect may result from an assumption that new technology, once available, is always eagerly sought and quickly adopted. In other words, we may believe that the problem of diffusion of innovation is largely a problem of communication.

As anyone knows who has fought in a technological controversy like the environmental movement, even technologists will resist change, including technological change. The introduction of new technology—new techniques, processes, or information—into an established problem-solving area often requires a great deal more than communication of innovation. Often it requires a revolution, a struggle of ideas. This revolution accompanies the deployment of technology and precedes the social revolution which that deployment may induce.

I want to look at the epistemology of internal technological revolutions, that is, at the way in which new knowledge comes to modify or replace existing practices. Technological revolutions such as the environmental crises, the natural food controversy, and the nuclear power plant controversy will be my chief examples. To begin with, we must take a general look at technological problem solving, which is often governed by broad policies or established procedures that simplify and give coherence to an area of technological practice. We shall ask how established policies come to be changed, for it is the overthrow of such policies which constitutes a technological revolution.

A. Evaluation Policy

Suitability and correctness are key concepts in applying technology. In deciding to use available technology to deal with a given problem, one must determine the suitability of that technology for the task at hand. If electric power is needed, then nuclear power generation may be considered. If product shelf life is low, then the availability of chemical means to retard deterioration may be investigated. This evaluation process is the heart of practical reasoning: one has a problem, and one considers ways of solving it. Correspondingly, when a solution technique has been chosen and planning gives way to design and implementation, there must be a way to determine when the task is correctly done. The right power plant must be designed and built; the proper additive must be added. These criteria guide the technical manager and the engineer at all levels, from the formulation of national policy to the design floor.

Determining suitability and correctness necessarily involves both values and facts. A task has been correctly done only when certain goals are achieved without consequences whose negative value is greater than the positive value of the goal—in short, when the good outweighs the bad. Benefit-cost analysis is, of course, a paradigm case of this kind of practical evaluation; however, judging the net value of a particular solution may not involve money. Qualitative factors are often, if not always, to be considered.

Factual knowledge is required to predict or estimate what will be achieved by a solution. Similarly, to predict or estimate the correctness of a proposed design or plan requires normative knowledge, because determining the net value of a proposed solution depends on deciding which factors to be considered and how they are to be valued.

For example, to evaluate a proposed dam, one must decide whether flood protection benefits are to be considered and, if so, exactly how these benefits are to be weighed. Similar decisions must be made for such factors as recreation potential, low flow augmentation, power generation, and water supply. Then there are social consequences, environmental impacts, economic effects, and esthetic concerns, as well as considerations of utility, economy, efficiency, and safety. Each of these factors must be evaluated and either dismissed as irrelevant or esoteric, or else included as relevant. Then both the magnitude and the value of each relevant factor must be worked out for thousands of detailed decisions. The manager or engineer must know which values count and how much each counts if he is to proceed at all, for without the means of determining these values he cannot know when a job is being done correctly or when the problem is solved.

In any relatively stable area of technological application these decisions are relatively unproblematic. Normally they will have been fashioned into an established policy so that only the details unique to each case need to be worked out. We shall term the set of such decisions the evaluation policy of the technological area in question. Thus we may speak of the evaluation policy of the food industry according to which new additives are evaluated as to utility, safety, benefit, and so forth. Similarly, until recently the recreational uses of reservoirs were not normally countable as good consequences of dams under Federal public works evaluation policy; even though recreation was a recognized consequence, it was just not figured into the benefit-cost ratio.

In a major technological area, evaluation policy may include a combination of scientific theory, engineering principles, rules of thumb, laws, court decisions, commonly accepted moral precepts, and in-house regulations, as well as professional standards and the accepted techniques of many disciplines. Where technological practice is stable, an evaluation policy is normally implemented in the form of standard decision-making procedures. The correct performance of a particular task is then a matter of good judgment within the standard procedures. For example, in the food industry whether a new additive is safe or not is determined by following certain standard test procedures. In the Corps of Engineers, the cost and benefits of a proposed dam are evaluated according to strict rules which specify what factors are to be examined, for instance material costs, anticipated flood protection, anticipated recreation uses, and land cost, as well as how each such factor is to be measured. These rules are built into recognized procedures—often codified in regulations or manuals—which are routinely employed to solve problems.

The totality of procedures in any given technological area may involve techniques from many different fields. For example, the evaluation of a proposed food additive may involve procedures drawn from medicine, chemistry, biology, operations research, marketing, industrial engineering, civil engineering, and so on, developed both within and outside the food industry. Usually no single person or group of people has the power to determine evaluation policy in a given technological decision. Most major decisions concerning the use of technology are based upon following hundreds, if not thousands, of standard practices, few of which can be changed or overridden by executive decision.

The evaluation policy defines good administrative and engineering practice at any given time; questions as to the correctness of the policy do not generally arise. Normal technological practice does not involve creation of new technology; rather it involves the artful application of well-understood and well-recognized decision-making procedures. These procedures provide a coherent conceptual framework within which technology is applied to the solution of problems. Such problems are often difficult, requiring creativity and skill for their solution, but the presence of a stable evaluation policy embodied in recognized decision-making procedures guarantees that what counts as a good solution is well defined and recognizable. The engineer or administrator who knows the accepted procedures also knows when he is doing his job well. He knows how to approach his problems, what is to be considered, and what ignored. Just how crucial to technical practice the presence of a stable evaluation policy and standard procedures are will become clear when we consider the consequences and underlying formal structure of a breakdown in evaluation policy.

Evaluation policy articulated in standard procedures plays a central organizing role in technological practice; such practices may be thought of as operationally defining the normative concepts which enter into the solution of society's problems.

Established evaluation policy may be incorrect, however. We may discover that our practices leave out factors which ought to be considered, or perhaps that we are weighing things improperly. The possibility of error and its discovery raises the question of how evaluation policy comes to be changed. This question will concern us for the rest of this essay.

Fortunately our task is made easier by existing work in the philosophy of science, particularly Thomas Kuhn's analysis of scientific revolutions.[1] This is so because accepted evaluation policy plays a role in organizing the use of technology analogous to the role which, according to Kuhn, is played by a scientific theory or paradigm in organizing explanation in the sciences. Let me briefly describe a somewhat Kuhnian model of paradigm replacement and then apply this model to revolutionary change of technological evaluation policy.

Kuhn states that the heart of a scientific discipline is a system of concepts which he calls a paradigm. These concepts may be built up from or depend upon several elements, including (a) the main theoretical laws of the field, (b) classic experiments which exhibit the concepts at work, (c) methodological principles which determine what work in the field should look like, (d) an ontology of theoretical entities whose presence in the world is supposed to explain the phenomena in question, and even (e) metaphysical principles.

Kuhn then asks how such concept systems, or paradigms, come to change. The accepted use of a paradigm in a field is called "normal science." As a paradigm is accepted because it promises to explain certain problems such as the motions of masses or the shapes of animals, the normal job of a scientist is to work out these explanations. The normal scientist uses the paradigm to explain phenomena: he does not test the paradigm. In fact, the paradigm is typically assumed to be true, or nearly so.

A lot of creativity is involved in normal science. Many new concepts were required, for example, to develop fluid mechanics out of Newtonian mechanics, and even the basic Newtonian concepts were considerably enriched. This process of enriching the basic concepts in a given field we shall call concept articulation. It is characteristic of scientific development that concepts are quite crude when they are first accepted, and their articulation is a long, difficult process. One only has to trace the history of concepts such as "force," "inertia," "natural selection," or "the positron" to see this.

If a scientist doing normal science fails to explain given phenomena, he does not thereby falsify the theory he uses. In the short run, he may only show his own inability rather than any weakness in the theory. Nevertheless, in the long run problems can arise for a theory and lead to its overthrow. These problems Kuhn calls anomalies; he identifies several kinds. A theory may slowly be undermined when prolonged attempts by the best people fail to produce an explanation of some phenomena. The failure of Newton's laws to account for mercury's motion is a classic case of this. New phenomena such as X-rays or the electron, which are incompatible with the ontological claims of a theory, may be discovered. As a theory is used over time, a body of unsolved problems may accumulate. These problems undermine confidence in the paradigm, and attention turns toward the possibility of a new theory. New knowledge sets the stage for a revolution, which occurs when another paradigm is developed and the old one overthrown.

The development of and controversy over a new theory is called "extraordinary science." This period is characterized by confusion. A new theory is usually crude at first and, because it offers promise rather than precision, it may seem a poor substitute for the old one. Also, the need for a new theory may be open to question; whether the old theory has failed or not is seldom clear-cut, and advocates of the old theory may plausibly deny that anomalies exist. They can often maintain that appropriate auxiliary hypotheses would save the old theory and that the new one therefore marks a mistaken direction. Holders of old and new theories have different basic concepts and, in a sense, speak different languages. For this reason, they often seem to talk past one another without being able to come to grips with their differences.

This, then, is the basic outline of Kuhn's account of conceptual revolutions in science: a recurring cycle of theory acceptance, concept articulation and normal science, anomaly accumulation, new theory development, and revolution characterized by confusion. Then the cycle begins again. To be sure, some of Kuhn's claims are questionable. For example, his views about the nature of scientific controversies seem to me to be mistaken. Lack of communication occurs not because no common observational basis is available but because of the complexity of the evaluation process. This complexity is discussed below for technological controversy, but applies equally to scientific controversies. On the whole, Kuhn's description is very compelling and, I believe, quite correct. This description underlies the theory of technological revolutions which follows.

B. The Structure of Technological Revolutions

The received evaluation policy in an area of technological application plays a role analogous to that played by an accepted paradigm in an area of scientific explanation. The evalution policy organizes problem solving and provides procedures whose application is assumed to yield the best available solution. Moreover, just as a scientific theory is more than merely a set of laws or equations, so the basis of engineering technique is more than merely a set of equations to be solved. Such tools are always embedded in a system of accepted procedures for doing the job correctly, and these procedures in turn embody a specific evaluation policy, that is, a system of judgments as to what factors are important, how each is to be measured, and how it is to be valued.

If evaluation policies are like paradigms then we may expect the radical change of such policies to resemble revolutionary theory change in science. This is in fact the case; the general pattern is as follows. Assume an evaluation policy operating with concomitant standard procedures. The object of such a policy is to guide us in transforming a present state of affairs into the best possible future state (here value plays that role of target which is played by truth in the scientist's efforts). In this situation any indication that the best possible state is being missed is an anomaly for the accepted evaluation policy. Anomalies may arise when repeated attempts using standard procedures fail to eliminate known ills. In the short run, we may blame extraneous factors or our own incompetence, but eventually the possibility of alternative modes of action must arise. This kind of anomaly is, of course, the counterpart of the failure of a scientific paradigm to yield an explanation for a particular phenomenon. A technological example is the persistence of pests in the face of massive applications of pesticides over the last decades.

Another kind of anomaly which is of particular interest arises when new scientific or technical knowledge enables us to see that our standard procedures do not evaluate all factors correctly. In particular, we may discover that some factor which has been ignored is really very important, now that we know more about it. Examples of this kind of anomaly are the discovery, as a result of advances in biochemistry, of long-term effects of pesticides, the discovery of side effects of food additives through advances in medicine, and the discovery of adverse ecological effects of dams through advances in ecology.

This kind of anomaly is particularly troublesome because the knowledge which supports it, that is, the knowledge that some heretofore ignored factor is important or that some factor is incorrectly measured, may lie outside the group or discipline in charge of the technology in question. The Corps of Engineers, for example, had little, if any, contact with ecologists prior to the mid-sixties when the environmental crisis began to grow. Also, the evidence for anomaly or misevaluation may be tentative, controversial, or merely qualitative. Indeed, revolutions may be fueled by very soft data, and this explains some of the seeming incommensurability of opposing views.

Those who appreciate the new evidence will argue for a new evaluation policy which takes the newly discovered consequences into account. Exactly what this new policy should be may not be clear, however, even to its advocates. It may be a long way from the initial perception of working procedures for evaluation of that factor. For this reason, the proposed new evaluation policy may seem speculative and visionary to those who actually employ the existing policy. These technologists may perceive that the proposed policy is not sufficiently articulated to support working procedures, and hence it does not seem at all to be a viable alternative. The lines may become drawn between them and those who advocate a new evaluation policy as a moral necessity on the grounds that normal problem solving could not proceed under the new policy.

At this point the issue may spread into the public or political domain. The revolutionary scientist or technologist, or a popularizer such as Rachel Carson or Paul Ehrlich, finding his words unheeded by the technologists who carry out existing policy, may appeal directly to some higher authority or even to the public and to the lawmakers, presenting the case as a moral crusade against the archaic and wanton practices of the established technological order. The chances of success are good for this line of attack because the lay public may not appreciate the differences between the crude new evaluation policy and the well-articulated established policy. The public may not see that, if the new policy is forced on the technological area in question, then normal problem solving will have to be suspended while the details of new standard procedures are worked out.

If the revolution succeeds by the public route, legislation may be passed or orders given which outline a required new evaluation policy. It is then up to those working in the technological area in question, together with those who possess the new knowledge, to work out new procedures for carrying out this policy. In the interim it is literally impossible to evaluate a proposed problem solution. Needless to say, this stage is one of confusion and unhappiness for the technologist; production slows or stops as the unfamiliar task of articulating new evaluation concepts is thrust upon the engineers and technical administrators. Disorder prevails until new evaluation procedures are worked out and normal problem solving is resumed.

C. The Revolution in Water Resources Management

I hope it is apparent by now that the process which I have outlined is going on at this very moment in a number of technological areas. One good example is water resource management; the Corps of Engineers is currently going through a revolution in dam-building policy. The immediate causes of this revolution are the National Environmental Policy Act of 1969 (Public Law 91-190) and the River and Harbor Flood Control Act of 1970 (Public Law 91-611).

During the 1960s, a number of books and articles appeared which argued that the Corps's procedures for evaluating the costs and benefits of flood control dams were inadequate. These writings charged that the Corps was incorrectly evaluating factors it routinely considered, such as flood damage, recreation, and water supply, and also that it was ignoring significant factors such as environmental and social impact. Many case studies were offered to show that such factors as siltation, loss of animal habitat, erosion, loss of fast water, and general ugliness were not being considered when proposed dams were evaluated. One must remember that many of these studies could not have been made prior to the rather recent development of relatively sophisticated ecological and engineering techniques.

The initial response of the Corps was indifference, fed by an awareness that the ecological considerations in question could not be evaluated by the simple quantitative techniques currently being used to figure cost and benefit. Environmental effects had always been judged to be "secondary effects," a conceptual category which tended to equate immeasurability with insignificance. As the Corps remained unmoved in the face of accumulating evidence, the case against it grew into a moral crusade.

All this changed with the passage of the National Environmental Policy Act (NEPA). This act required that evaluation of any proposed civil works project with significant environmental impact include preparation of an environmental impact statement. With the passage of the act, the Corps's standard procedures for evaluating proposed projects became obsolete. It soon became apparent that while NEPA required environmental evaluation it gave no guidelines for that evaluation. The Corps was now in the position of having to develop that evaluation policy which it had been blamed for failing to follow but which had, in fact, never existed. The size of this task slowly became apparent; project after project stalled for lack of agreement as to what constituted adequate environmental analysis, and successive drafts of environmental impact statements were rejected.

The case was further complicated by the passage of the River and Harbor Flood Control Act of 1970. This act extended the concept of "significant effect" of a project to include social as well as ecological effects. According to Section 122 of the act, the Corps is now required

to assure that possible adverse economic, social, and environmental effects relating to any proposed project have been fully considered in developing such a project, and that the final decisions on the project are made in the best overall public interest, taking into consideration the need for flood control, navigation, and associated purposes, and the cost of eliminating or minimizing such adverse effects as the following:

1. Air, noise and water pollution:
2. Destruction or disruption of man-made and natural resources, aesthetic values, community cohesion and the availability of public facilities and services;
3. Adverse employment effects and tax and property value losses;
4. Injurious displacement of people, businesses and farms; and,
5. Disruption of desirable community and regional growth.

One did not have to be an ecologist or sociologist to realize that some of these factors, like "community disruption," were going to be very hard to evaluate. Many people at all levels in the Corps found themselves involved, not in their normal work of planning, design, and construction of projects, but in evaluation policy formulation. Needless to say they did not like it. This was a time of severe questioning and absolute confusion, as proposal after proposal was rejected as being inadequate. Lower-level administrators and managers were asking for guidelines from above, while higher-level reviewers were looking for suggestions from below. The Corps slowly turned to the ecology groups for advice, but the ecologists were unfamiliar with the detailed problems of dam ecology. It turned out that no one knew how to predict specific environmental effects in specific cases. It became apparent that literally thousands, perhaps hundreds of thousands, of specific points of fact and value would have to be isolated and settled before planning could resume in a normal way.

I do not want to give the impression either that all this confusion is bad or that there is any alternative to an overall conceptual revolution in such cases. The job of articulating a new Corps evaluation policy is being done, slowly and with much blind groping. Mechanisms for increased public contact with the planning process are now being developed. The next phase should see the growth of procedures for dealing with specific environmental factors. This phase should include the development of simple quantitative relationships when these are reasonable together with a body of rules of thumb and exemplary cases for qualitative issues. Presumably some factors will be judged intractable, and the category "secondary effect" will be revived. Then the Corps will settle down to (enlightened) normal technological problem solving, until a new body of anomalies is built up and the cycle begins again.

In terms of concepts, the problem can be described as follows. Consider the concepts "cost of a dam" and "benefit of a dam." Until recently these concepts were clearly defined by normal evaluation procedures; that is, the procedures for planning, designing, and constructing a beneficial dam were clearly understood in their day-to-day application. However, new knowledge, derived especially from ecology, made it increasingly clear that certain effects of dams—particularly bad ones—were not being taken into account when cost and benefit were computed. This new knowledge constituted an anomaly for the existing evaluation concepts. The initial result was the overthrow of the old concepts and procedural definitions of cost and benefit and the replacement of these by new concepts. But new concepts, like new theories, are unarticulated; that is, the definition or application of these concepts in day-to-day procedures had yet to be worked out. In the interim, the concepts of cost and benefit were literally undefined, with the result that no one knew what to say or do in any particular case.

From an engineering point of view, this was a strange crisis, but, from the point of view of philosophy of technology, it was a familiar and indeed a necessary feature of rational concept revision and procedural change. Progress would probably be impossible without this kind of confusion. By understanding that this is a "normal" conceptual revolution, we can better deal with it. We know, for example, that new day-to-day evaluation procedures to figure cost and benefit will be hammered out, but that they will have to be quantitative and simple because benefit-cost evaluation must be quantitative and government procedures must be simple. Those aspects of environmental impact which can be approximated by simple, quantitative evaluation procedures must be simple and will be taken into account. Some other qualitative aspects will be taken care of by precedent and rule of thumb, while still other impacts will be ignored because we cannot now figure them in.

This is a social revolution of the first magnitude, but even social revolutions have epistemological constraints. Our analysis cannot supply new procedures—that requires creative concept development by the engineers and environmentalists themselves, which they are not trained to do, but do anyway.

D. The Revolution in Good Food

The "good food problem" is another case in which new knowledge has led to an attack on established procedures for evaluating technology. Many people today believe that most of the food sold in grocery stores is unnatural and unhealthy, if not dangerous. Why do they believe this? Not, I think, because it is true. Rather, the case is like that of Paracelsus and the rise of chemical pharmacology in the sixteenth century.

Paracelsus used known poisons such as mercury and arsenic as medicines. When challenged, Paracelsus is supposed to have said, "Speak to me not of poisons, speak to me only of doses." [2] The transition from "poisonous substance" to "poisonous dose" was a conceptual revolution and one which required great technical sophistication. Rather than dividing the world into poisonous and nonpoisonous stuffs, the empirical medicine practiced by Paracelsus required that each application of a stuff be analyzed and its effects predicted. This development depended upon and was no doubt motivated by a growing skill in quantitative analysis and scientific insight. No wonder the Galenic doctors, whose evaluation policies made no provision for such distinctions, were horrified; the empirics were clearly poisoners.

With the development of highly sophisticated food chemistry in the last few decades, together with comparable advances in medical science, we face the same kind of conceptual problem that troubled the Galenists. Development of food chemistry has presented us with an ever-increasing array of potentially useful food additives, from preservatives, flavoring, and enriching agents to create new foods all the way to coagulants, lubricants, and antifoaming agents to facilitate production. Until recently our criterion for judging the safety of this kind of potentially useful food additive was basically that it seemed to have no detectable harmful effects under standard laboratory and field test conditions. These tests were pretty crude, and lots of new additives passed. However, as medical science advanced, more and more substances were found to be capable of producing adverse effects on laboratory animals. One old procedure for identifying "bad" chemicals for use in foods was to test a massive dose on a rat; any detectable adverse effect was good grounds for rejection. That procedure is inadequate for dealing with the new chemistry of food. We have become so much better at producing and detecting effects that everything turns out to be bad.[3] The result has been a growing crisis in evaluating additives as dangerous or safe. As this crisis has become public, the conviction has set in that any chemical alteration of foodstuffs is harmful, and that we must return to completely natural foods to avoid being poisoned. The actual problem, however, is that our new technology cannot be adequately evaluated by old procedures. The result is a kind of instability which grows as our old procedures give more and more unacceptable results.

The conflict between natural food advocates and those who scoff at the natural food reform movement may be seen as a conflict between people who are just using the established standards for good food and those who sense that those standards have been superseded, but don't have any new standards. This conceptual confusion signals a conceptual revolution. The dispute stems from a breakdown of the standard procedures for rational decision making. For one group it is obvious that we are poisoning ourselves; for the other side this is ludicrous. Resolution of this problem for us will require just what it required for Paracelsus. We need to be able to distinguish poisonous from nonpoisonous doses and to extrapolate more precisely from tests on laboratory animals. What is needed, however, is more profound than merely supplementing our stock of knowledge. What is needed is replacement of one concept by another.

It will not be easy to accept the concept that chemicals, like saccharin, which cause cancer under certain circumstances in rats are "safe enough" for use in our food. Note the concept shift from "safe" to "safe enough." Nor will it be easy to decide between accepting a drug which can cause cancer and one which can cause heart disease. Our new understanding of the subtle effects of chemicals will make this kind of decision unavoidable. To make such decisions will require new modes of thought.

The articulation process may in general require, or precipitate, new theories in the sciences. This happens, I think, because the crisis directs attention to new problems which the existing theories cannot deal with. The environmental impact problem is directing research into new areas in ecology, sociology, economics, and pollution engineering. The good food problem is responsible for new directions in chemistry, ecology, and medicine. Asking new questions in all these areas will ultimately affect physics and mathematics. In this way conceptual issues ebb and flow, circulating throughout the intellectual world.

E. The Issue Tree

Technological revolutions occur because new knowledge, particularly new science, enables us to realize that our accepted evaluation policy is incorrect—that we are missing the best possible solution. This realization may dawn slowly, and perhaps only through public action, because technical managers do not want to stop problem solving just to work out a new evaluation policy. If a conceptual revolution does occur, then production will drop as confusion sets in. This is a change-over cost which must be paid. Then in the chaos of conceptual articulation a new evaluation policy and day-to-day operating procedures are hammered out, and work gradually resumes.

Let us now take a closer look at the process of technological revolution in which a new evaluation policy is articulated. A number of major technological areas are currently in this phase of revolution. As an example, I shall consider nuclear power. I believe that the controversy over nuclear power in recent years has taken a very definite and typical form and that, moreover, the controversy is such that the articulation of a new evaluation policy is its result.

To make this point I must introduce the concept of an *issue tree.* An issue tree locates and relates the points at issue in an evaluation policy dispute in such a way that all points of disagreement are represented.[4] A practical decision rests upon the risk- or cost-benefit evaluation of each proposed solution. This evaluation may be quantitative or not; we may just consider "risk" or "cost" to mean "bad stuff" and "benefit" to mean "good stuff." If a single course of action is being considered, we may assume that a decision rests upon evaluation of total cost, total benefit, and a cost-benefit criterion which says how much greater benefits must be than costs to justify action.

The cost-benefit criterion is always debatable, as are total costs and total benefits. Total cost will usually be a combination of several subcosts, each of which is actually established separately. For each subcost three basic questions can arise. The first of these is the question of *relevance*. Is the factor in question actually a cost of the proposed action? Not every consequence of an action is a cost; some may be benefits while many others may well be of neutral value. For example, while it may be apparent that the presence of a nuclear power plant will have a substantial impact on the social and economic life of a nearby community, it may not be clear that this impact is, all in all, a cost. Under traditional nuclear power evaluation policy the factor of social impact has not usually been considered a cost, but this omission has become controversial as traditional policy has come to be questioned.

Second, if it is a cost, what is its *magnitude?* Even if a cost is considered relevant, how its magnitude is to be determined may be controversial. It is universally accepted, for example, that if radiation from a nuclear power plant causes cancer and leukemia this is a genuine cost of the plant, perhaps even a prohibitive cost. It is far from clear how the magnitude of such effects is to be estimated. This question has been a major source of controversy in the nuclear power issue.

Finally, there is the question of *valuation*. When the magnitude of a relevant factor has been determined, this magnitude must be converted into a value, that is, weighed against all other factors. If low-level radiation can be shown to cause leukemia, then we shall have to decide what difference this makes, vis-à-vis such factors as cheap energy or conservation of fossil fuels.

The questions of relevance, magnitude, and valuation also apply, of course, to possible benefits. A proposed action such as general use of nuclear power may have ten or twenty potential major costs, for which there are thirty to sixty possible points of disagreement. Moreover, major costs such as thermal pollution or radiation-induced disease may be further divisible. Thermal pollution has many bad effects, for example, and if enumeration is carried to this finer degree there may well be hundreds of separable costs. The same is true for benefits, though usually these are fewer because an action is typically proposed to produce a few clearly defined benefits.

We may have about fifty possible major points of disagreement. We can expect actual disagreement on most of these, if the issue is a hot one, but that is not all. As a cost, for example, is challenged, evidence will be offered on both sides. The debate then spreads to the level of the evidence for the claimed relevance, magnitude, and valuation for each cost and benefit. There may be a dozen major items of evidence for each claim about each cost or benefit. In the nuclear power case, for example, debate has raged over the evidence for or against various death rates to be anticipated from leukemia, cancer, and so on if nuclear power is used.

In this way a controversy may very quickly come to involve hundreds of specific issues, grouped about the major cost and benefit claims (see Figures 1, 2, and 3). Not only are there possible hundreds of items of evidence for the relevance, magnitude, and valuation of the costs and benefits, but for each item of evidence we can ask three questions (basically the same three as we asked about costs). We can ask: Is it really evidence (relevance), how reliable is it (magnitude of credibility), and how much does it count (confirmation value)? In the nuclear case, debate has already shifted to the relevance, credibility, and value of the laboratory experiments which are claimed as evidence for a certain predicted incidence of leukemia due to radiation. At this point the debate is already very complex, but in cases such as nuclear power it has gone even further, for, as questions are asked about the evidence, these questions are then debated in terms of evidence for or against that evidence. In this way it is possible for debate to work down to, say, the evidence for the relevance of certain evidence as evidence for a certain cost's having a certain magnitude, or to the evidence for the confirmation value of a certain evidence as evidence for a

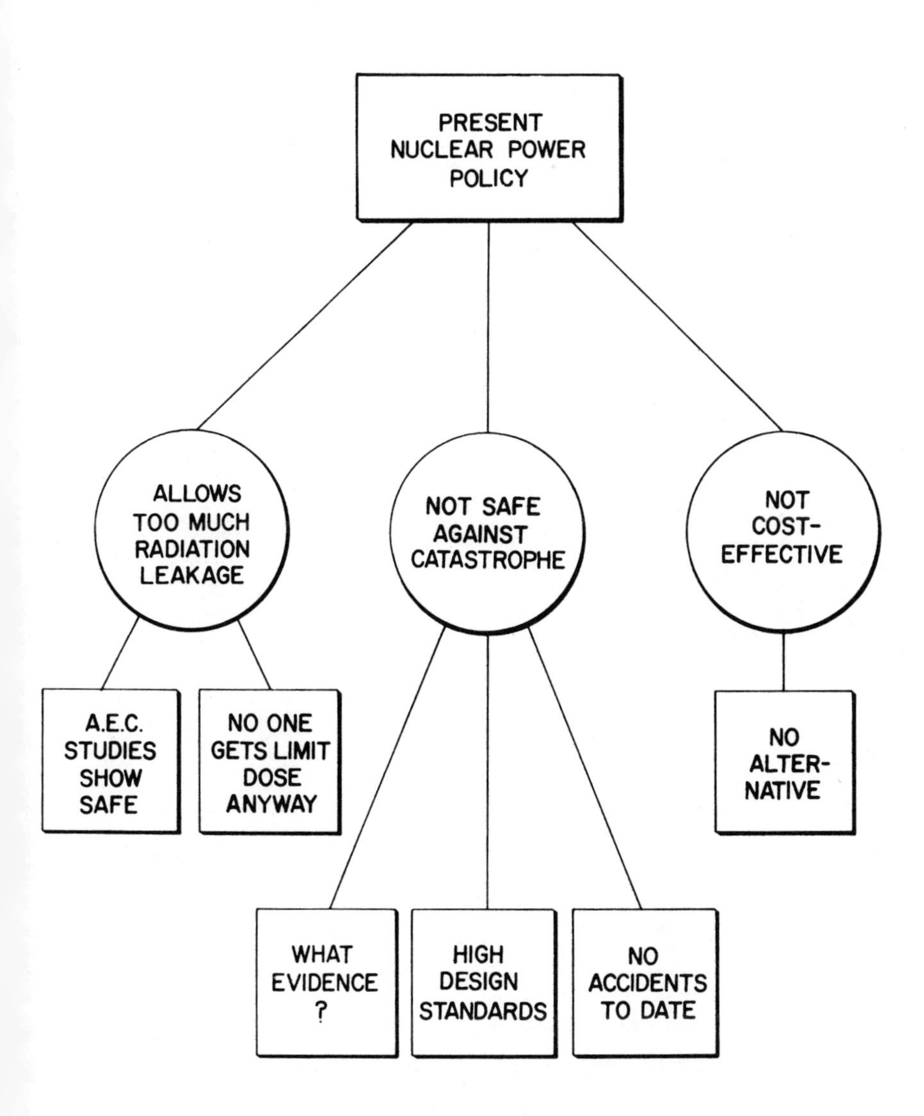
PRESENT
NUCLEAR POWER
POLICY
ALLOWS
TOO MUCH
RADIATION
LEAKAGE
NOT SAFE
AGAINST
CATASTROPHE
NOT
COST-
EFFECTIVE
A.E.C.
STUDIES
SHOW
SAFE
NO ONE
GETS LIMIT
DOSE
ANYWAY
NO
ALTER-
NATIVE
WHAT
EVIDENCE
?
HIGH
DESIGN
STANDARDS
NO
ACCIDENTS
TO DATE

Figure 1

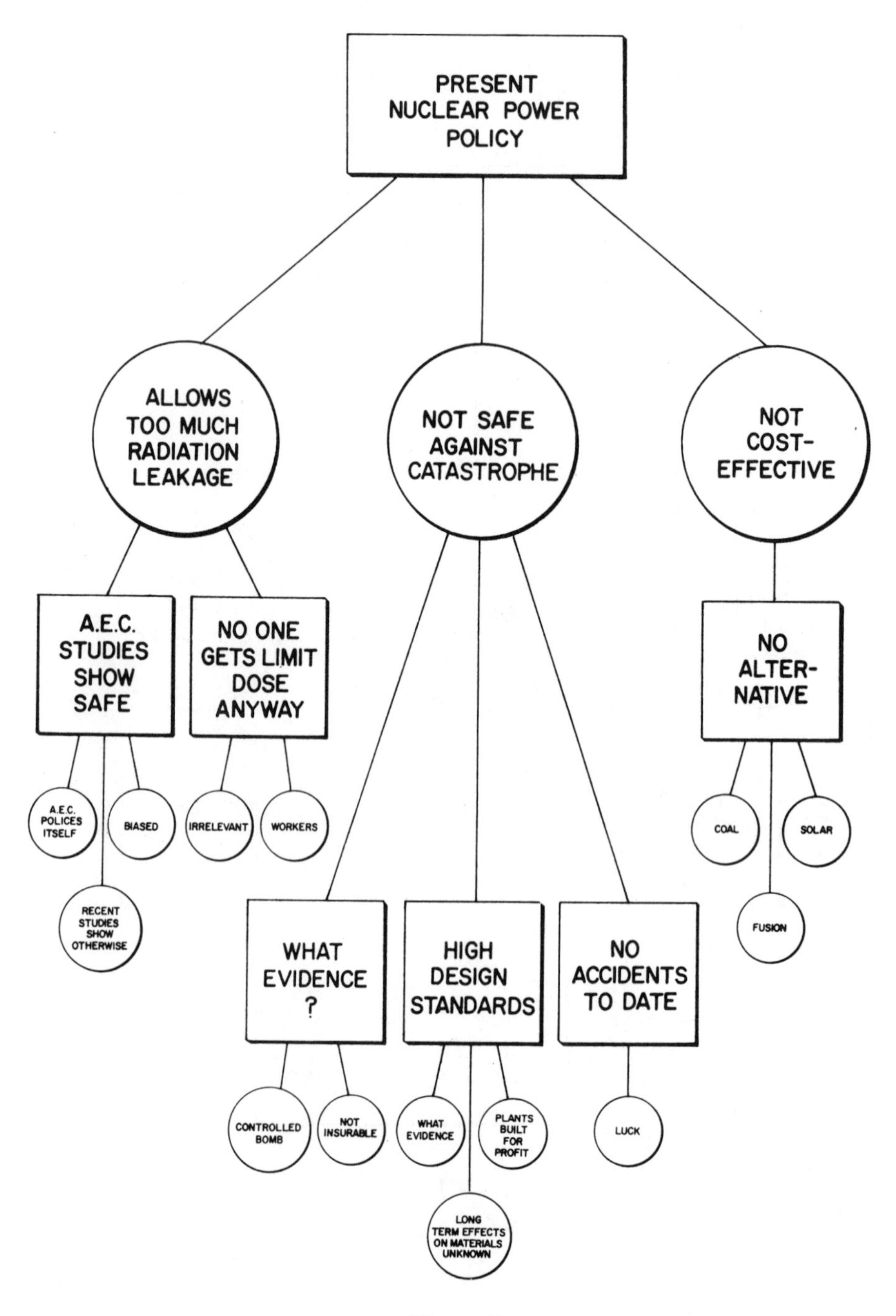
PRESENT NUCLEAR POWER POLICY
ALLOWS TOO MUCH RADIATION LEAKAGE
NOT SAFE AGAINST CATASTROPHE
NOT COST-EFFECTIVE
A.E.C. STUDIES SHOW SAFE
NO ONE GETS LIMIT DOSE ANYWAY
A.E.C. POLICES ITSELF
BIASED
RECENT STUDIES SHOW OTHERWISE
IRRELEVANT
WORKERS
WHAT EVIDENCE ?
HIGH DESIGN STANDARDS
NO ACCIDENTS TO DATE
CONTROLLED BOMB
NOT INSURABLE
WHAT EVIDENCE
PLANTS BUILT FOR PROFIT
LONG TERM EFFECTS ON MATERIALS UNKNOWN
LUCK
NO ALTER-NATIVE
COAL
SOLAR
FUSION

Figure 2

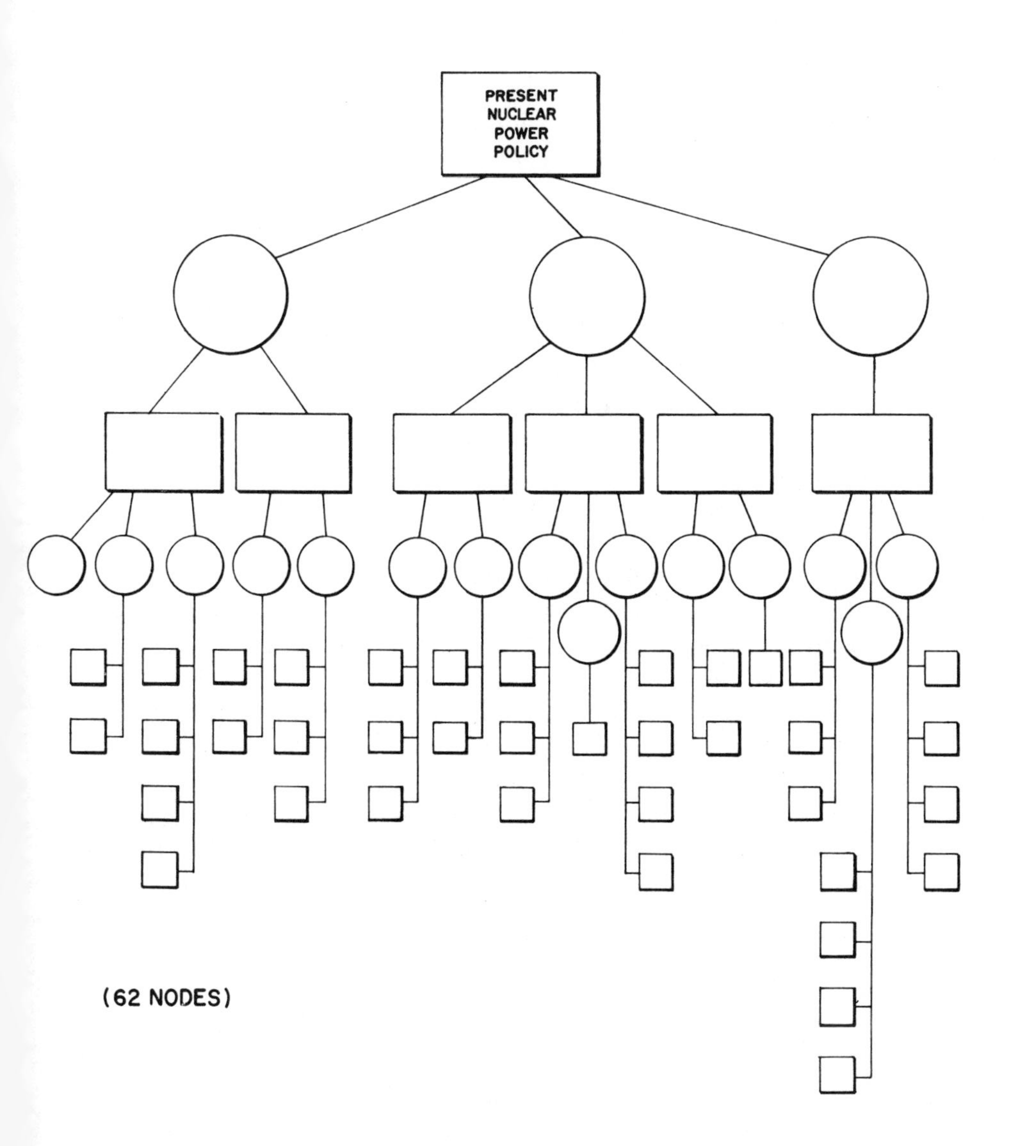

Figure 3

certain claimed benefit's being in fact a benefit, and so on. If debate spreads this far, there are potentially thousands of points of disagreement possible in the controversy. These points are arranged hierarchically, in a tree. Few of these points can be decided clearly. Most of the questions at issue in the nuclear controversy actually involve such subtle issues as the methodological soundness of certain experiments, the legitimacy of certain extrapolations, or the credibility of certain speculations.

It is possible to work out a controversy like the power plant issue as a tree structure. We see which moves have been made and which have not. We see the untried lines discovered and pursued as time goes by, with debate spreading down to the subterranean depths of such issues as the evidence for the magnitude of some particular claimed cost or benefit. Of course, this kind of issue analysis does not solve the problem. That is a job for the people involved. We only make clear the structure of the debate, though often this clarity helps resolve the issues.

One characteristic of technological revolutions is the role that new knowledge may play in generating a revolution. In order to attack traditional cost and benefit evaluation, one must be prepared to offer evidence. Correspondingly, new evidence for previously unsuspected or immeasurable consequences is required to support proposals to change existing evaluation policy. It is by working down through the issue tree, offering and defending one's evidence, that proposed evaluation policy changes are evaluated. The core of most current technological controversies is the problem of revising existing evaluation policies in the light of new knowledge about the consequences of our actions, including new techniques for predicting these consequences.

A second characteristic is that technological revolutions generate intense speculation. As controversy over evaluation policy develops, the antagonists seek to develop all possible lines of attack and defense in the tree. Every claim is carefully scrutinized for weakness, and no hypothesis with a shred of plausibility is left uninvestigated. As the level of evidence is reached, questions of fundamental scientific and engineering methodology are raised. Here our hypothesis-raising abilities are especially important, and intense scientific speculation is required to uncover and develop all possibilities. In addition, as new consequences come into view, the question of the magnitude of cost or benefit value to be assigned is taken up and debated, so that eventually even our most basic values are questioned. A typical controversy is a complex combination of factual and normative issues, with a tree structure.

Many of the possibilities raised at this stage are qualitative suggestions of possible phenomena. It is these qualitative claims and the insistence that they be taken seriously that technological managers initially reject as unworkable. How, they ask, are we to include aesthetic, environmental, ecological, or sociological consequences in our cost-benefit analyses? They may not stay to hear the answer until compelled to do so by law. The technologist appreciates the detail and precision of the established policy, and he must discover at least the promise of such detail and precision in his opponent's proposals if he is to take those proposals seriously. This detail may be largely absent in the initial stages of debate. Moreover, as discussion proceeds and the issue tree is fleshed out, the qualitative and tentative nature of many claims may fail to close debate. In this stage rational disagreement is possible, and where one stands may be a matter of one's own background information or intuition, organizational loyalty, or even party affiliation. In actual practice there may be few people with the fortitude to fully explore their opponent's case, where thousands of points are at issue.

The confusion cannot last, however. This is an important point. Development of the issue tree focuses attention on crucial issues, which are then explored. As initially qualitative hypotheses are proposed, they are investigated until hard data settle the question one way or another. In this way the issue tree guides the research, which in turn resolves the issues. If an objection to established policy is sustained, then the research which finally settles the case will also provide the basis for the new procedures needed to implement the changed policy. Thus, for example, the claim that nuclear power development under existing regulations would lead to a given incidence of leukemia can only be finally substantiated by determining the magnitude of this effect in specific cases. Should we actually become able to determine this magnitude, then we will finally be in a position to incorporate this effect in our evaluation policy.

The criticism of established policy and the development of the issue tree guide that research which ultimately terminates in a new working evaluation policy. The issue tree, which first expanded wildly to the confusion and consternation of all concerned, finally contracts as issue after issue is settled, until a consensus is reached. As this contraction occurs, a new evaluation policy comes into focus. The initial articulation phase is over and normal decision making resumes. All known costs and benefits have been explored, needed research pinpointed, and a new evaluation policy put on a firm foundation.

When many of the controversial points cannot be decided rationally in a reasonably short time, but a decision must be made, one finds sincere but decisive disagreement between experts. When several thousand points are at issue and there is room for reasonable disagreement on most of them, the result can be a complete polarization of opinion. At this point, the standard procedures for professional judgment have broken down. All benefit-cost analysis is shaky. When it is actually criticized by experts, it often dissolves completely into a morass of possibilities and intangibles. At this stage decision making shifts to authority. It seems clear that one of the main functions of administrative structure is that it enables decisions to be made even when opinion is not unanimous.

How the nuclear power plant issue is settled will depend upon who is in power in government and who is in power in the engineering and scientific communities. What environmental factors the Corps of Engineers has to consider is ultimately for Congress and the Corps to decide. In short, politics and administration are necessary because reasonable men of good will can fail to agree. The issue tree model shows that a single issue can include literally thousands of points of possible conflict. The amazing thing is that we usually do agree. This shows the iron grip which accepted standards usually have on any profession. Normal problem-solving work cannot proceed without such self-imposed control. Thus it is vital that there be a mechanism to force rapid articulation of new evaluation policies and rapid acceptance of these policies once they are articulated.

In the long run, however, convergence to a new equilibrium position is a matter of truth, not power.[5] Anomalies cannot be legislated away, in either technical practice or social being. As long as we can recognize the bad, we shall strive for the good.

Notes

1. Thomas Kuhn, *The Structure of Scientific Revolutions* (Chicago: University of Chicago Press, 1962).
2. The story is now considered apocryphal, but it still captures the conceptual issue. See J. G. Jacobi, *Paracelsus: Selected Writings,* rev. ed. (New York: Pantheon, 1958).
3. See, for example, James S. Turner, *Chemical Feast: Report on the Food and Drug Administration* (New York: Grossman, 1971).
4. We have applied the issue tree concept to many practical issues of policy and regulatory reform. See David E. Wojick, *Issue Analysis: An Introduction to the Use of Issue Trees and the Nature of Complex Reasoning* (Pittsburgh, Pa.: David E. Wojick Associates, 1975), and *Planning for Discourse: A Manual for the Management of Complex Processes, Based upon Issue Analysis Measures* (Fort Belvoir, Va.: Institute for Water Resources, 1978).
5. See *Back to Normal,* a report to the St. Paul District, Corps of Engineers, U.S. Army, on the management of engineering issues (Pittsburgh, Pa.: David E. Wojick Associates, 1978).

Mario Bunge

15

Philosophical Inputs and Outputs of Technology

Technology is often considered soulless, aphilosophical, or even antithetical to philosophy. This paper contends that such an image of technology is erroneous and that:

1. Far from being aphilosophical, let alone antiphilosophical, technology is permeated with some of the philosophy it has inherited from pure science along with scientific methods and theories—as exemplified by its reliance on the philosophical principle that we can get some knowledge of reality through experience and reason, and even improve on it.
2. Far from being philosophically passive or sterile, technology puts forth a number of philosophically significant theories, such as automata theory, and important (though perhaps mistaken) philosophical views, such as pragmatism.
3. Far from being ethically neutral, like pure science, technology is involved with ethics and wavers between good and evil.

In other words, this paper proposes the thesis that technology has a philosophical input and a philosophical output and, moreover, part of the latter controls the former. If this is true, then technology is not cut off from culture nor is it a detachable part of culture; technology is instead a major organ of contemporary culture. This being so, the philosopher must pay it far more attention than before; he should build a fully developed philosophy of technology related to but distinct from the philosophy of science.

Tasks of the Philosophy of Technology

The concern of the philosophy of technology—one of the underdeveloped areas of philosophy—is the investigation of the philosophy inherent in technology as well as of the philosophical ideas suggested by the technological process. Some of the typical problems in the philosophy of technology are these: (a) Which characteristics does technological knowledge share with scientific knowledge, and which are exclusive of the former? (b) In what does the ontology of artifacts differ from that of natural objects? (c) What distinguishes a technological forecast from a scientific forecast? (d) How are rule of thumb, technological rule, and scientific law related? (e) Which philosophical principles play a heuristic, and which a blocking, role in technological research? (f) Does pragmatism account for the theoretical richness of technology? (g) What are the value systems and the ethical norms of technology? (h) What are the conceptual relations between technology and the other branches of contemporary culture?

Where are we to search for the philosophical components of technology? Clearly not among the products of technology—cars, drugs, healed patients, or victims of technological warfare—which are about the only technological items the anti-technological philosopher is acquainted with. We must search for philosophy among the ideas of technology—in technological research and in the planning of research and development. We are likely to find them here, as philosophy is found in every department of mature thinking. Indeed mature thinking is always guided (or misguided) and controlled (or exhilarated) by methodological rules as well as by epistemological, ontological, and ethical principles. Just think of the problems posed by the design of any new product. Is the relevant scientific knowledge reliable, and is it likely to be sufficient? Will the new product be radically new—that is, will it exhibit new emergent properties—or will it be just a rearrangement of existing components? Shall we design the product so as to maximize performance, social usefulness, profit, or what?

Since the philosophical components of technology must be searched for among technological ideas, we had better start by recalling what the loci of these ideas are. Moreover, since there is some uncertainty about what "technology" includes, we should enumerate the branches of technology as we understand it.

Branches of Contemporary Technology

We take technology to be that field of research and action that aims at the control or transformation of reality whether natural or social. (Pure science, if it is experimental, also controls and transforms reality but does so only on a small scale and in order to know it, not as an end in itself. Whereas science elicits changes in order to know, technology knows in order to elicit changes.) We discern the following branches of technology;

Material	Physical (civil, electrical, nuclear, and space engineering) Chemical engineering Biochemical (pharmacology) Biological (agronomy, medicine)
Social	Psychological (education, psychology, psychiatry) Psychosociological (industrial, commercial, and war psychologies) Sociological (politology, jurisprudence, city planning) Economic (management science, operations research) Warfare (military science)
Conceptual	Computer sciences
General	Automata theory, information theory, linear system theory, control theory, optimization theory, and so forth

This list is not exhaustive, and some technologists may feel ill at ease with the bed fellows I have chosen for them. The list is intended to be only a partial extensional definition of "technology." It includes the miscellany I have called "general technology" because its theories can be applied almost everywhere regardless of the kind of system; we shall see later in the paper that it constitutes the great contribution of technology to metaphysics. On the other hand, the list does not include futurology, because the latter is just long-term planning and hence is part of social technology.

Let us now locate the areas of maximal conceptual density regardless of subject matter: there we must cast our net. To this end we must take a brief look at the technological process.

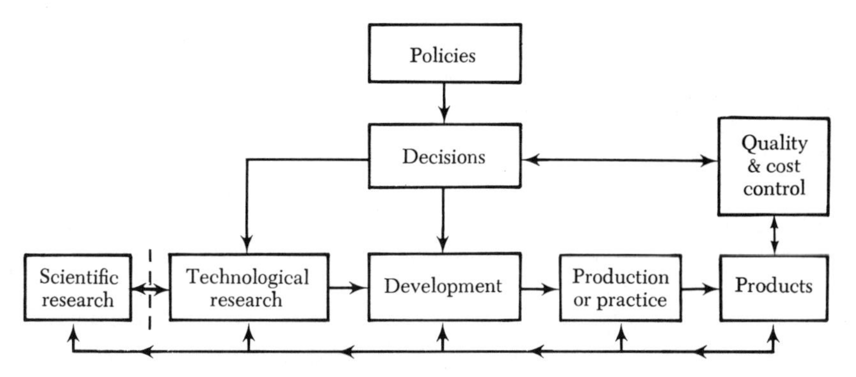

Figure 1. Flow Diagram of Technological Process. The first stage, scientific research, is occasionally missing or completed at a scientific institution —hence the dotted vertical line. The end product of a technological process need not be an industrial good or a service; it may be a rationally organized institution, a mass of docile consumers of material or ideological goods, a throng of grateful if fleeced patients, or a war cemetery.

Technological Research and Policy

A technological process exhibits the stages shown in Figure 1.

Most technological ideas are found in two of the stages or aspects of a technological process: policy and decision making (largely in the hands of management) and research (in the hands of investigators). In any high-grade technological process, such as one taking place in a petroleum refinery, in a hospital, or in an army, managers as well as technological investigators (but not technicians and blue- and white-collar workers) employ a number of sophisticated conceptual tools—belonging, for example, to organic chemistry or operations research. If they are innovative or creative, policy makers and investigators will try out or even invent new theories or procedures. In sum, technology is not alien to theory, nor is it just an application of pure science; it has a creative component, which is particularly visible in the design of technological policies and in technological research.

Consider technological research for a moment. Methodologically, it is no different from scientific research. In either case, a research cycle looks schematically like this: (1) spotting the problem; (2) trying to solve the problem with available theoretical or empirical knowledge; (3) if that attempt fails, inventing hypotheses or even whole hypothetico-deductive systems capable of solving the problem; (4) finding a solution to the problem with the help of the new conceptual system; (5) checking the solution, for example by experiment; (6) making the required corrections in the hypotheses or even in the formulation of the original problem. Besides being methodologically alike, both kinds of research are goal-oriented; however, their goals are different. The goal of scientific research is truth for its own sake; that of technological research is useful truth.

The conceptual side of technology is neglected or even ignored by those who equate technology with its practice or even with its material outputs. (Curiously enough, not only idealist philosophers but also pragmatists ignore the conceptual richness of technology. Hence neither of them can be expected to give a correct account of the philosophy inherent in technology.) We must distinguish the various stages or aspects of the technological process and focus on technological research, as well as on the design of technological policies, if we are to discover the philosophical components of technology.

Before we face our specific problem we shall make one more preliminary investigation—this time into the conceptual relations among technology and a few other branches of culture, both alive and dead.

Near Neighbors of Technology

Nothing, especially not technology, comes out of nothing. Hence nothing, especially not technology, can be understood in isolation from its kin and neighbors. Modern technology grows out of the very soil it fertilizes, industrial civilization and modern culture. The distinction between civilization and culture is particularly useful for understanding the nature of technology. One can have some modern industry without modern culture, provided one imports technological know-how and does not expect great technological innovations. One can have scraps of modern culture without modern industry—provided one is willing to put up with a one-sided and rickety culture. No creative technology, however, is possible outside modern civilization (which includes modern industry) and modern culture (which of course includes modern technology).

In particular, modern technology presupposes not only ordinary knowledge and artisanal skills but also scientific knowledge, hence mathematics. Technology is not a final product, either; it shades into technical practice—the practice of the general practitioner, the teacher, the manager, the financial expert, or the military expert. Things are not completely pure in or around technology; besides its artistic and philosophical components, one occasionally finds traces of pseudoscience and pseudotechnology. Table 1 shows some of the nearest neighbors of technology. To complete the picture, add mathematics, crafts, arts, and humanities, as in Figure 2, p. 280 below.

Having sketched a map of technology and having listed some of its neighbors very schematically, we are now in a position to try to explore the philosophy inherent in technological research and policy making .

PROTOSCIENCE	SCIENCE	TECHNOLOGY	TECHNICAL PRACTICE	PSEUDO-TECHNOLOGY
Ancient & medieval physics & astronomy	Modern physics & astronomy	Physical engineering	Engineering practice	Astrology
Ancient & medieval mineralogy & part of alchemy	Chemistry	Chemical engineering	Chemical engineering practice	Alchemy
Ancient & medieval natural history	Biology	Agronomy, medicine	Agrotechnical & medical practices	Homeopathy, chiropractic, Lysenkoism
Philosophy of mind (partly)	Psychology	Psychopathology	Drug & behavior therapy	Psychoanalysis, graphology
	Economics	Economic & financial planning	Economic management	Economic miraclemanship
		Computer science	Computation & control	GIGO computeering

Table 1. The nearest neighbors of technology.

The Epistemology of Technology

Technology shares with pure science a number of epistemological assumptions. We mention only the following: (1) there is an external world; (2) the external world can be known, if only partially; (3) every piece of knowledge of the external world can be improved upon if only we care to. These assumptions belong to epistemological realism. The classical technologist was not only a realist but usually a naive realist, in that he took our representations of reality for more or less accurate pictures of it. The modern technologist, involved as he is with constructing sophisticated mathematical models of things and processes, is still a realist but a critical one. He realizes that our scientific and technological theories are not pictures but symbolic representations that fail to cover every detail (and sometimes the very essence) of their referents. He knows that those theories are over-simplifications and also that they contain many concepts—like the proverbial massless piston—which lack real counterparts.

However, the critical realism of technology is tempered and distorted by a strong instrumentalist or pragmatist attitude, the normal attitude among people intent on obtaining practical results. This attitude is obvious from the technologist's way of dealing with both reality and the knowledge of it. For him, reality, the object of pure science, is the sum total of resources (natural and human), and factual knowledge, the aim of pure science, is chiefly a means.

In other words, whereas for the scientist an object of study is a *Ding an sich*, for the technologist it is a *Ding für uns*. Whereas to the scientist knowledge is an ultimate goal, to the technologist it is an intermediate goal, something to be achieved only in order to be used as a means for attaining a practical goal. It is no wonder that instrumentalism (pragmatism, operationalism) has such a great appeal both to technologists and to those who mistake technology for pure science.

Because of his pragmatic attitudes, the technologist will tend to disregard any sector of nature that is not or does not promise to become a resource. For the same reason he is prone to push aside any sector of culture unlikely to be instrumental for achieving his goals. This is just as well as long as he is open-minded enough to tolerate whatever he disregards.

The pragmatic attitude toward knowledge is reflected, in particular, in the way the technologist treats the concept of truth. Although in practice he adopts the correspondence conception of truth as adequacy of the intellect or mind to the thing, he will care for true data, hypotheses, and theories only as long as they are conducive to the desired outcomes. He will often prefer a simple half-truth to a complex truth. He must, because he is always in a hurry to get useful results. Besides, any error made in neglecting some factor (or some decimal figure) is likely to be overshadowed by unpredictable disturbances his real system may undergo. Unlike the physicist, the chemist, or the biologist, he cannot protect his systems against shocks other than by building shock-absorbing mechanisms into them. For similar reasons, the technologist cannot prefer deep but involved theories when superficial ones will do. However, unless he is a pseudotechnologist, he will not shy away from complex and deep theories if they promise success. (For example, he will employ the quantum theory of solids in the design of solid state components and genetics in obtaining improved varieties of corn.) The technologist, in sum, will adopt a mixture of critical realism and pragmatism, varying these ingredients according to his needs. He will seem to confirm first one and then another epistemology, while actually all he intends to do is to maximize his own efficiency regardless of philosophical loyalties.

The technologist's opportunistic conception of truth is just one—although major—epistemological component of technology. We shall now cite two specific items of epistemology that have taken part in technological developments, one in education, the other in artificial intelligence. It is well known that Pestalozzi's educational techniques were based on the slogan of British empiricism, "No concept without a percept." Likewise the philosophical basis of Dewey's educational techniques was the pragmatist thesis, "No concept without an action." The philosophy underlying artificial intelligence studies contains one major ontological hypothesis, "Whatever behaves like an intelligent being is intelligent," and a batch of epistemological hypotheses, among them

"Every perception is the acceptance of an external stimulus" and "Some spatial patterns are perceptible and discrete."

There is more to the epistemology of technology, but we must hurry on to the metaphysics of technology.

The Metaphysics of Technology

Technology inherits some of the metaphysics of science and has in turn produced some remarkable metaphysics of its own. We shall list without discussion a few examples of each.

Here are some of the metaphysical hypotheses inherent in both scientific and technological research:

1. *The world is composed of things,* that is, it is not simple, and it is not made of ideas or of shades of ideas. (Were this not so, we could not get things done by cleverly manipulating things—people among them. Mere wishes or incantations would suffice.)
2. *Things get together in systems* (composed of things in more or less close interaction), and *some systems are fairly well isolated from others.* (Otherwise we would not be able to assemble and dismantle things, nor would we be capable of acting upon anything without at the same time disturbing everything else.)
3. *All things, all facts, all processes, whether in nature or in society, fit into objective stable patterns (laws).* Some of these laws are deterministic, others are stochastic, and all are objective. (Otherwise we would not need to know any laws in order to transform nature and society: ordinary knowledge would have sufficed to bring forth modern technology.)
4. *Nothing comes out of nothing and nothing goes over into nothingness.* There are antecedents or causes for everything, and whatever is the case leaves some trace or other. If this were not so, there should be no need to work and no worries about energy.
5. *Determination is often multiple and probabilistic rather than simple or linear.* (If this were not so, we would be unable to attain most goals through different means, and there would be no point in searching for optimal means or in calculating probabilities of success.)

So much for the metaphysics that takes part in technological research and policy making. Now let us look at some of the metaphysical outputs of contemporary technology. While some of them are loose though important theses, others are full-blown ontological theories. Among the former we point out the following:

1. With the help of technology man can alter certain natural processes in a deliberate and planned fashion.
2. Thanks to technology man can create or wipe out entire natural kinds, thus increasing the variety of reality in some respects and decreasing it in others.
3. Because artifacts are under intelligent control or are endowed with control mechanisms which have not emerged spontaneously in a process of natural evolution, they constitute a distinct ontic level characterized by properties and laws of its own—whence the need for elaborating a technological ontology besides the ontologies of natural and of social science.

As for the metaphysical theories evolved by contemporary technology, they belong in what I have called general technology. They are high grade (though mathematically often simple) general theories such as automata theory, the general theory of machines, general network theory, linear system theory, information theory, control theory, and optimization theory. They qualify not only as technological (or scientific) theories but also as ontological theories for the following reasons. First, they are concerned with generic traits of entire genera (rather than species) of systems: they are cross-disciplinary theories. (Think of the variety of applications of automata theory and control theory.) Second, those theories are stuff-free (independent of any kind of material), hence independent of any particular physical or chemical law. (They focus on structure and behavior rather than on specific composition and mechanism.) Third, those theories are untestable without further effort, if only because they issue no predictions. (They can be made to issue projections and thus become testable upon conjoining them with items of specific information concerning the concrete systems they are applied to.)

In sum, whether they like it or not technologists have built a conceptual building which houses all of the metaphysics of science plus some distinctly technological metaphysics. Metaphysics, banned from philosophy departments, is alive and well in the schools of advanced technology.

The Value Orientation of Technology

To the scientist all concrete objects are equally worthy of study and devoid of value. Not so to the technologist: he partitions reality into resources, artifacts, and the rest—the set of useless things. He values artifacts more than resources and these in turn more than the rest. His, then, is not a value-free cosmology but one resembling the value-laden ontology of the primitive and archaic cultures. One example should suffice to bring this point home.

Let P and Q be two components or properties of a certain system of technological interest. Assume that, far from being mutually independent, Q interferes with or inhibits P. If P is desirable (in the eyes of the technologist) then Q will often be called an impurity. Unless the impurity is necessary to obtain a third desirable item R (such as conductivity, fluorescence, or a given color), the technologist will regard Q as a disvaluable item to be minimized or neutralized. To the scientist Q may be interesting or uninteresting in some respects, but never disvaluable.

This value orientation of technological knowledge and action contrasts with the value neutrality of pure science. True, social science does not ignore values but attempts to account for them. However, to pure science nothing is either pure or impure in an axiological sense, not even pollutants. In pure science valuation bears not on the objects of study but on the research tools (e.g., measurement techniques) and outcomes (e.g., theories). One lunar theory may be better (truer) than another, but the moon is neither good nor bad. That is not so for the space scientist and the politician behind him. Whereas the technologist evaluates everything, the scientist *qua* scientist evaluates only his own activity and its outcomes. He approaches even valuation in a value-free fashion.

The value orientation of technology gives the philosopher a splendid opportunity to analyze the valuation process in concrete cases rather than setting up a priori (or else conventional) "value tables." It can even inspire him to build realistic value theories, where valuation appears as a human activity, largely rational, done in the light of definite antecedent knowledge and definite desiderata. As a matter of fact, technology has already had an impact on value theory; utility theory (the theory of subjective value), though originally proposed as a psychological theory, has recently been revived and elaborated in response to the needs of managers. One may also think of a theory of objective value even more closely in tune with technology—one defining value as the degree of satisfaction of an objective need.

We turn now to a few other instances of the impact of technology upon philosophy.

Technology as a Source of Inspiration for the Philosophy of History

We have seen that technology is both a consumer and a producer of philosophical ideas. In addition, it can inspire or suggest interesting new developments in the philosophy of action, in particular ethics, legal philosophy, and the philosophy of history. Let us look into the last.

A number of historians are applying mathematics to problems in history. Here are a few examples of the mathematization of history: (a) cleansing historical data (such as chronologies) with the help of mathematical statistics; (b) finding historical trends or quasi-laws in a number of socioeconomic variables (notably by the French historians of the *Annales: Economies, Societes, Civilisations*); (c) building mathematical models of certain historical processes, such as the expansion and decline of empires; and (d) studying certain historical events and processes in the light of decision theory. This last approach, suggested partly by management science, is legitimate with reference to deliberate decisions affecting the life of entire communities. The passing of important new legislation, the launching of a war, the call to a nationwide strike, and the outbreak of a planned revolution are occasions for the application of decision theory. Indeed, all the necessary components are there or can be conjectured: the decision makers who are supposed to maximize their expected utilities, the goals, the utilities of them, the means or courses of action considered by the decision makers, and the probability of attaining a given goal with a certain means.

The philosophy of history can acquire a whole new dimension in the light of decision theory, provided, of course, it is not employed to resurrect the great hero theory of history. Certainly, important areas of historiography, such as the anonymous history studied by historical demography, historical geography, and economic history, remain beyond the decision-theory approach. However, in an increasingly technological society, rational (but, alas, often wicked) action, based on carefully designed policies, plays an increasingly important role and can therefore be partly understood with the help of decision theory.

Technology as a Source of Inspiration for Ethics and Legal Philosophy

Other fields of the philosophy of action that technology can fertilize are ethics and legal philosophy, by teaching them to spell out norms as grounded rules or even as conclusions of arguments. Thus, instead of issuing blind commands of the form "Do *x*," or blind ethical norms of the form "You ought to do *x*," the technologist will proceed as follows. He will propose and test grounded rules of the form "Do *x* in order to get *y*," on the basis of the knowledge that doing *x* does in fact bring about *y* either invariably or with a certain probability. By stating explicitly the ground for a rule of action, one kills three birds with one stone: (a) one breaks the fact/norm barrier, (b) one transforms moral decision making into a rational activity, and (c) one dispenses with the logic of norms.

This proposal, even if feasible, does not allow us to build a value-free ethics. This would be impossible, because moral decision making is as value-oriented as technological policy design. What technology can teach us is, rather, to render values explicit so as to be able to examine them critically instead of receiving them uncritically. In other words, it is impossible to translate a normative sentence into a value-free declarative sentence without loss. On the other hand, it is possible to spell out a norm into a pair law sentence-value sentence, in this way: "Do *x*" or "You ought to do *x*" may be construed as short for "There is a *y* such that *x* brings about *y* and (you value *y* or there is a *z* such that not doing *x* brings about *z* and you disvalue *z*)." The command (or the norm) and its expansion, though not logically equivalent, are related in that the former is just an abbreviation of the latter.

For example, "Do not cheat" can be expanded into "(Any) cheating does (some) harm and you do not want to do any harm." But the same norm can also be expanded into "(Any) cheating jeopardizes your credit and you want to keep your credit in good standing." This ambiguity is to be blamed on the norm itself and not on its rational translation. In any event, a norm, when grounded and formulated in the declarative mode, appears as a consequence of a set of premises. And at least one of these premises is a law statement while at least one other is a value judgment. Consequently, the handling of norms requires only ordinary logic (instead of the logic of norms) and value theory. In other words, we can reconstruct normative science without norms, but with values.

(Superficially, ordinary logic would seem to suffice. Thus, in the case of the injunction not to cheat because it causes harm and harm is undesirable, we would seem to have just an instance of *modus tollens,* namely: $C \rightarrow H$ & $\neg H \therefore \neg C$. However, the H occurring in the first premise differs from that occurring in the second: the latter is not really H but rather "H is valuable." Likewise, the conclusion is a value statement. A task of value theory is to compute the value of the conclusion in terms of the values occurring in the premises. But we cannot go into this here.)

What holds for ethics holds for legal philosophy: here, too, norms are profitably expanded into complex statements or construed as consequences of sets of premises. For example, "Murderers must be put away" somehow follows from "Murder endangers the social structure and we value the social structure." However, the same norm also follows from premises in a different field, e.g., "Murderers are sick people and it is disvaluable to leave sick people at large," as well as from premises inspired in still other value systems. The advantage of such expansions is obvious: they force the law giver to lay bare the grounds of positive law—which is often cruel, unfair, or even absurd—and invite him to ground legal technology on sociology and psychology.

In sum, technology suggests that we replace every authoritarian set of imperatives with a grounded set of rules—rules based on laws and value judgments. In this way, whatever was implicit or even concealed can be analyzed, criticized, reconstructed, and systematized. Technology can thus act as a methodological model for the normative sciences, in particular ethics. Unfortunately, far from having served as a moral model, technology is in need of some ethical bridling. This deserves another section.

The Dubious Morals of Technology

Knowing is a good in itself. (Even knowing how to inflict pain may be valuable, as it can assist us in avoiding the act of inflicting pain.) However, there are ways and means of knowing, and some of them may be morally objectionable, such as torturing and killing people in order to find out more about fear. Hence scientific research gets somewhat involved with ethics. In practice, a few rather obvious strictures usually suffice to keep it unsoiled. There are, of course, uncertainty zones, but they can be bounded. For example, in research into fear mild torturing might be condoned provided it is done with the free consent of the experimental subject and it can be safely predicted that it will not be traumatic. In short, pure science needs only a mild external ethical control. As a matter of fact, scientific research has built into it an ethical code of honesty, responsibility, and hard work that can inspire other human activities.

Things are different in technology. Here not only some of the means and ways of knowing may be impure, but also the entire technological process may be morally objectionable for aiming exclusively at evil practical goals. For instance, it is wicked to conduct research into forest defoliation, the poisoning of water reservoirs, the maiming of civilians, the manipulating of consumers or voters, and the like, because the knowledge gained in research of this kind is likely to be used for evil purposes and unlikely to serve good purposes. It is not just a matter of an unexpected evil use of a piece of neutral knowledge, as is the case with the misuse of a pair of scissors: the technique of evil doing is evil itself. The few valuable items it may deliver are by far outnumbered by its negative output. Try to find a good use for the stocks of lethal germs accumulated for chemical warfare, for example, or for plans for the rational organization of an extermination camp.

Technology can then be either a blessing or a curse. That it is always a blessing, if not in the short run then in the long run, is a tenet which has been preached by a number of progressive philosophers since the dawn of the modern period. Other philosophers claimed instead that technology is a curse, but they did so for the wrong reasons—because they were against social progress and cultural expansion. It is only very recently that most of us have come to realize that technology itself can in fact be wicked and must therefore be checked. We have learned that, while accelerating advance in some respects (such as the size of the GNP), technology is also accelerating our decline in other respects (such as the quality of life) and is even jeopardizing the very existence of the biosphere.

Of course, there is nothing unavoidable about the evils of technology. Except for isolated cases of unexpected bad side effects, technology could be all good instead of being half-saintly and half-devilish. It is up to the policy makers to have the technological investigator produce good or evil technological items. It is up to the technologist to take orders or to disobey them. In any event, technology is by its nature morally committed one way or another, and it needs some ethical bridling.

The Ethics of Technology

Every human activity is either explicitly controlled or criticizable by some behavior code which is partly legal and partly moral. In particular, the technological process has usually been guided (or misguided) by the following maxims:

1. Man is separate from and more valuable than nature.
2. Man has a right (or even the duty) to subdue nature to his own (private or social) benefit.
3. Man has no responsibility toward nature: he may be the keeper (or even the prison warden) of his brother, but he is not the nanny of nature.
4. The ultimate task of technology is the fullest exploitation of natural and human resources (the unlimited increase in GNP) at the lowest cost without regard for anything else.
5. Technologists and technicians are morally irresponsible; they are to carry on their task without being distracted by any ethical or aesthetic scruples. The latter are the exclusive responsibility of the policy makers.

These maxims constitute the core of the ethics of the technology that has prevailed heretofore in all industrial societies, regardless of the type of social organization. Certainly those maxims are not justified by technology itself: rather, they justify boundless exploitation of the natural and social resources. Moreover, they have not evolved within technology or science but within certain religions, ideologies, and philosophies.

In recent years we have come to distrust these maxims or even reject them altogether because we have started to realize that they condone the dark side of technology. As yet, we have not offered an alternative ethical code. It is high time we attempted to build alternative ethics of technology, ones with different desiderata and based on our improved knowledge of both nature and society, which were largely unknown at the time the old code was formulated, toward the beginning of the seventeenth century. If we wish to keep most of modern technology while minimizing its evil components and negative side effects, we must design and enforce an ethical code for technology that covers every technological process and its repercussions at both the individual and the social levels. Such a code should consist of the following components: (1) *An individual ethical code* for the technologist *qua* investigator. This should include the ethics of science, namely the set of ethical norms securing the search for truth and its dissemination. It should also take into account the peculiar moral problems faced by the technologist bent on attaining noncognitive goals. These additional norms should emphasize the personal responsibility of the technologist in his professional work and his duty to decline taking part in any project aiming for antisocial goals. Such moral imperatives, or rather grounded rules, should be consistent with (2) *a social ethical code* for technological policy making, research, and development of practices, disallowing the pursuit of unworthy goals and limiting any technological processes that, while pursuing worthy goals, interfere severely with further desiderata. This social ethical code should be inspired by the overall needs and desiderata of society rather than being dictated by any privileged group within it. Otherwise it would be unfair, and it might not be enforceable.

Such a two-tiered ethical code would make impossible, or at least reprehensible, the "Dr. Jekyll–Mr. Hyde" type of scientist who deserves both the Nobel prize for his contributions to elementary particle research and a hanging verdict for designing diabolical new means of mass murder. There would be no toleration of double ethical standards today if there were not two ethical codes, one for the pure scientist and the other for the impure technologist. If we are to keep technology in check, we need a single ethics of technology covering its whole wide spectrum, from knowledge to action.

Conclusion: The Centrality of Technology

Nobody denies that technology is central to industrial civilization. What is sometimes denied is that technology forms an essential part of modern intellectual culture. Indeed, it is often held that technology is alien or even inimical to culture. This is a mistake, one which betrays a total ignorance of the intellectual richness of the technological process, in particular of the innovating one. The mistake has obnoxious consequences, for it perpetuates the training of scholars with a traditional (preindustrial) cast of mind and conceptual equipment, contemptuous and afraid of whatever they do not understand about modern life. When they wield power in governmental or educational institutions, such people try to isolate the technologist as a skillful barbarian who must be kept in his modest place as the provider of material comfort. By behaving in this way, those scholars in fact deepen the gaps among the various subcultures and miss the chance of contributing to steering the course of technology along a path beneficial to society as a whole.

Like every other culture, ours is a complex system of heterogeneous interacting components. Some of them are already past their creative prime, others are blossoming, while still others are just budding. The creative components of our culture are some of the humanities, mathematics, science (natural and social), technology, and the arts. Modern technology is both an essential component and the youngest of all. Perhaps this is why we do not fully realize how central it is to our culture. In fact, instead of being an isolated component, technology interacts strongly with every other branch of culture. (On the other hand, art hardly interacts at all with mathematics.) Moreover, technology and the

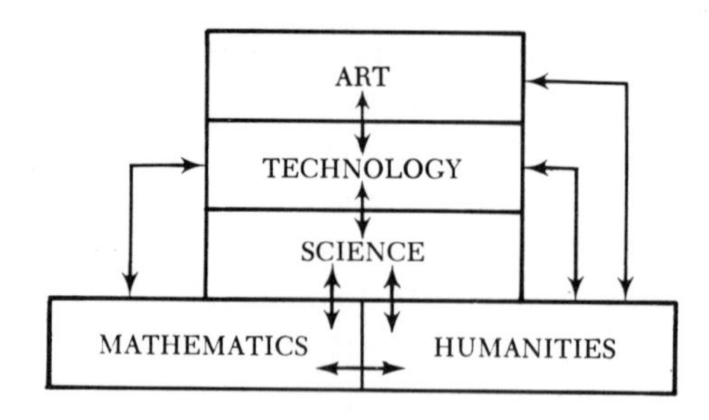

Figure 2. Flow Diagram of the System of Contemporary Culture. The noncreative components have been discarded.

humanities (in particular philosophy) are the only components of living culture that interact vigorously with all the other components (see Figure 2). In particular, technology interacts fairly strongly with several branches of systematic philosophy: logic, epistemology, metaphysics, value theory, and ethics.

Not only does technology interact with every other living sector of contemporary culture, in particular philosophy, but it overlaps partially with some of them. Thus, architecture and industrial design are at the intersection of technology and art; much of physics and chemistry is as much engineering as it is science; applied genetics is hardly distinguishable from pure genetics; and even some of metaphysics is at the intersection of technology and philosophy, as was discussed above.

Like science, technology consumes, produces, and circulates philosophical goods. Some of these are the same as those activated by science; others are peculiar to technology. Thus, because of its emphasis on usefulness, the epistemology of technology has a pragmatist streak and is therefore coarser than the epistemology of scientific research. On the other hand, the metaphysics and the ethics of technology are richer than those of science.

Because of the conceptual richness of technological processes, and because of the multiple contacts between technology and the other creative components of modern culture, technology is central to that culture. We cannot ignore the organic integration of technology with the rest of modern culture if we wish to improve the health and even save the life of our culture. We cannot afford to ignore the nature of technology, let alone despise it, if we want to gain full control over technology in order to check its dark side.

We must then build up all the disciplines dealing with technology, not least of them the philosophy of technology—the more so since it is often mistaken for the philosophy of science. The history, sociology, and psychology of technology tell us much about technologies and technologists, but only the philosophy of technology makes it its business to tell us what the methodological, epistemological, metaphysical, and ethical pennants of technology look like.

Werner Koenne

16

On the Relationship between Philosophy and Technology in the German-speaking Countries

The school reform carried out in 1809 by Wilhelm von Humboldt (June 22, 1776–April 5, 1835) is of fundamental importance for understanding the relationship between philosophy and technology, the role which technology and engineers played in public life and in society, and the development of the dispute between philosophy and technology in the German-speaking countries.[1] This school reform brought the first national order into Prussian education and—with the increasing power of Prussia among the small German states after the Napoleonic wars—it became the model for the entire German educational system. Humboldt's ideas determined both the essential features of this reform and its educational goal. The contradiction between this educational system with its claim for universal validity and the technological transformation of all spheres of life due to intrinsic dynamics is the fundamental factor underlying the German position in the discussion of technology.

Having completed his studies of law and classical languages, Humboldt became a civil servant in 1790,[2] but resigned after no more than one year to live with his wife "for his own education and happiness." His decisive intellectual development took place between 1802 and 1809, when he was the Prussian ambassador in Rome. During that time he developed his ideal view of man on the basis of antiquity, as Winkelmann (1717-86) had done forty years earlier.[3] This ideal view of man became the educational goal of his reform during his activity of only sixteen months in the Prussian Ministry of Education (February, 1809–April, 1810).[4]

That was the time of the Prussian defeat by Napoleon and also the time of national rebirth.[5]

The reformed Prussian school system became a model, and in the course of the nineteenth century the entire German school system was oriented toward this model.[6] This school reform had a number of extremely positive aspects. It divided the educational process into three stages: elementary education, high school education, and university education, the first two stages being no more than a preparation for university education. Technical and vocational schools were largely excluded. Thus Humboldt was the creator of the *Humanistisches Gymnasium* (Latin grammar school), the dominant educational institution throughout the nineteenth century, graduation from which was virtually *the* prerequisite for a university education. The educational aim in these grammar schools was the development of man according to the model of classical Greece, to bring forth a man who in his idealized form might have lived in ancient Athens and expressed himself in works of art and philosophy. This goal was to be achieved, above all, through teaching Latin and Greek. With this idealized view of man as the aim of education, any relationship to practical work and social reality was largely suppressed. The goal was the development of personality for its own sake. Very little attention was given to social problems and to the technological developments which were already recognizable at the time.[7] Thus a kind of arrogant isolation was created, one which separated those who were educated according to idealistic ancient humanism from a world which had developed quite differently. Generations of "educated" Germans thus did not become Greeks but did develop a rather disturbed relationship vis-à-vis technological and social reality. It must not be overlooked, however, that on the basis of this education a number of fascinating ideas were developed which had a considerable influence on the further social development of man. Philosophers, of course, primarily went through this education. Familiar with Latin and Greek which they had learned in grammar school, they could read the original texts of Plato and Aristotle. On the other hand, they showed a relatively disturbed view of the natural sciences and technology and of social change. In the nineteenth century, contacts with problems which might have arisen in the field of constructional technology were largely avoided.

The University of Berlin, founded in 1809, gave professorships to the philosophers Fichte (1762-1814) and Schleiermacher (1768-1834) among others, and later on to Hegel (1770-1831) and Schelling (1775-1854). The attractiveness of this university was correspondingly great, and its opinion of science was decisive. Originally, the intended goal was a global unity of knowledge, but this was immediately impeded by nonrecognition of technology and by attempts to exclude the natural sciences. The low opinion of craftsmanship which the Greeks had held became even stronger when the ideal of an autonomous man became the guiding image. The fundamental ideas of the Enlightenment gained little ground in Germany. The philosophy of German idealism had immediately countered the empirical critical approach to the facts of nature and society with a comprehensive view of the world derived from preexisting totality. The role attributed to the natural sciences and technology in this world view—the role of an institution providing the necessities of life—did not correspond to the actual development of technology.

The autonomous development of technology and the increasing importance of the natural sciences for the Prussian state in its efforts to hold its stand against other states resulted in an increasing influence of technology and the natural sciences on all spheres of human life. The number and training of engineers had to be regulated in the interest of the state. Polytechnic schools were established. Toward the end of the nineteenth century these developed into schools of technology and often had to fight with the universities for their promotion and qualification rights. It can easily be understood that both for psychological reasons and because of insufficient knowledge about technology this discipline, which was fought against and was not recognized as a science, was hardly deemed worthy of the attention of philosophy. The spirit of the school reform and the role attributed to philosophy at the universities practically prevented any genuine confrontation with technology.[8]

The more the world became shaped by technology, the more difficult and unbearable this position became. In his efforts for recognition the technologist was therefore deserted by the philosophy of that time. His need for an adequate system of values consequently forced him to work out his own system. This view of the world and its system of values could not, of course, live up to the criticism of a trained philosopher, and initially were not relevant to the social situation either. As far as value and moral systems were concerned, it was the churches and the legislative branch of the state that were competent. The idea that scientific discoveries and technological possibilities might change the meaning of morals and ethics was simply unthinkable, and in some ways is not even recognized today. The result was that philosophy first of all felt that it had to protect a higher point of view. The possibility of a positive investigation into the problems of technology and its implications was barred to the individual philosopher because of his education. To a certain extent the technologists were willing—for the same reasons—to accept this position of refusal. One must not overlook the stabilizing role of mutual recognition, the need for adjustment in academic life, and self-continuity through the recruitment of new professors.

The establishment of societies was important for the further development of a self-understanding among technologists, as well as for the development of technology. In Austria, the ÖIAV (Association of Austrian Engineers) was formed in 1848 (the year of the revolution). In Germany, several associations were formed, the most important of which is the VDI, founded in 1855. These societies did not concentrate on the engineer as a person so much as on technology itself and on its recognition.[9] It is surprising that these associations were not interested in political or social fights but rather in fundamental recognition by society of the role and performance of modern technology. The interests of these societies were hardly economic or professional in nature; in promoting technology they worked in the public interest. This was the so-called *Gründerzeit* from 1870 to 1914.

The most attractive social system for engineers was one willing to make efficient use of their services for the benefit of all. In their demand for participation, the engineers were opposed by the lawyers and officers with their monopoly of leading positions in government.[10] Basically, Germany and Austria at that time were organizations of civil servants and military officers ruled by a rather inflexible aristocratic elite.

The social importance of the fight for recognition and equality of other types of schools with the grammar schools in Germany, which is still going on today, can only be understood in connection with the sociological situation created by Humboldt's school reform.[11]

In this fight for social recognition the technologists were aided by the wars of 1870-71 and 1914-18. In the German-French war of 1870-71 there were already battles of materiel which were a challenge to organization and technology. World War I, however, was of decisive importance for technology and for engineers in Germany for these reasons: (a) There had been no previous war in which technology had played such a decisive role; in preparing for war and during the war itself the engineer had become indispensable. (b) Aware of their indispensability, the engineers wanted to demonstrate their idea of maximum benefit for the public during and after the war. (c) The importance of this performance aroused cultural critics who in turn thought technology was the cause of everything negative. (d) The defeat of 1918 (as in 1945) seemed to open up possibilities for changing the existing social system into one that would recognize the role of technology.

In the twenties, antitechnological cultural critique reached a certain climax.[12] The big cities, the consequences of World War I, inflation, and efforts to rationalize industry seemed to reflect the "devilish character" of technology. In the field of classical education a polarization developed; it was expressed in pairs of concepts such as culture and civilization, nature and technology, soul and technology, creative individual and anonymous mass, organic and mechanical, irrationality and rationality.[13] This antitechnological thinking was surprisingly close to antidemocratic thinking and was mostly connected with a criticism of the Weimar Republic; in fact, it was certainly one of the reasons for the Republic's decline. This relationship has been repeated in the antidemocratic and antitechnological attitude of the "New Left" in Germany in discussions during the past decade. Its roots lie in the roles of philosophy and technology in the intellectual life of Germany. The claim of the philosophy of German idealism for absoluteness, like the similar claim of Marxism, is in contradiction to the critical attitude of science and technology, which instead stems from the age of Enlightenment. Thus it may be that negative reactions against complex structures of democratic systems and against technological relations, which crop up under different names, have a common cause. The fact that the wave of criticism in the past decade began in university departments which were far from technological-constructive or scientific-analytical thinking might be regarded as evidence of this.

Among engineers, the attack in the twenties first of all aroused the advocates of technology, who, forced by their own education and the language of their opponents, advanced onto philosophical grounds in their efforts to defend technology. This led to attempts—sometimes rather strange and idealistic—to define the nature of technology. Dessauer is the outstanding example here.[14] In all these efforts, including those after World War II, the defenders of technology were formed by the system of values and concepts of the educational system introduced by Humboldt. Recognition of technology, therefore, depended on underlining its intellectual importance, by which greater value could be attributed to it. It has to be borne in mind that, independent of all these efforts at classification, individual people continued to carry out their technological activities both functionally and operationally.

The analogy of machine and state developed by engineers contrasts with that defensive position. According to the mechanical model, politics should be determined by the balance of powers. The idea of technocracy connected with this model came to acquire two meanings: on the one hand the rule of technology or of the system produced by it, and on the other hand the rule of technologists or experts. One-sided exaggeration of socio-technological movements is a characteristic of this model, which was much more strongly developed in the United States. The role of the engineer and consequently the role of technology was regarded as that of a mediator in the sense of Aristotle. Thus the technologist in his role and in his task within society was again bound by the views of a philosophy which had no contacts with technology. This view of himself, which can be traced back to the educational system, is the reason why the German technologist has shown and still shows a tendency to work for ideologies which offer him the expected role. This was true for the engineers under Hitler, and the same applies to the engineers in East Germany today. In both systems they have been given recognition.

Development after World War II was basically identical to that after World War I. Engineers had profited from their social role both during the war and afterward in the period of reconstruction. With the rise of technology, criticism of technology also set in. The predominating philosophical school in the German-speaking countries in the immediate postwar years was existential philosophy, which was basically alien and hostile to technology. The leading personalities are Karl Jaspers [15] (1883-1969) and Martin Heidegger [16] (1889-1977). Jaspers's attitude towards science, which he considers to be the foundation of technology, is a negative one.[17] What he wants to do among other things is to destroy belief in the scientific comprehension of the world. He calls upon man to be himself. The analogy to Humboldt's view of man is obvious. Technology is regarded as a neutral means of shaping the external conditions of life. Thus technology should be restricted and must not threaten what is genuinely human.

Heidegger's claim for *Gelassenheit* (calmness) vis-à-vis technology reflects a similar attitude. Both opinions reflect the social position and role of philosophy. In his ontology Heidegger, however, also regards technology as a special constellation of man and being. To him, the danger is that this relationship between man and being that is required by technology will prevent actual access to truth and to the *Ereignen des Seins* (occurrence of being). Although Heidegger's attitude toward the problem of technology is certainly distinctive, it is by and large characterized by the traditional German differentiation between culture and civilization. Another characteristic feature is that when he speaks of technology he always seems to think of heavy engineering of the late nineteenth century. Thus his entire discussion about the problem of technology, its values and importance, is tied to the specific situation of educated Germans produced by the dominant schools of the time and by the social values of the nineteenth century.

The development of philosophy itself and the confrontation with science and technology which started from the neopositivism of the *Wiener Kreis* (Vienna Circle) constitutes a certain contrast. The logical empiricism of the Vienna Circle was contradictory to the prevailing philosophy of that time. It was not a continuation of the German school philosophy of the nineteenth century. With its emphasis on science and its hostility toward metaphysics, it was rather a belated child of the Enlightenment which had not fallen on fertile ground in Germany. Thus it was close to science and technology from the very beginning. One of its main goals in the thirties was a uniform theory of the natural sciences and their methods of cognition. Owing to well-known political events, this philosophy did not have the opportunity to develop further in Germany. Again the reason lies in the social position of philosophy in Germany. As a result, the logical positivist position was nonexistent in philosophical discussion of technology after the war, which shows once again that after World War II technology was evaluated in keeping with the traditions of German education.

After 1945, there was a tendency in the Federal Republic of Germany to introduce the humanities in the schools of technology. Efforts like those in Berlin to organize entire study programs on the basis of a combination of technology and the humanities, particularly philosophy, have been abandoned since then. This combination, however, proved to be fruitful in the opposite direction; that is, science and technology had an influence on philosophy. Technology and science did not orient themselves along the lines of philosophy, but instead philosophy became increasingly familiar with the peculiarity and the problems of technology. In the meantime, the social position of technology had become considerably stronger than that of philosophy. For these reasons, topics of philosophy started to shift, which is probably not in keeping with the original intentions of the integrators, with their motives rooted in Humboldt's educational system. A fact which seems to confirm this is that the development of technological and scientific departments at the universities was not promoted to the same extent.

The discussion about university matters triggered by the New Left in the sixtiees again focused attention on the role of the university within society. The arguments of the New Left—and this is not surprising after what has been said previously—showed a certain alienation and hostility toward the social system's dependence on technology and toward technology itself. Thus old arguments are often repeated in new terminology. The demand that the technological possibilities should constitute a part of independent, higher systems shows that these systems in turn assume a religion-like, nonempirical character and thus proves to be a genuine offspring of German idealistic philosophy. For a number of idealistic reasons, however, it seems that this time the defenders of technology are no longer willing to accept this way of thinking and its jargon. They use philosophical arguments rooted in the Enlightenment and in logical empiricism, and thus a contrast between the theory of science and the leftist philosophical position has developed.[18] At the same time, during recent years the dialogue between the theory of science and technology has increased. This position has obviously been made possible by a parallel change in education as a result of an overall reformation of the school system as well as by changes in social and economic conditions. One cannot yet say which of the two forces will have a greater influence on the institutes of education in the future. After some initial successes of the idealistic trend it seems now that the theory of science is gradually advancing.

Of course, these positions have their effect on the relationship between technology and values. The difference lies mainly in the field of ontology. The humanistic and idealistic school maintains an ontology of being, while on the technological-scientific side there seems to be an ontology of function. In the idealistic school, efforts are made to develop an overall system of being; the important thing is the nature of things and of values. In this view there are recognizable absolute value differences of being, and it becomes necessary, in this absolute value system, to also allocate a position to technology and its results. One such absolute system was that of high scholasticism, which was gradually dissolved by nominalistic concepts. The opposing idea of functional value pertaining to a thing or an activity not per se but only in relation to other things and activities was already present among the nominalists of the fifteenth century. As is well known, this paved the way for the development of science and technology. The history of this dissolution process is the history of the development of science and technology. At the same time it is also the history of a new philosophy, which even today has not yet acquired self-awareness although its basis is no longer the ontology of being derived from the Greeks, or rather from Aristotle. This difference with all its radical consequences still remains to be thought out.

This functional-value thinking is also reflected in the ideas and methods of technology assessment. However, there is danger that an ontological value system may again be imposed on this functional method. It will probably be one of the most important future tasks in the relationship between philosophy and technology to work out the underlying philosophical basis of a functional approach. As a result, the fundamental content of two thousand years of European thinking will probably be questioned; this may be compared to the fundamental changes which modern technology has wrought upon the lives and social systems of individuals and of mankind. Because of the powerful position of philosophy in society and politics which, in my opinion, is particularly characteristic of Europe and the German-speaking countries, this discussion will probably take place mainly there. In those countries, school systems, education, and social values are subject to a tradition which is probably contradictory to the thinking of the engineer and the scientist. The task facing us is to articulate this contradiction, to bring it to light, and to make it part of global thinking.

Notes

1. The term "German-speaking countries" essentially stands for Austria and Germany; statements relating to the time after 1945 refer mainly to the Federal Republic of Germany.

2. As a reporter at the Supreme Court of Berlin. He married June 29, 1791.

3. Winkelmann was in Rome from 1755 on; in 1757-58 he became librarian for Secretary of State Archinto; from 1763 on, president of the antiquities and documents of the Vatican library. His interpretation of the nature of Greek art *(edle Einfalt und stille Grösse)* became definitive for subsequent times.

4. During that period Humboldt was director of the Department of Culture and Education of the Ministry of the Interior.

5. Following the Peace of Tilsit in 1807, German reforms began under Stein and Scharnhorst.

6. It brought the first uniform regulation of the school year, tuition, transfer provisions, examinations, and so forth.

7. Here are a few dates illustrating the development of technology at

that time: Pascal in 1642 and Leibnitz in 1673 had built the first calculating machines; in 1735 pig iron was produced from iron ore by means of coke; in 1776 (more than thirty years before the school reform) Watt's first steam engine was in operation at Wilkinson's iron mill; in 1805 the first automatic loom was put into operation; in 1807 Fulton's steamboat was running; in 1826 Ressel invented the ship screw; and in 1835, the year of Humboldt's death, the first railway line in Germany ran from Nürnberg to Fürth.

8. When the phenomenon of "technics" later entered the picture, it was mainly looked at from the viewpoint of Aristotle's concept. It has been proved that the modern expression *Technik* in the German-speaking countries is not to be understood as a word borrowed from Greek. Rather, it goes back to the late Latin expression *ars technica.* This was used by Caspar Schott (ca. 1666) in his book *Technica curiosa sive mirabiliae artis libris XII comprehensa* as the description of mechanical machines. E. Heyde, "Zur Geschichte des Wortes 'Technik'" (Concerning the history of the word "Technik"), *Humanismus & Technik,* vol. 9, no. 1 (1963).

9. One is justified in seeing in this movement a certain reciprocal position to the predominance of philosophy and human sciences. It would be an enlightening study for those interested in the role and position of technics to pursue the motives and consequences of these associations.

10. In 1877, Max Maria von Weber characterized the position of the technician in the following way. Just as there are social climbers in society, so in the society of nations there are climbers, too. In neither one are they looked upon favorably; they are often feared, yet everywhere they are slowly and with difficulty achieving recognition and equality. Such a climber on an international plane is the professional class of the technician. The centuries-old professions, those of teachers, soldiers, and farmers, don't quite know in which rank to place them. University sciences look upon them as intruders. For government they are an uncomfortable and new element in the mechanism of the state. All consider them, at least deep down in their hearts, a necessary evil. "The Position of the German *Techniker* in Civil and Social life," in *Populäre Erörterungen von Eisenbahn-Zeitfragen* (Vienna, Budapest, Leipzig: 1877), ch. 6, p. 5.

11. For example, one of the issues was admission to a university education without knowledge of Greek and Latin. To this day, a knowledge of Latin is required for study in the faculty of philosophy (which includes the natural sciences) at an Austrian university.

12. Examples can be found in Thomas Mann's *Zauberberg* (1924), whose hero Hans Castorp is an engineer, in Fritz Lang's film *Metropolis,* or in Oswald Spengler's *Der Untergang des Abendlandes* (Munich: C. H. Beck, 1923), p. 1187, and *Der Mensch und die Technik* (Munich: C. H. Beck, 1931).

13. The first concept stands for the higher value that is to be defended. These pairs of concepts repeatedly crop up in different forms.

14. In his book *Arguments about Technics,* F. Dessauer describes technics as follows: "Technics is the existing reality of ideas, existing through the final forming and processing of resources given by nature."

15. Karl Jaspers, *Die geistige Situation der Zeit* (Berlin-Leipzig: De Gruyter, 1931); Jaspers, *Von Ursprung und Ziel der Geschichte* (Frankfurt

am Main: Fischerbücherei, 1955); Jaspers and Kurt Rossmann, *Die Idee der Universität,* (Berlin-Gottingen, Heidelberg: Springer, 1961).

16. Martin Heidegger, *Identität und Differenz* (Pfullingen: Neske, 1957); Heidegger, *Sein und Zeit,* 7th ed. (Tübingen: Niemeyer, 1953); Heidegger, "Die Frage nach der Technik" (The arts in the technological age), series of lectures (Darmstadt: Wissenschaftliche Buchgesellschaft, 1956; Heidegger, *Gelassenheit* (Pfullingen: Neske, 1959).

17. See also Jaspers's *Ursprung und Ziel der Geschichte,* p. 82.

18. The dispute between K. Albert and K. Popper and the Frankfurt School is one of several examples.

Frances Svensson

The Technological Challenge to Political Theory

In a recent volume entitled *Philosophy and Technology,* a number of scholars examined the question whether an independent philosophy of technology exists. Without reformulating all the arguments pro and con, it must be stated that *this* analysis is based on the assumption of a distinct field of inquiry called the philosophy of technology. The primary task of this philosophy, to paraphrase Winch, must be the explication of technological reality as such and in general.[1] Politics, from an essentially technological (in Ellul's sense) point of view, "represents the set of means by which man puts to use the forces inherent in his social organization".[2] Technology, from the same perspective, becomes "the set of means by which man puts the forces and laws of nature to use, in view of improving his lot or modifying it as may be agreeable to him." [3] The working out of the political implications of technology, then, must involve an analysis in the valuational mode of the extent to which, through certain kinds of political decisions and decision processes, man "improves his lot." Since this obviously incorporates both assumptions about man's nature and a concept of "improvement" grounded in social aspirations, ideas which are beyond the ken of empirical analysis and of formal logic, it requires the exercise of political philosophy in the classic mode. This is the technological challenge to political theory.

I. Jacques Ellul

According to Jacques Ellul, perhaps the most publicized of the philosophers of technology to date, the "conjunction of state and technique is by far the most important phenomenon of history."[4] This conjunction carries with it a valuational process deriving, in his view, largely from technology; certainly, as he states, this conjunction "is not a neutral fact."[5] Skolimowski, in his analysis, suggests the concept of technological progress as the key to a philosophy of technology; he goes on to describe this mode of progress as "the pursuit of effectiveness in producing objects of a given kind."[6] Ellul seems to be operating with the same basic assumption about the operations of technology, for he emphasizes "efficiency" as the hallmark of the technological operation. This concept of efficiency, whether operationalized as bureaucratic rationality or cost accounting or program planning/budgeting, provides a thrust within social institutions which transcends the dictates of those ideologies that, in the absence of political philosophy, have been left in command of the field. Thus, Ellul asserts, "The structures of the modern state and its organs of government are subordinate to the techniques dependent on the state."[7] As a result of this process, technique comes to assume the role of ideology; it dictates the mode of operation and therefore the process of valuation. In effect, "The state is no longer in a position to reject the most efficient means possible."[8] It is to this sort of development that Bell points when he introduces the prospect of an "end of ideology" in the contemporary world. What Bell and others fail to examine is the nature of a social system (whether universal or local) in which essentially mechanistic technological processes, founded in an assessment of effectiveness and thrust forward by a technological imperative of open-ended expansion, usurp the human function of choice.

Ellul himself addresses the problem of the role of political philosophy, but from a negative point of view. First, he notes, technology removes many traditional questions of political theory from consideration: "Technique puts the question, not whether a given state form is more just, but whether it permits more efficient utilization of techniques. The state is no longer caught between political reality and moral theories and imperatives. It is caught between political reality and technical means."[9] From the point of view of technology, then, political theory is asking the wrong questions—even if we assume that it continues to ask any at all. From the point of view of contemporary political philosophy, in turn, there is every reason to acquiesce in this assumption of irrelevance, not merely passively, but actively, by applying what life-force it can muster in these empiricist times to quantitative means analysis, abjuring the whole issue of the ends involved. Ellul sees this situation as inevitable, apparently, for he rejects the whole notion of the continuing relevance of autonomous political thinking: "Things happen today in the political sphere without the benefit of the minutest theory. There is no longer any question of a state in the classic sense. To think otherwise is a laughable error. . . ." [10] If, however, it is true, as Ellul goes on to predict, that a "monolithic technical world is coming to be," [11] it would seem that the error of assuming the possibility of a continuing less-than-total political form (of whatever nature) is not only not a laughable error; it may be the key to the generation of alternatives to technological hegemony.

Ellul himself attests to the fact that such an alternative is needed. One of the most critical factors in the situation is the fact that "at present there is no counterbalance to technique." [12] As a result of its unchallenged progress through the human sphere, technology assumes a hitherto unprecedented aspect: "the character of technique renders it *independent of man himself*." [13] The social world which emerges from this matrix is an highly deterministic one, in which, "the prodigious increase in our means of action makes it impossible for us to claim any control whatever over those means. Rather, they control us." [14] It is also a world characterized by "a definitive primacy of action over thought, meditation, choice, judgment." [15] Finally, it is a world in which the means determine the ends. There are a number of grounds on which Ellul's logic might be contested. For example, Ellul would seem to be guilty of what Dewey referred to as the fallacy of the "means-end continuum." This is contained in the idea that all ends are means to further ends and that therefore the concept of "means" has limited utility at best and suffers from suppressed premises.[16] However, the most critical failing of Ellul in this context is his apparent resignation in the face of the technological imperative. Although he concludes his study of the impact of technology on man's social relations by suggesting two possible sources of salvation—the discovery of new technical instruments for the management of technology and the discovery of "a new end for human society in the technical age" [17]—he undermines any inspirational role he might have played in the recruitment of a new generation of political philosophers at the outset. "It is vanity," he states in referring to the technological society to be, "to pretend it can be checked or guided." [18] Humanism, however manifest, he dismisses as "merely a pious hope with no chance whatsoever of influencing technical evolution." [19] It seems highly problematical whether man can survive Ellul's technological society as human—or whether he ought even to try.

II. Victor Ferkiss

In *Technological Man,* Victor Ferkiss, a political scientist, offers a fundamentally very different view. To him, technology promises not merely the fulfillment of man's nature, but the possibility of precipitating a quantum jump in his very potential: "Humanity today is on the threshold of self-transfiguration, of attaining new powers over itself and its environment that can alter its nature as fundamentally as walking upright or the use of tools." [20] Dismissing the "mistaken" views of critics like Ellul, Ferkiss asserts that the greatest danger facing man in the present era "lies not in the autonomy of technology or in the triumph of technological values but in the subordination of technology to the values of earlier historical eras and its exploitation by those who do not understand its implications and consequences. . . ." [21] Man, in other words, is not keeping pace with his instruments in the evolutionary race. The thrust of Ferkiss's analysis, then, is toward the isolation of those social institutions which have retarded man's progress toward the realization of his potential in the technological sphere. That this evolution is desirable seems obvious, for it represents progress: "Industrial man, whatever his characteristics, is evolving into something different and superior—technological man." [22]

Technological man is a new ontological construct, one which Ferkiss acknowledges is at present "more myth than reality." [23] It is a construct in which he sees great promise, however. Indeed, Ferkiss posits—although with a very different valuation—much the same type of technological juggernaut (or "existential revolution") to which Ellul is committed. He contends however that the challenge of this social transformation can be met: "The race's only salvation is in the creation of technological man."[24] The nature of this man remains somewhat obscure. Ferkiss asserts that "technological man will be man in control of his own development within the context of a meaningful philosophy of the role of technology in human evolution."[25] He also asserts, "Technological man, by definition, will be possessed of the world view of science and technology, which will themselves provide a standard of value for future civilization."[26] Therein lies the rub. Is it possible for man to express his humanity within the context of a philosophy of technology dominated by the technological imperative? Is it possible for a theory of valuation founded in the mechanistic pursuit of effectiveness to assure man the possibility of the agonies and glories of choice? If commitment to a world view represents at base an act of faith, then perhaps an act of revelation is needed to convince the skeptic of the validity of this position.

Ferkiss shares with Mesthene and others a belief that technology can generate answers to all the problems it raises. He carries this belief to an extreme position when he supports the idea that "a culture that permits science to destroy its values without permitting science to create new ones is a culture that destroys itself."[27] This can only be an act of blind faith. On what possible grounds can the leap be made to the assumption that the source of destruction is—must be—the source of resurrection? By what logical process will science generate the values and the process of valuation which satisfy human needs? How indeed can Ferkiss arrive at his own formulation of technological man other than by committing the heresy of political philosophizing?

There is a reason why Ferkiss is willing to look to science to perform the valuational task, particularly in the social sphere: the "existential revolution" arises in a context of "political default."[28] This is because "the political system simply has been unable to structure the issues and to relate them to the decision-making process in such a way as to enable the popular will to be expressed."[29] Its failure is a failure of rationality; the irrationality of contemporary political processes "makes the emergence of technological man impossible."[30] The image with which we are left is one of man "bewildered and impotent, a prisoner of his most primitive atavisms and the plaything of the fates."[31] This does seem to be a harsh indictment of present social reality, and in fact one from which we have been proclaimed to have escaped by the triumph of the empirical world view. In fact, it is difficult to see how this value assertion could be effectively supported by the science from which Ferkiss would henceforth derive values and valuations. Be that as it may, it does point to the basic fact which can never be forgotten in the examination of the elucidation of philosophies of technology—they are all embedded in a world view and cannot be understood apart from it.

The image of technological man remains curiously obscure in Ferkiss's work, as does the "meaningful" philosophy through which man will assert his control of the technological processes. Three basic principles underlie the new philosophy: "the new naturalism, the new holism, and the new immanentism." [32] Roughly stated, the new naturalism asserts the integral relationship between man and nature; the new holism stresses the interconnectedness of all phenomena; and the new immanentism emphasizes the self-creative principle inherent in life systems. This is a promising beginning for a new philosophy. That it happens to be in principle incompatible with the operations of technology as presently expressed is, however, something that Ferkiss fails to confront, perhaps because philosophy in the classic mode is not a primary interest of Ferkiss. It is fine to assert that technology can accommodate itself to these directives under the guidance of man; it is something else entirely to develop a philosophy which incorporates these principles within its very structure, which in fact is grounded in them. That technology will be a very different one from the current, ruthlessly exploitative type. The taming of the technological imperative will not be accomplished by the manipulation of a few words on paper or the bravura assertion of the idea that "in the era of absolute technology, freedom and identity must take on new meanings or become meaningless." [33] The latter probability is altogether too likely.

C. B. Macpherson, in a recent article,[34] attempts to move toward an examination of the conception of man which underlies the conception of the possibility of "freedom." In describing the "liberal, individualist concept of man as essentially a consumer of utilities" and the "concept of man as an enjoyer and exerter of his uniquely human attributes or capacities," [35] however, he does not push the ontological frontiers back far enough. Underlying assumptions of man as consumer or enjoyer are even more basic assumptions about the nature of the world and man's relationship to it. Technology as we know it today has emerged in a metaphysical context in which the world is mechanistic, subject to lawful operation, immutable in its basic nature, passive in its relationship to man, exploitable—above all, instrumental. The image of man which emerges within this system is arrogant, willful, fey, exploitative—man as the master, philosophically, of all he surveys. If, as Ferkiss asserts in contradistinction to Ellul, we may not only master the technological forces but indeed further realize ourserves through them, it will not be an easy task. The principles which Ferkiss outlines as the core of a new philosophy happen to be in direct contradiction to the whole weight of our experience with technology over the last several centuries, and to the kind of men we have become. We have met the enemy, and it is us.

III. The Inadequacies of Political Philosophy

The interface between technology and society is the critical social frontier of the present age. Its impact reaches not only values conceived as properties, but more important the process of valuation itself: technology carries with it profound consequences for human society, including political structure. The task of political philosophy has been classically conceived to encompass two basic concerns: the description of existing and possible forms of political organization and the prescription of those forms which would most comprehensively meet the demands of man's pursuit of self-realization, however conceived. Indeed, it used to be said that political theorists and philosophers were "the professional custodians of political values." [36] If, however, political theory can be said to be that which political theorists do (and so it has increasingly become), then today it reflects along with its parent disciplines an "expedient apathy concerning its limits." [37] Far from exploring the parameters of the society/technology interface, political philosophy is scarcely able to discern it. The majority of professional activity directed toward the problem of technology by political scientists focuses on the question of science-technology policy and on the political role of science-technology "experts" and institutions within the framework of an empiricism uninvested with philosophic analysis.

Thus, in an age preoccupied with political crisis which many observers relate integrally to an emergent technological crisis, political theory "has become disengaged from political facts. Even worse, it has become disengaged on principle. . . ." [38] Reduced to mere description through abnegation of valuational-prescriptive responsibility, political theory has become a travesty of its former self, its adherents turning to "historical, textual analysis, 'rediscovering some deservedly obscure text or reinterpreting a familiar one' or to 'using low-powered logic on traditional technical problems in an essentially quixotic way.' " [39] As a consequence of the retreat from prescriptive statements, technology is viewed within political science in a descriptive mode, with its impact on human society outside the purview of contemporary political theory.

Political philosophy in general reflects themes dominant in philosophy at large. As philosophy came to accept the canons of logical empiricism, social and political philosophy fell under its rules as well, particularly after World War II. The result—a full-scale retreat from prescriptive statement, on the basis of a new concern for justification—resulted in not only prescriptive paralysis, but also something of a state of *reductio ad absurdum* in the descriptive sphere. The fundamental inadequacy of assuming the empiricist world view as the source of an image of man arises from the fact that within this mode of analysis man's nature is reduced solely to observable behavior. His needs, his wants, his aspirations become merely mechanistic responses to physical stimuli. He becomes Skinner's prototype: ultimately, he becomes man in the image of machine. Yet it is man in the image of man that political philosophy has always sought—at least until its castration by the forces of rigid empiricism. It is man transcendent—man whose nature strives toward spiritual as well as physical gratification—that must contend with the technological age. Why? Because the threat technology run amok poses to man is the suppression of man's opportunity to express his humanity.

It is not necessary to assume totalitarian technological regimes in the style with which we are familiar (e.g., Nazi Germany, Stalinist Russia) to support this assertion. Indeed, Ellul argues forcefully that these represent merely transitional stages in the emergence of a truly "total" state system in which "nothing useless exists," "there is nothing arbitrary," and "only technical necessity counts." [40] In this new world, marching to the beat of the technological imperative, there is no room for the agonies and glories of choice. There is therefore no room for humanity. To the extent that we must acquiesce to an image of man in order for it to emerge triumphant, we still possess the capacity to reassert ontological alternatives and to embed them in social philosophies which will enhance and nourish alternative visions. In political philosophy as in life, to choose may well be a gamble; to abstain, on pseudo-principle or otherwise, is suicide.

One of the major tasks of political philosophy, as noted above, has been the problem of prescribing the conditions of man's self-realization. Because empiricism provided no access here, the task has largely been abandoned. However, in blocking the path of ethical inquiry, empiricism itself violates the essential requirement of research and understanding. It may well be that the reassertion of the prescriptive task in political philosophy through a critical examination of the society/technology interface will produce sharp controversy. It would almost certainly produce an increase in the number and kinds of valuations that could be brought to the decision-making process in contemporary society, particularly in the critical area of technology assessment as a political function. There are those who would assert that we are overburdened with choices in our present situation and who would deplore the introduction of more conflicting perspectives. Those who hold such views may be faulted in at least two ways. First, in assuming that there are already a sufficient number of alternatives available in the evaluation process, they fail to grasp one of the most significant facets of contemporary, technologically oriented society: the emergence of artificial needs generated by a technology of self-enhancing production and of marginally distinct synthetic "goods" and "services" available for selection through a superficial "choice" process. The political theme of wings of both major political parties in the United States during the late sixties and early seventies, most succinctly expressed in the slogan "A choice, not an echo," is but one manifestation of the emergence of the problem of meaningful choice in the political context.

A more basic criticism of this position of fearing undue choices, however, lies in a challenge to the assumption that a multiplicity of choices implies disorder, chaos in the decision process, and ultimately social breakdown. Indeed, this very perception of "choice" as a social operation is posited on a conception of "progress" uniquely attuned to the technological era. According to this perspective, "rational" movement is toward increasing simplicity, centralization, and unitary (monolithic?) structure; in other words, toward decreasing the choice options in evaluation and assessment (see Junger and Marcuse on this theme). Much of the thrust of technology assessment under this conception, thus, is directed toward eliminating complications in application rather than toward exploring potential applications. To acquiesce in this vision is to accept, in effect, an ethical statement. An empiricist worthy of the name should at least be willing to pursue an investigation of such an alternative thesis as Whitehead's, which states in connection with the problem of choice, "Life tedium is the fatigue derived from the thwarted urge toward novel contrast." [41]

IV. How the Technological Imperative Has by Default Become the Essence of Political Evaluation

It has been asserted that technology carries within itself a so-called technological imperative, the thrust of which is to pursue technological progress without end. This imperative is reinforced by the principles inherent in the philosophy of science, which must incorporate Pierce's famous dictum: Do not block the road to inquiry. In the case of technology, this means, in effect: Do not interfere with the working out of the potential inherent in technological development. As these developments dovetail they form purposive matrices. Some, like Junger, saw in this development an ultimately centralizing process, "an ever more concentrated involvement and assimilation which tries to wield the technological arsenal into one gigantic instrument despite its specialization." [42] To the extent that the theory of valuation as developed in contemporary political philosophy defers to the empiricist legacy and defines itself out of the sphere of rational discussion, and therefore out of discussion entirely, a vacuum is created into which the implicit valuational processes of technology can seep. There are indeed such processes.

The concept of effectiveness which underlies the technological process manifests itself in a number of implicit directives in the social sphere. Thus, for example, such values as standardization, automation, homogenization, concentration, and systematization emerge as instrumental in the pursuit of the technological imperative. Why? Because effectiveness/efficiency of production involves the elimination of unnecessary duplication of effort and reduction of diversity. The old cliché about the reduction of man to the status of a "standardized part" has more than a grain of truth in it. The urban concentration, the homogenizing process of modern education, the reduction of the human role through automation, and so forth are all hallmarks of the present age which have attracted attention in various sectors. As Marcuse notes: "Indeed, what could be more rational than the suppression of individuality in the mechanization of socially necessary but painful performances; the concentration of individual enterprises in more effective, more productive corporations; the regulation of free competition among unequally equipped economic subjects; the curtailment of prerogatives and national sovereignties which impede the international organization of interests." [43]

These "values" have become major components of the valuational process itself. That is "good" (socially, in this case) which contributes to technological efficiency. There is a suppressed premise at work here. The goal of open-ended technological development is assumed to have been accepted, whether by decision or by default. The seal on the bargain is the output of technological production: material goods. The byproduct is the materialization of values. That which cannot be expressed quantitatively, that which cannot be produced "efficiently" through technology, cannot be taken into account in the valuational process any longer. And, and this is the great indictment of man in the modern age, it is not missed.

The technological imperative as principle might well be extended to the realm of valuation when it is understood that this technological progress is dominated by a particular mode of activity, "the pursuit of effectiveness in producing objects of a given kind."[44] The concept of effectiveness becomes the open-ended "end" against which activity, human as well as technological, is measured. This occurs because the technological imperative is a purposive one. One of the basic questions of philosophy is the relationship between purpose and function. Far from fulfilling the limited functions assigned to it by man in a social context, technology has come to manifest purpose, and therefore to entail goals, in ways which are somehow beyond the ken of direct human intervention.

Technology moves along axes of development which derive from an inherent logical structure. While attention has been focused on the nature of that internal structure, and particularly on the relationship between the modes of progress of technology and science (see Bunge, Feibleman, Jarvie, and Skolimowski in *Philosophy and Technology*), little attention has been paid to the problem of the basic purposiveness of the technological process, the sense in which it seems committed to goals. The technological imperative, to pursue technological progress, is in fact a goal, but a goal in the style of a contemporary theory of valuation which no longer permits the philosopher to come to grips with ethics as a living system. The new goal is rather an attempt to avoid the whole issue of appraisal in valuation theory; it is a "non-goal," because it is completely open-ended. It points toward everything, but actually to nothing at all but continuing activity. It attempts to posit nothing but the value neutrality of objectified activity. It has already been emphasized that this is an indefensible position.

Reducing man to his quantifiable measure, empiricism leaves to its adherents, including political philosophers of the present period of "normal" social science, no program for coming to grips with man's transcendent nature, his immaterial needs, wants, and aspirations. Empiricism as a guide to valuation leaves us with no end in the realm of values but mere survival. This survival can be calculated along only two axes: maximization of the life-span of the individual and maximization of the size of the human community at large. In both cases, valuation concentrates on the satisfaction of physical needs to the often total neglect of spiritual ones. In a world in which these positions represent the most fundamental ethical directives, the emergence of a constellation of materialized and materialistic values, moral, aesthetic, and so on, is hardly surprising. It is in large measure to the technological imperative that such values, and valuation processes, must be traced. By introducing the idea of the preeminence of the mode of effectiveness and efficiency as the measure of value in technological operations, the technological imperative seems to provide a standard of valuation particularly suited to the era of the empiricists. In failing to challenge the human validity of the ethical stance which has accompanied the technological imperative, political philosophy has defaulted on its task. It has also defaulted on its humanity.

Conclusion

"On the day when crime dons the apparel of innocence," stated Camus in *The Rebel*, "it is innocence that is called upon to justify itself." [45] In an age beset by crises of social and political life, the pseudo-innocence of political philosophy in its pursuit of the value neutrality ostensibly prescribed by current paradigms of social science cannot be supported. Technology carries with it a tremendous baggage of values and valuations deriving from the technological imperative. Standardization, systematization, concentration, homogenization, perhaps ultimately sterilization (intellectual and emotional, if not necessarily physical) may well be entailed by that imperative, at great cost to man as we know and have traditionally valued him. The task before the political philosopher is not merely the bystander's description of the technological millennium; it is the human being's call to active participation in the prescription of social and political structures and processes. The methodology which must accompany this reassertion of the rightful role of political philosophy is one which has ancient roots in speculative philosophy. It must draw its strength from the rigor with which the Greeks, long before the emergence of the arbitrary and philosophically indefensible dichotomies of the recent intellectual past, applied the principles of *episteme* and *doxa*. The risks of this sort of philosophy, both individual and social, are perhaps grave, as history suggests; but then, choice, that most human of contexts, holds out the possibility of agony as well as glory. It could not be otherwise. The death of political philosophy, like the abortion of philosophy of technology, has been heralded in the public forum. Let us hope that, as in the case of Mark Twain's death, the situation has been greatly exaggerated.

Notes

1. Peter Winch, *The Idea of a Social Science* (London: Routledge and Kegan Paul, 1958), p. 8.
2. Nathan Rotenstreich, "Technology and Politics," in Carl Mitcham and Robert Mackey, eds., *Philosophy and Technology* (New York: The Free Press, 1972), p. 151.
3. Ibid., p. 151.
4. Jacques Ellul, *The Technological Society* (New York: Vintage, 1964), p. 233.
5. Ibid., p. 247.
6. Henryk Skolimowski, "The Structure of Thinking in Technology," in *Philosophy and Technology*, p. 45.
7. Ellul, *Technological Society*, p. 271.
8. Ibid.
9. Ibid., p. 277.
10. Ibid., p. 279.
11. Ibid., p. 428.
12. Ibid., p. 301.
13. Ibid., p. 306.
14. Jacques Ellul, *The Political Illusion* (New York: Vintage, 1967), p. 238.
15. Ibid.
16. Abraham Kaplan, *The Conduct of Inquiry* (San Francisco: Chandler, 1964), p. 395.
17. Ellul, *Technological Society*, p. 430.
18. Ibid., p. 428.
19. Ibid., p. 430.
20. Victor Ferkiss, *Technological Man* (New York: New American Library, 1969), p. 29.
21. Ibid., pp. 35-36.
22. Ibid., p. 73.
23. Ibid., p. 202.
24. Ibid.
25. Ibid.
26. Ibid., p. 203.
27. Ibid., p. 205.
28. Ibid., p. 159.
29. Ibid.
30. Ibid., p. 164.
31. Ibid.
32. Ibid., p. 207.

33. Ibid., p. 31.
34. C. B. Macpherson, "Democratic Theory: Ontology and Technology," in Mitcham and Mackey, eds., *Philosophy and Technology*, pp. 161-70.
35. Ibid., p. 161.
36. Thomas Thorson, *The Logic of Democracy* (New York: Holt, Rinehart and Winston, 1962), p. viii.
37. David Easton, *The Political System* (New York: Alfred Knopf, 1960), p. 93.
38. Alfred Cobban, "The Decline of Political Theory," *Political Science Quarterly*, vol. LXVIII (Sept., 1953), p. 331.
39. Thorson, *Logic of Democracy*, p. 19.
40. Ellul, *Technological Society*, pp. 287, 288.
41. Alfred North Whitehead, *Process and Reality* (New York: Harper and Row, 1957), p. 270.
42. Ernst Junger, "Technology as the Mobilization of the World through the Gestalt of the Worker," in Mitcham and Mackey, eds., *Philosophy and Technology*, p. 277.
43. Herbert Marcuse, "The New Forms of Control," in Albert H. Teich, ed., *Technology and Man's Future* (New York: St. Martin's Press, 1972), p. 74.
44. Ibid., p. 45.
45. Albert Camus, *The Rebel* (New York: Vintage, 1956), p. 4.

David Edge

18

Technological Metaphor and Social Control

There's a "Peanuts" cartoon in which Lucy and Charlie Brown are leaning on the wall, discussing life:

> "There's an old song, Charlie Brown, that says life is like a railroad."
> "I've never been on a train."
> "Have you ever been out to the airport?"
> "I've seen the airport, but I've never flown on a plane. . . . I took a trip on a bus once. . . . Is life like a trip on a bus?"
> "Forget it. . . . "

In 1969, during the big student demonstrations in Washington, the BBC showed a TV news sequence (I think it originated at CBS) that consisted mainly of edited clips of "vox pop" interviews with middle-aged bystanders who talked about the demonstrations as a healthy "letting off of steam" by the young, a "safety valve to relieve society's pressures," and so on. However, the final clip was one of the student leaders addressing the crowd: "You've heard them all say we're the safety valve—but I say they're dead wrong. We're the explosion!"

A slightly modified version of this paper was published in *New Literary History*, vol. VI (1974-75), 135-47. This paper was prepared while the author held a joint appointment as Senior Fellow of the Society for the Humanities and Senior Research Associate of the Science, Technology, and Society Program at Cornell University. The author is indebted to his colleagues in the Science Studies Unit, Edinburgh University, for introducing him to many of the ideas expressed, and many of the sources cited, in this paper.

In recent years, there has been a notable increase in interest, among philosophers of science, in the cognitive functions of metaphor.[1] The process whereby we construe an uncertain, obscure, or puzzling area of experience in terms of one both familiar and apparently (in at least some respects) similar, the displaced pattern acting as a metaphorical redescription of the unfamiliar, has been shown to be central to many key scientific innovations.[2] Technological devices are significant components of our familiar, everyday world: they therefore share in this process, forming the literal basis of metaphors which give implicit, tacit structures to our thought and feeling and that "fill our consciousness." Indeed, many of the most influential of the theories of modern science have an explicit origin in such "technological metaphor." [3] (The pervasive metaphors of cybernetics, reconceptualizing the brain and society in terms of the behavior of computers and other electrical networks, offer one striking example: how often do you hear the term "feedback"?) We would do well to explore the extent of the dynamics of this process by which our imagination comes to be "dominated" by those very devices which we devise in order to "dominate" and "control" our environment and human society.

Writers on the cognitive functions of metaphor tend to stress two aspects of the process. First, that the successful metaphor does not merely provide answers to preexisting questions, but rather, by radically restructuring our perception of the situation, it creates new questions, and, in so doing, largely determines the nature of the answers. Judith Schlanger, for instance, in discussing the technical origins of metaphors which are central to recent advances in molecular biology, writes that "the concept of cell regulation establishes the field for which it sets the boundaries and is the coordinator." [4] That concept, like the cybernetic metaphor of brain function, derives from the "hardware" of control technology, and it is hardly surprising that, when authors come to discuss the inadequacies of such metaphor, they do so in terms derived from the metaphor itself—talking, for instance, of the brain's impenetrable "black boxes" or of "the mysterious nature of human encoding and decoding." [5] This is not a matter of human weakness. We cannot think about the matter otherwise, because otherwise there is no "matter" to think about. As Schlanger comments, "The cybernetic analogue provokes and instigates its own theoretical elaboration."

Second, there is a realization that, in some important sense, we do not choose a successful metaphor—rather, it chooses us. We are "seized by it." The metaphor acts to eliminate confusion, to structure chaos: its action involves, as Schon notes, "a transition from helplessness to power." Moreover, those on whom the metaphor acts in this way are driven to take it literally. Not for them the detachment implicit in the realization that their activity is essentially metaphorical: they "take it seriously" and proceed to spell out their new perception in direct detail, imposing the categories determined by the metaphor. In Douglas Berggren's terms, they have lapsed from metaphor to "myth."

These two features are strongly reminiscent of the action of a "paradigm," as T. S. Kuhn characterizes it.[6] And, just as Kuhn's later work is moving in the direction of a sociological analysis of the basis of scientific activity, so we might profitably enquire as to the social forces which may determine the use of metaphor, predisposing us to assimilate it into our imagination and to take it seriously. I have three suggestions to make about how we might approach this problem.

The first starts from the simple observation that successful metaphors tend to have ambiguous associations. (Religious symbols, of course, are the classic instance.) We've already noted one such metaphor: the "safety valve." It is only activated when the pressure reaches dangerous proportions, but the noisy outrush of steam signifies that the danger is under control. This metaphor can sometimes reassure the Silent Majority, and sometimes alarm them! (In this it is like that other notorious technological metaphor, the "swinging pendulum.") These confused associations arise from an ambivalence in our experience of the technology itself. I must confess that when our pressure cooker lets off steam I want to rush for the door! The most deeply rooted form of this metaphor, which equates a human skull with a steam boiler, reflects a critically ambivalent attitude to the violent expenditure of human energy (especially by the young) and to enthusiasm in general. Such behavior seems to be both necessary and dangerous. If the connotations of danger can be attributed to the prevalence of spectacular boiler explosions in the mid-nineteenth century,[7] and hence to the ignorance at that time of the second law of thermodynamics and of basic principles of metallurgy, would this alter our feelings?

There is a related ambivalence at the heart of the railway metaphor, which can account for much of its exploration in modern literature.[8] It concerns the tension between the throbbing, hissing, fiery primal energy of the engine, vigorous and eager to be off the leash, and the exact, geometrical, purposeful discipline of the rails. The powerful imagery generated by this tension gives the metaphor much of its symbolic energy. (It has some interesting corollaries: in American folksong, for instance, the fireman is usually depicted as a randy, undisciplined, hedonistic, drunken character, with a girl at every stopover, while the engineer is "straight," happily married, and sober. One stokes the fire, while the other keeps to the rails!) One recalls the uneasy truce that Lovejoy has called "the ethic of the middle link"—the view of the human predicament so widespread and influential throughout the late eighteenth century, which saw men as forced to accept "an unsatisfactory but nevertheless unavoidable compromise between their animal nature and their rational ideals." The railway steam engine symbolizes this "middle link ethic" to perfection. Life is, indeed, like a railroad, and man like a railway engine: in the face of this competition, buses haven't a hope. What technology has provided us with, then, one might say, in this relatively brief cultural interlude, is merely a fresh way of sharpening our appreciation of both poles of this essential tension. Alternatively, you could argue that this tension, if it really exists, is not all that important. The visions and ideals of the Enlightenment thinkers may have been delusory and related only to their own, unique social experience; the problem they posed may be of no general consequence. In focusing our attention on this tension, and in acting as the vehicle by which an eighteenth-century dilemma has been vividly transported into the twentieth century, technology may have led us all off the rails.

The use of railway metaphor within the emerging class structure of America illustrates another ambiguity.[9] To the middle class, and putative middle class, the railway symbolized the endless possibilities of exploitation, the journey of adventure to those far frontiers of rich potential where they would find their rewards. It had a romantic quality, positive associations. To the underprivileged, as those of you who are familiar with traditional Negro blues will know, the metaphor had negative associations. The railway station was a place of separation from your lover. The railway recurs as a symbol of desolation and despair; it appears as an impersonal fate, inexorably tearing apart the securities and transitory comforts of personal existence. If it offers any hope, it is the rather attenuated one of escape from a living hell on earth to somewhere (Chicago?) where life might be slightly more tolerable. Only in the Gospel song does it offer, metaphorically displaced, the endless frontier of an escape to Jesus.

Two widely differing class (and racial) groups, two radically different perceptions—but each supported by the ambiguous connotations of one central metaphor. And what is true of the particular case of the railway is true, *a fortiori,* of the general manifestation of the paradigm technological metaphor—Society as Machine. To those in power, it expresses liberation and hope: to those exploited, repression and despair. The ambivalence lends life to the metaphor, even today. It is one that a wide spectrum of class interests can still take seriously.

This centrality of an ambiguous metaphor can be supported (often inappropriately) by conflict. To take a related example: there is a lively debate these days in Britain (and I gather elsewhere, too) over the content of the mass media. Those in control of the media justify their policies in grand, liberal terms—they are enlightening, educating, fostering humane values, and so on —and a vigorous pressure group maintains a continuous crossfire —the media are corrupting, a major cause of increasing crime and violence, morally degenerating, and so on. Both groups have a vested interest in believing that the media are influential. In the midst of this debate, the view (supported by the bulk of empirical research) that the media have relatively little effect on people's beliefs, values, attitudes, and behavior doesn't get much air time! When it does surface—as in the recent U.S. presidential commission report on pornography—it is found to be politically embarrassing and has to be shouted down.

Central to this debate, and lurking within the Society/Machine metaphor, is the notion that the action of society is centrally controlled, coherent, and manipulative. Controversy can keep this alive, too. Those in power like to think that the course of events has been coherent and guided by humane, rational policy, and they justify their actions in these terms. Their critics also make the "coherence" assumption, but see the policies of those in power as mistaken, foolish, or actively evil.[10] In such a climate, the lack of control in society is not acknowledged. Inappropriate images and metaphors can be perpetuated.

The second approach to the social analysis of metaphor recognizes the potential of metaphor to alter feelings and attitudes toward oneself and others, and toward the natural world. Here, for instance, is Owsei Temkin's account of the development of the metaphor of the body as a machine:

> Descartes' metaphor of the body machine proved most fruitful in many respects. In the first place, it made room for a more active attitude toward the body. Galen had imagined the human organism to be so perfectly constructed that an improvement was not even thinkable. Besides, nature was constantly at work to protect and cure. . . . But a machine has only a certain number of regulations, which in many cases may prove insufficient to restore the damage. One of the consequences of the Cartesian concept . . . was a difference in the evaluation of the healing power of nature and of medical interference. The Galenists upheld the healing power of nature whereas many Cartesians tended to stress its limitations. Boyle, for instance, who followed Descartes in the metaphor of the human machine, argued elaborately that many natural reactions in disease were not beneficial but harmful and that the physician, therefore, had to combat rather than encourage them—this, in spite of Boyle's belief that the human body had been fashioned by God with infinite wisdom. Once this belief weakened it could be asked whether the body was a good or a bad machine. Thus, Helmholtz, in considering the eye as an optical instrument, found it so full of defects that he for one would have felt justified in returning it to the optician who had dared to sell it to him. And perhaps it is not by chance that the period of the nineteenth century which made the most fruitful applications of the metaphor of the body machine also became interested in "dysteleology." By this theory Haeckel designated organs which were useless, and dysteleology found its practical culmination in the removal of the healthy appendix as an altogether

useless and dangerous part. In the days of Galen this would have been rank heresy.[11]

This is an account of the slow, steady working out of the affective implications of a metaphor. As the metaphor begins to "bite," cognitively, it brings with it attitudes appropriate to its literal referent. A similar account could be given of the development of the metaphor of the Universe as a Machine (specifically, the Universe as a Clock, and its attendant Deist theology of God as Clockmaker). Both these accounts run parallel to the socio-economic history of Western Europe, the development of technological means to harness energy efficiently and to increase agricultural production, and the associated growth in population and wealth. As Lynn White has argued, in several controversial essays,[12] this development involved a radical change in attitudes toward nature. Cooperation gave way to exploitation. He cites the diffusion of the heavy plough in Northwest Europe as one technological innovation which made such dominance possible and traces the shift in attitudes via, for instance, a succession of calendar motifs, which steadily abandon impersonal, natural symbols for the seasons and substitute pictures of human activities—ploughing, threshing, and so forth. As White puts it, "Man and Nature are now two things, and Man is the master." The idea of the universe as a clock offers a pervasive metaphor, with impeccable cognitive credentials, within which such attitudes and feelings come to be seen as perfectly natural and are hence given social validation and support. The ambiguity of the symbolism and myths of the Judeo-Christian religious tradition allows an appropriate reinterpretation of the sacred texts.

We are, of course, now talking about religious perspectives. James Fernandez has recently offered an analysis, in similar terms, of certain tribal systems of religious symbols and rituals.[13] He talks of a "metaphoric strategy," by which we can alter our position in "quality space." Metaphoric strategies "involve the placing of self and other pronouns on continua." The religious believer, for instance, enters a ritual in a depressed relationship to his social and physical environment. The ritual starts by metaphorically equating him with, say, a worm, with its earth-bound, submissive connotations. Thus identified, the ritual then transposes him metaphorically to, say, an eagle—dominant, free. Fernandez comments: "We need to become objects to ourselves, and others need to become objects to us as well. . . . The shift in feeling tone—of adornment and disparagement—may be the dominant impulse to metaphor. . . . People undertake religious experiences because they desire to change the way they feel about themselves and the world in which they live. . . . We come to understand these operations only if we study metaphoric predications upon pronouns as they appear in persuasion and performance. The strategy of emotional movement in religion lies in them."

The third approach to the social basis for the diffusion of metaphor includes aspects of the first two and is, to my mind, the most powerful and suggestive. It emphasizes the role of metaphor (and of institutionalized forms of knowledge in general) in establishing and reinforcing moral and social control.

Charles Rosenberg has recently illustrated the way in which scientific and technological metaphors can be used to add authority to moral injunctions.[14] He discusses the popularity, in America in the late nineteenth century, of the metaphor of the human nervous system as an electrical (telegraph) network, around which flows a (limited) amount of a fluidlike "nervous energy." This metaphor, when allied with thermodynamic notions (notably the conservation of energy), produced a potent source of moral homilies on the virtues of moderation, adding the force of "science" to controling moral notions of the time. It was popular precisely because it did so. I would guess that much of the popularity of the Skull/Boiler metaphor, and of the Energy/Restraint tension exhibited by the railway engine, could be analyzed in similar terms. (Rosenberg also shows how the electrical metaphor for the nervous system "helped to express the ambivalence of many Americans toward progress, toward urbanization, toward the treacherous fluidity of American life"—and, again, the analysis could serve equally well for aspects of the railway metaphor.)

Metaphors of this kind, with similar moral implications, are, of course, still with us. Konrad Lorenz, for instance, writes in his classic work, *On Aggression:* "I believe . . . that present-day civilised man suffers from insufficient discharge of his aggressive drive."[15] Lorenz has been vigorously attacked by many of his scientific colleagues for just this kind of "metaphorical sloppiness," which encourages the reader to draw out moral and social implications where they do not "properly" exist and to call on Lorenz's authority to support "controversial" political and ethical positions.[16] But, in the perspective I am suggesting, the authority of the scientist, in society at large, rests on what he says: if people find it congenial, he will be believed; if not, his authority vanishes. Of course, attempts to reverse this trend, and to derive a new ethic from science, are not new[17]—attempts, that is, to take the scientific community as an overriding authority and to wield that authority in order to "persuade" people to alter their moral beliefs and behavior—but the verdict on all such attempts seems to echo Stephen Toulmin's verdict on Evolutionary Ethics: "The support given by Evolution to ethics serves as a source of confidence in our moral ideas, rather than as an intellectual justification for them." (Still less, of course, as a reason to *change* them!)

The most comprehensive statements of the position I am describing here can be found in the tradition in social anthropology that stems from the work of Emile Durkheim.[18] Mary Douglas's article "Environments at Risk" is particularly apposite. She discusses the problem faced by those ecologists associated with the environmental conservation movement in "being believed." She notes that they portray the earth as a unified system—the ecosystem. (In our terms, this is to say that they are propounding a fresh metaphor for the world, and man's place within it—with all the appropriate cognitive and affective implications.) They argue that the integrity of that system, and ultimately the fate of the human race, is at risk, due to the activities of the human race itself—or, at least, of identifiable portions of it, the exploiters and polluters. This picture, of course, is presented by these scientists with clear moral intent: the object is to persuade us to modify our desires and values and to act so as to restrain our inclinations toward material advancement. But, as Mary Douglas argues, drawing on a range of anthropological material, "No one can impose a moral view of nature on another person who does not share the same moral assumptions." So, to put it crudely, if the moral consensus is toward restraint, then these ecologists are likely to be listened to, and their advice acted on; but if there is no such consensus, they will not be generally believed, and their metaphorical world view will not gain any wide popularity.

In her book *Natural Symbols,* Mary Douglas elaborates the notion that "the view of the universe, and a particular kind of society holding this view, are closely interdependent. They are a single system. Neither can exist without the other." She suggests a typology with which to classify four relatively distinct kinds of social experience, and, with a wealth of anthropological detail, she shows that these can be plausibly related to four distinct kinds of cosmological belief. The details need not concern us here, except to note two points. First, that modern Western society contains people whose social experience (in Mary Douglas's terms) are of distinct kinds. (She is herself centrally concerned with the plight of the orthodox, Friday-fasting, working-class, "Bog Irish" immigrants in the Catholic Church in London, ministered upon by articulate, mobile, middle-class, symbol-blind, and uncomprehending priests—and with the rationally unbridgeable cosmological chasm between them.) Second, that the related cosmologies differ markedly in their metaphysical and moral coherence (and hence in the extent to which they lend themselves to a unified system/organism/mechanism metaphor) and also in the degree to which the universe and the powers (if any) which control it are seen as benign, or neutral, or malevolent. This scheme, when combined with the other aspects I have mentioned, seems to me to offer a powerful potential source of illumination on both the optimistic, "rational" assurance of those who sponsor and manage the modern technological enterprise and the pessimistic, "irrational" passion of those whose social experience and cosmology force them to view that enterprise as more impersonal than humane, more sinister than benign.

One final, rather gloomy, meditation. Mary Douglas writes that "credibility depends so much on the consensus of a moral community that it is hardly exaggerated to say that a given community lays on for itself the sum of the physical conditions which it experiences." [19]

Technology now has a central role in determining "the sum of the physical conditions which our society experiences," and it is itself expressly designed and promulgated in order to accommodate consensus views of how things should rightly proceed. If it lends itself to ambiguous symbolic reference, carries affective implications, reflects moral patterns, and evokes cosmological echoes, it will therefore tend to generate its own metaphorical force. And that force—embracing, as it does, critical perceptions of ourselves, our society, and our environment—will be essentially conservative.

With that thought in mind, consider the kind of cartoon image with which we are all now familiar, in the pages of *The Ecologist* and elsewhere: "Spaceship Earth" is plunging through smoke and grime, its structure slowly disintegrating and no one at the controls, but carrying a complacent complement of passengers. Or this remark by Ralph Lapp, in his book *The New Priesthood:* "We are aboard a train which is gathering speed, racing down a track on which there are an unknown number of switches leading to unknown destinations. No single scientist is in the engine cab, and there may be demons at the switch." [20]

You don't need me to tell you how prevalent such images are![21] The sense of "crisis" is conceived of in terms of a technology which enshrines an extreme form of centralized control. Society (and the world) is "out of control" like a crashing, disintegrating aircraft without a pilot. This metaphor brings with it the corollary that order, sense, purpose will be restored by mending the plane and by putting in a competent, well-meaning pilot (or, as Lapp would have it, "a scientist in the engine cab"). One common reaction to our present social problems (such as environmental pollution) is for people to say that it demonstrates that our existing centralized institutions of political control are defective and then to proceed to attempt to shore up and strengthen those centralized "controls." Technological metaphor tends to confirm this conservative reaction, since we devise technologies specifically to "fit," to serve, and to extend our preconceived notions of "control" as we have institutionalized them. The very form of our technology necessarily spells out this assumption, refining it, in sophisticated forms, to a high art. It may be that the dialogue between the conservation movement and the government agencies is unwittingly perpetuating inappropriate metaphors.[22]

A common view of the relationship of man and technology is

contained in this remark by Juergen Schmandt: "Technology is seen as an all-consuming monster which has so enslaved man that the normal relationship between man and his tools—that of master and servant—has been turned upside down." [23] This notion of the "normal relationship" seems closely tied to metaphors deriving from simple tool-users, such as carpenters. A carpenter uses tools selectively for a preconceived purpose; the tools lie passively until used; their use affects neither the carpenter nor his preconceived goal. Here, indeed, is a "master/servant" metaphor: the tools are "commanded." When such a metaphor fails to do justice to the situation, the only possible modification is to stand it on its head (in electronic parlance, it defines a "flip-flop," with only two stable states), hence the power of the Sorcerer's Apprentice myth. But how normal is this kind of tool using? Perhaps the experience of the artist is (or could be) more typical: as he struggles with his materials, he is forced to reconceive his goals (and himself); he neither dominates, nor is dominated by, his materials. The model of artistic creation, with its dialectic symbiosis, seems more congruent with the data of cultural anthropology, the experience of great scientists—and, indeed, with the interaction of men with computers. The popularity of the simpler, "cleaner," more authoritarian and dualist alternative may reflect some of modern society's hidden structure and its deeper fears. However, this is no reason for taking it as a model with which to shape the way in which we attempt to control technology. Rather, it can be seen as a challenge so to change society that people will be predisposed to adopt more humane and creative (and less fear-ridden) alternatives. Technology itself, by epitomizing other styles of control, can provide bases for such alternative metaphors: it can also change society in such a way as to make those alternatives seem "appropriate."

Mary Douglas senses parallel possibilities and dangers. In *Purity and Danger,* she argues that all pollution taboos in primitive societies can be interpreted as ways of preventing the fundamental schemes of classification in those societies from being infringed: the taboos "keep the categories pure" and protect them from danger. Ritual cleansing and the expulsion of polluters represent the essentially conservative reenactment and reinforcement of traditional categories and principles. In "Environments at Risk," she cites the example of the Eskimo girl who, by persistently eating caribou meat after winter had begun, broke a fundamental taboo of her tribe. She was, by unanimous decision, banished, to freeze to death. Thinking of the conservation movement, Mary Douglas comments: "Are we going to react, as doom draws near, with rigid applications of the principles out of which our intellectual system has been spun? It is horribly likely that, along with the Eskimo, we will concentrate on eliminating and controlling the polluters. But for us there is the recourse of thinking afresh about our environment in a way which was not possible for the Eskimo level of scientific advance. Nor is it only scientific advance which lies in our grasp. We have the chance of understanding our own behaviour."

With the remark that "our own behaviour" must include the metaphorical activity I have been describing, both within science and in society at large, this might be an appropriate note on which to close. However, since I opened with the mass media, let me close with them, too. *The Listener* of 1 March 1973 contained this item:

High Heaven.

When John Mortimer said, in a BBC1 *Sunday Debate,* that he thought of morality "as a kind of civic sewage system," with "no further pretension than the basic protection of each other," Robin Day intervened: "You don't think human beings are entitled to be looked at on a higher level than sewage?" "No, no," said Mortimer. "That's not what I said, Robin. I wasn't talking about human beings, I was talking about ethical systems. Human beings have the value they give themselves. If they give themselves freedom, free choice, self determination, then they're as high as the gods. If they give themselves the value of trams which run along the predestined lines of the moral and ethical views imposed by authoritarian states or authoritarian religions, that is all they are."

Trams! Good grief.

Notes

1. See, for instance, Max Black, *Models and Metaphors* (Cornell University Press, 1962); D. A. Schon, *Displacement of Concepts* (Tavistock Publications, 1963: reprinted, 1967, by Tavistock Social Science Paperbacks, under the title *Invention and the Evolution of Ideas*); D. Berggren,

"The Use and Abuse of Metaphor, I and II," *Review of Metaphysics,* vol. 16, 237-58, 450-72, and "From Myth to Metaphor," *Monist,* vol. 50 (1966), 530-52; M. G. Hesse, "The Explanatory Function of Metaphor," in Y. Bar-Hillel, ed., *Logic, Methodology and Philosophy of Science* (North Holland Publishing Co., 1965), pp. 249-59, and Hesse, *Models and Analogies in Science* (Notre Dame Press, 1966), which contains a useful bibliography. For an extensive annotated bibliography on metaphor, see Warren A. Shibles, *Metaphor* (The Language Press, 1971).

2. See, for instance, Owsei Temkin, "Metaphors of Human Biology," in R. C. Stauffer, ed., *Science and Civilization* (University of Wisconsin Press, 1949), pp. 169-94; Eduard Farber, "Chemical Discoveries by Means of Analogies," *Isis,* vol. 41 (1958), 20-26; Harvey Nash, "The Role of Metaphor in Psychological Theory," *Behavioral Science,* vol. 8 (1963), 336-45; Robert M. Young, "Darwin's Metaphor: Does Nature Select?," *Monist,* vol. 55 (1971), 442-503; and John C. Marshall, "Minds, Machines and Metaphors," *Social Studies of Science,* vol. 7, no. 4 (Nov., 1977), 475-88. An interesting early reference is D. Fraser Harris, "The Metaphor in Science," *Science,* vol. 36 (1912), 263-69. Also relevant are D. C. Bloor, "The Dialectics of Metaphor," *Inquiry,* vol. 14 (1971), 430-44, and "Are Philosophers Averse to Science?" in D. O. Edge and J. N. Wolfe, eds., *Meaning and Control* (Tavistock Publications, 1973), pp. 1-30; and Martin Landau, "On the Use of Metaphor in Political Analysis," *Social Research,* vol. 28 (Fall, 1961), 331-53. Schon (*Displacement of Concepts,* noted above) discusses both scientific and technological innovation.

3. See especially Karl W. Deutsch, "Mechanism, Organism and Society: Some Models in Natural and Social Science," *Philosophy of Science,* vol. 18 (1951), 230-52. This paper is particularly interesting since, after an elegant and concise introduction on the historical role of technological metaphor in the development of scientific theory, Deutsch proceeds to expound another (the cybernetic metaphor). See also my own "Technological Metaphor," in Edge and Wolfe, eds., *Meaning and Control,* pp. 31-59. Other examples are discussed in works cited in footnote 2.

4. Judith Schlanger, "Metaphor and Invention," *Diogenes,* vol. 69 (1970), 21.

5. See, for instance, C. C. Anderson, "The Latest Metaphor in Psychology," *Dalhousie Review,* vol. 38 (1958-59), 176-88. Deutsch's paper closes with a short discussion (pp. 250-52) in which he analyzes the differences between his metaphorical referents in terms of the metaphor itself.

6. In *The Structure of Scientific Revolutions* (University of Chicago Press, 2nd ed., 1970). See also his "Reflections on My Critics," in I. Lakatos and A. Musgrave, eds., *Criticism and the Growth of Knowledge* (Cambridge University Press, 1970) and "Second Thoughts on Paradigms," in F. Suppe, ed., *The Structure of Scientific Theories* (University of Illinois Press, 1972).

7. See B. Sinclair, *Early Research at the Franklin Institute: The Investigation into the Causes of Steam Boiler Explosions: 1830-1837* (Franklin Institute, 1966); J. G. Burke, "Bursting Boilers and Federal Power," *Technology and Culture,* vol. 7, no. 1 (1966), 1-23; and a review by Burke of Sinclair, *Technology and Culture,* vol. 9, no. 2 (1968), 230-32.

8. See T. R. West, *Flesh of Steel* (Vanderbilt University Press, 1967); and H. L. Sussman, *Victorians and the Machine* (Harvard University

Press, 1968). Thomas Carlyle's essay, "Signs of the Times" (1829), is a classic and historic commentary on the effects of the machine metaphor.

9. Leo Marx, *The Machine in the Garden: Technology and the Pastoral Ideal in America* (Oxford University Press, 1964) is relevant here.

10. Alasdair MacIntyre accuses Herbert Marcuse of inappropriately perpetuating just such a notion in this way, in *Marcuse* (Fontana Modern Masters, 1970), esp. pp. 71-72. MacIntyre uses the example of the debate over the Vietnam war.

11. MacIntyre, *Marcuse,* fn. 2, pp. 180-82.

12. Notably *Medieval Technology and Social Change* (Oxford University Press, 1962) and *Machina ex Deo* (M.I.T. Press, 1968). See also the essay by Richard A. Underwood, "Toward a Poetics of Ecology: A Science in Search of Radical Metaphors," in Richard E. Sherrell, ed., *Ecology, Crisis and New Vision* (John Knox Press, 1971).

13. "Persuasions and Performances: Of the Beast in Everybody . . . and the Metaphors of Everyman," *Daedalus,* vol. 101, no. 1 (Winter, 1972), 39-60. See also Robert P. Armstrong, *The Affecting Presence* (University of Illinois Press, 1971).

14. "Science and American Social Thought," in D. Van Tassel and M. G. Hall, eds., *Science and Society in the United States* (Dorsey, 1966), pp. 137-84. An excerpt can be found in Barry Barnes, ed., *Sociology of Science* (Penguin Modern Sociology Readers, 1972), pp. 292-305. See also Charles E. Rosenberg, *No Other Gods: On Science and American Social Thought* (Johns Hopkins University Press, 1976).

15. K. Lorenz, *On Aggression* (Methuen, 1966; University Paperback ed., 1967), p. 209. Lorenz is here quoting from an earlier (1955) paper of his entitled "On the Killing of Members of the Same Species."

16. For a particularly vigorous attack, along these lines, on Lorenz's book, see the review by S. A. Barnett in *Scientific American,* vol. 216, no. 2 (Feb., 1967), 135-38. Some similar ideological touches can be found in Deutsch's paper cited in footnote 3 above.

17. The attempts to derive an ethic from evolutionary theory are particularly well documented and scrutinized. See A. G. N. Flew, *Evolutionary Ethics* (Macmillan, 1967); Stephen Toulmin, "Contemporary Scientific Mythology," in A. MacIntyre, ed., *Metaphysical Beliefs* (SCM Press, 1970); and A. Quinton, "Ethics and the Theory of Evolution," in I. Ramsay, ed., *Biology and Personality* (Blackwell, 1965).

18. In Britain, two major exponents are Mary Douglas and Basil Bernstein. For Douglas's views, see her *Purity and Danger* (Routledge & Kegan Paul, 1966) and *Natural Symbols* (Barrie & Rockliff, 1970) (both books now available in Penguin paperback); also "Environments at Risk," *Times Literary Supplement,* 30 Oct. 1970, pp. 1273-75, since reprinted in J. Benthall, ed., *Ecology, the Shaping Enquiry* (Longmans, 1972), and in M. Douglas, *Implicit Meanings: Essays in Anthropology* (Routledge and Kegan Paul, 1975), pp. 230-48. For Bernstein, see first his essay in Michael F. D. Young, ed., *Knowledge and Control* (Collier-Macmillan, 1971) and other bibliographies in that volume. See also S. B. Barnes, "On the Reception of Scientific Beliefs," in Barnes, *Sociology of Science,* fn. 14, pp. 269-91.

19. "Environments at Risk," p. 1274.

20. Ralph E. Lapp, *The New Priesthood: The Scientific Elite and the Uses of Power* (Harper and Row, 1965), p. 29. The entire paragraph from which these two sentences are taken extends the metaphor very ingeniously.

21. There is a fairly extensive psychoanalytical literature on this topic. See, particularly, Robert W. Daly, "The Specters of Technicism," *Psychiatry: Journal for the Study of Interpersonal Processes,* vol. 33, no. 4, (Nov., 1970), 417-31; and Bruno Bettelheim, "Joey: A 'Mechanical Boy,'" *Scientific American,* vol. 200 (March, 1959), 116-27. This literature is, I find, somewhat unhelpful in discussing the topic with those involved, since it tends to encourage people to "distance" themselves from the issue, as something which "only affects the abnormal." For most people, the reaction to Daly's remark that "given contemporary symbols of power, efficacy, and heroic human action, can one wonder that troubled persons employ these symbols of power in their neurotic constructions" is an immediate "Not me, Lord! I'm not troubled or neurotic!"

22. In the same way, and for similar reasons, as the debate over the mass media. See also MacIntyre's remarks on Marcuse cited in footnote 10.

23. Juergen Schmandt, "Technology and Man: Who is in Control?," unpublished manuscript, Program on Technology and Society, Harvard University, p. 2.

Henryk Skolimowski

19

Philosophy of Technology as a Philosophy of Man

On Change

The ancient Chinese had a saying, "I curse you to live in an age of change." For better or worse, Western civilization has almost continuously, particularly during the last century, lived in a time of change. We have reversed the Chinese view and persuaded ourselves that change is not a curse but a blessing, and the agent of "progress." We also believe that no change means no progress and, ultimately, stagnation and unfulfillment. "Change" is identified with external change, the most visible type of change and, indirectly, of progress. Change is an implicit presupposition in our metaphysics of progress, which is really the metaphysics of movement: we always go forward even if we go nowhere. We are constantly and deliberately causing change as the vehicle of progress.

Nevertheless, our understanding of the concept of change itself is curiously superficial and tenuous. Within the scope of our empiricist epistemology (bearing our concept of progress), the concept of "change" is conceived as linear, homogeneous, almost mechanical. In actuality, if change is to be conceived in such a way as to bring about *qualitative* progress, it must be comprehended as dynamic and nonlinear. Whitehead understood this very well. Most empiricists as well as prophets of technological progress are lacking this understanding. They do not see, to begin with, that the concept of change is a bewildering philosophical problem.

It has been pointed out during this symposium that technology flourished in China in the fourteenth century, that is, before the Western Renaissance and before our scientific revolution. This has two important implications. First, it is not true (as we so often contend) that the scientific revolution is a necessary prerequisite for the flourishing of technology. Second, it is not true (as we so often contend) that when technology flourishes it always brings about external change. In fourteenth-century China technology flourished, but there was no need for change because technology perfectly met the expected demands. Technology remained in a subordinate position as a dependent tool, not as the spearhead of progress. From the Western point of view, to say "there was no need for change" implies backwardness, stagnation, and decay. Actually, Chinese civilization was doing quite well at the time. We must realize that the difference in the outlook on change between the fourteenth-century Chinese mind and the modern occidental mind is not a small peripheral issue, but that it signifies different casts of mind and, indeed, different world views.

We in Western civilization consider change to be simply an agent of progress. Since progress is a necessary part of our conception of life, change is regarded as belonging to the substance of our life. Change, like perpetual motion, must go on and on, regardless of human costs. When we afford the luxury of pausing for a moment of reflection, we realize at once how strange it is for a human being to be caught in a perpetual lunatic motion called change. Change has become the driving force of our civilization. We do not question it because it has become identified with progress and progress until recently has been a sacred taboo: one cannot be against progress. Although we praise ourselves for being rational through and through, our Western metaphysics with its concealed myths of progress and of change is neither less mysterious nor more rationally justified than the metaphysics of other civilizations which we deplore as mysterious and irrational. Only history will tell where the ultimate wisdom lies, in ancient Chinese culture which deplored change or in Western culture which has hailed change and has chosen it as a major modality of its existence.

Actually, history is already beginning to tell us. We are becoming increasingly aware that the period of explosive material growth is coming to an end. This definitely means the end of the period of incessant external change, thus the end of further change as a vehicle of progress, and thus indirectly the end of the myth of progress. We are clearly heading toward a steady state in one form or another. In this steady state, whatever its ultimate form, we shall have to quietly abandon the myth of progress.

Philosophy of Technology as a Philosophy of Man

Leaving aside the concept of change, let us ask ourselves what is the reason for the sudden emergence of the philosophy of technology. Why are we so concerned with the *future* of technology? Paradoxical though it may sound, it may be that the philosophy of technology and our discussions about the future of technology have little to do with technology as such.

The emergence of the philosophy of technology reflects a delayed recognition of the importance of technology in making and breaking our civilization. Our increasing awareness that Western civilization may be breaking up has made us search for causes and links previously ignored. In the phenomenon of technology we find a focal point; here many paths converge. In this convergence is the main configuration of the network through which our civilization operates. Paths converging on technology include such concepts as "progress," "nature," "invention," "rationality," and "efficiency." The philosophy of technology is, in other words, the philosophy of our culture. It is the philosophy of man in a civilization which has found itself at an impasse, which is threatened by excessive specialization, fragmentation, and atomization, and which is becoming aware that it has chosen a mistaken idiom for its interaction with nature.

In our present discussion we are examining the ideas and ideals that have sustained our whole civilization and are built into its foundations. At stake is the viability of our entire intellectual equipment—the scientific-technological world view—for coping with the ecosystem and for supporting our further growth as human beings and as humane societies. Debates about the future of technology in this context are not about technology but about the future of a civilization and perhaps of mankind. We are little interested in the future of the technology of cutting tools, for example, or basket weaving; but we are vitally interested in the future of technology conceived as one unified phenomenon, because this phenomenon is interlinked in an extraordinary variety of ways with human society and has become a factor determining the future of society, as well as a co-defining component of this society. Technology has become, to use a Heideggerian term, a part of the Being of man. As long as its influence on us and on our progress and development was or was thought to be beneficial, we welcomed this symbiosis of man and machine. With the technological euphoria of the 1920s and 1930s, Le Corbusier, Buckminster Fuller, and other progressive thinkers called the house "a machine for living." This euphoria is now gone. We are examining our entire technological heritage.

A culture threatened by excessive instrumentalization finds a variety of reactions and outlets. Voyages beyond its restrictive bonds may be launched by means of drugs or by excursions into oriental philosophies. The emergence of the philosophy of technology is another reaction of a culture in peril of being strangled by excessive instrumentalization, for, to repeat, debates over the nature of technology are the discussions over the future of man. Whether we like it or not, whether we are well prepared or ill prepared, we are participants in a seminal debate over the fate and vicissitudes of a civilization. We are the custodians of a civilization.

I am in full agreement with Spengler, who asserts that "*technics is the tactics of living;* it is the inner form of which the *procedure* of conflict—the conflict that is identical with Life itself—is the outward expression." He continues, "*Technics is not to be understood in terms of the implement.* What matters is not how one fashions things, but what one does with *them.* . . . Always it is a matter of *purposive activity,* never of *things.*" [1] In a similar vein, Ortega y Gasset considers technology to be "the system of activities through which man endeavors to realize the extra natural program, that is himself."

In spite of its enormous bulk, technology offers us insufficient and in many ways defective techniques as tactics of living. Our tactics of living are simply wrong. In order to improve our tactics of living, in order to redirect the course of technology so that it provides new tactics of living, we must create a new kind of knowledge, a new discipline within which we can effectively rethink our present dilemmas. For lack of a better term, this new discipline could be called the philosophy of technology.

Thus the philosophy of technology must not be conceived as a mere scholastic discipline. Of course, a scholastic approach does give comfort and esthetic pleasure to those engaged in it, and these satisfactions are important. However, let us be aware above all that the philosophy of technology emerged as the result of a critical appraisal of the state of our civilization. It did not evolve to provide analytically minded philosophers with an arena in which to perform their marvelously effective analytical pirouettes. Our civilization has already produced too many technicians. Let us not delude ourselves that more technicians of a linguistic-analytical variety or any other variety will solve the problems of technology. It is our responsibility as philosophers, thinkers, historians, engineers, and enlightened citizens to meet those problems which we, as a civilization, have originated.

The Need for a New World View

A philosophical analysis of technology is nothing new under the sun. Aristotle was already engaged in it. In more recent times Spengler, Heidegger, Ortega y Gasset, Mumford, and others have contributed their share. We find that the phenomenon of technology is more complex and more far-reaching, however, than their analyses would seem to suggest. Contemporary philosophers by and large have ignored the problem. Mario Bunge is one of the salutary exceptions. I accept Bunge's conceptual analyses of technology, and particularly his expilicit claim that technology contains in itself a metaphysics, an epistemology, an ethics, but I find Bunge's program (though rigorous and yielding some illuminating results) to be fundamentally limited and in a sense essentially defective.

Bunge accepts a certain conceptual matrix within which he analyzes the metaphysical underpinnings of Western technology. This conceptual matrix, however, is a part of the Western world view shaped by scientific and technological rationality. Conceptual analysis of technology within a world view profoundly influenced by science and technology is bound to be partial and inadequate. To comprehend the course of our civilization as influenced by science and technology we seek a perspective above the partial and partisan world view of the scientific-technological *Weltanschauung.* Bunge's conceptual analyses of technology are at least partly self-referential: they use conceptual components of the world view they wish to appraise and possibly transcend.

Every metaphysics leaves out certain aspects of the world in order to concentrate on other aspects. No metaphysics is totally inclusive so as to give us perfect and complete knowledge. Concepts and categories, and in general the language of a given metaphysics, are not incidental but intrinsic features of it, in the sense that they serve to articulate the world in a specific way, as the given metaphysics conceives of it. A given language is specific to a given metaphysics. If we accept the language (its categories, concepts, and distinctions), we inadvertently accept the world view embedded in this language. One reason for our difficulties in overcoming the limitations of our present world view is that we use the language this world view has originated. This is also one of the reasons why Bunge's program is deficient in the long run, that is, when we seek an alternative life-style springing from alternative technology rooted in an alternative world view. By employing distinctions, concepts, and categories characteristic of the present empiricist-scientific world view, Bunge accepts the limitations of this view as embedded in these categories and distinctions.

Although we should not abandon conceptual analysis of technology within the present system, there is a much more important additional task awaiting us. We need to develop alternative world views, alternative metaphysics, as the basis for reflection on technology vis-à-vis society and civilization. The way toward understanding the metaphysical roots of technology, the way toward creating an alternative humanistic technology, must lead through the creation of an alternative metaphysical world view, which alone will enable us to grasp sharply and clearly the ramifications and consequences of present technology for a future humane society. Constructing alternative metaphysics is the most exciting intellectual task of our times, with profound and far-reaching practical consequences.

Where do we go from here? Nowhere, unless we rethink our entire intellectual heritage. We are at the juncture where philosophy has frequently found itself in the past: we are reassembling the conceptual pieces of the present civilization. In reassembling the pieces we are forced to concentrate on one of them, the phenomenon of technology. To reassemble these pieces will require much more than reshuffling the present categories of knowledge. What we are confronted with is not a little puzzle that can be easily resolved within the existing conceptual apparatus, but a major dilemma which will require an alternative set of conceptual structures and even an alternative view of knowledge. I hope it will not be taken as entirely facetious or outlandish to suggest that philosophers, thinkers, engineers, and historians will have to go back to school in order to rethink the present predicament, to schools in which alternative ways of viewing reality will be opened up for them. Philosophers, futurologists, and all other people who are concerned with the future of technology and thus the future of our culture should be sent to Indian reservations on which alternative world views are still adhered to, cultivated, and incorporated into alternative life styles; they should be sent to oriental societies and cultures in which alternative ways of interpreting the world are still viable and form an alternative basis of knowledge and of life styles.

We may not be able to rethink our predicament on our own, for our system of knowledge is self-referential. This system permits the existence of only those phenomena which support the claims of the system and excludes or minimizes phenomena and occurrences which seem to undermine the stability of the system. Technology in the present scheme of things is a part of our secular world view. Our affection for it and attachment to it are not a stupid infatuation with superficial gadgets. They are rather the residue of a long intellectual tradition, a residue of our longing for freedom via the instrument we have created and perfected. This instrument paradoxically has been endowed with more than merely instrumental functions: it was conceived as an instrument of liberation, as the vehicle of freedom, as a Noah's ark of hope, of prosperity, and of progress. All these longings and transcendental yearnings are built into our notion of technology. For this reason not only technocrats and simple-minded technicians with vested interests in technology, but also Nobel Prize winners and people of superior knowledge and exquisite refinement such as Sir Peter Medawar, unblinkingly insist that what has been spoiled by technology *will* be cured by technology and can *only* be cured by technology. These extravagant claims may seem naive when coming from sophisticated and sharp minds which in other contexts exhibit ruthless critical acumen. However, it is not naiveté that prompts Sir Peter to defend present technology so staunchly; it is rather his extraordinary commitment to the present world view. Nobel Prize laureates epitomize the feelings and disposition of the epoch that created them no less than ordinary farmers who admire the power of their tractors. The power of the myth of technology is so great and so dangerous precisely because it has pervaded the recesses of our Western mentality. Technology has become our physical and mental crutch to such a pervasive and perverse degree that even if we realize it devastates our natural and human habitat, our immediate reaction is to think about another technology which will mend it all. Technology is a state of Western consciousness. When we think "technology" we invariably think "control" and "manipulation." Our "most efficient" ways of dealing with present dilemmas result in further undermining our civilization, for these most efficient ways consist of further manipulation and fragmentation, the processes which are at the core of our troubles.[2]

We must understand that any attempt to humanize the present system by injecting more human values into it is doomed to failure, for the system is extremely resistant to such cosmetic operations. The present order of Western man, out of which have grown our ways of life, is based to a large degree on quantitative instrumental values. In terms of these values most, if not all, social and political assessments are made. The structure is exceedingly complex and its various parts beautifully support one other (see Fig. 1).

It is not sufficient to decorate the present quantitative system with some intrinsic human values by injecting human considerations here and there. As long as the quantitative instrumental basis remains unchanged and channels its imperative via descrip-

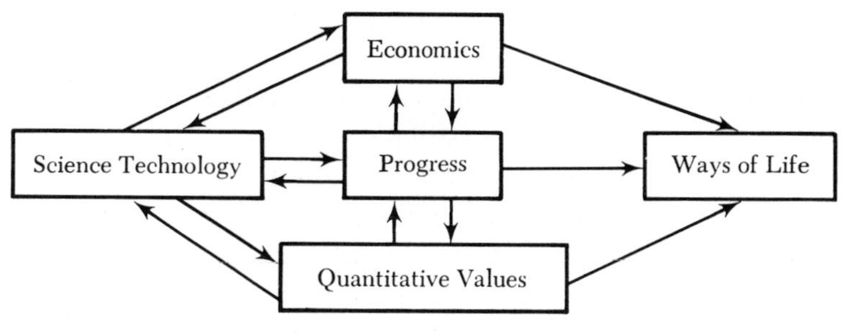

Figure 1.

tive science, via industrial profit- and efficiency-oriented technology, via economics geared to free enterprise, the order of things will remain the same, and we shall have to endure the same quantity-ridden way of life which is imposed on us now. Lechtman and Steinberg support this view when they say, "It is convenient to blame technology rather than to see it as symptomatic and expressive of value systems and orientations that characterize our world view." (p. 138 above).

Let me mention parenthetically that the prevalent approaches to values cultivated by the majority of contemporary philosophers are reductionist approaches, which indirectly serve the cause of the quantitative world view. At least four reductionist approaches are prominent nowadays:

1. *Linguistic analysis*, prevalent among analytic philosophers (especially exemplified by the late P. L. Austin), insists that we must study first the structure of moral utterances and from this study derive insight into the nature of moral phenomena. Unfortunately, the exercise of studying moral utterances has become an end in itself. We have a proliferation of types of moral utterances, but no comprehension of moral phenomena.
2. *Formalist approach*, prevalent among logicians (especially exemplified by von Wright), is only a more rigorous application of the linguistic approach, for it insists that we must study first the *logical* structure of moral utterances and then derive insight into the nature of moral phenomena. So far this has been a merely formal exercise which resulted in a proliferation of logical systems without corresponding comprehension of moral phenomena.

3. *Simple-minded technocracy,* prevalent among optimistic technocrats (especially exemplified by E. Mesthene), equates material choices with all human choices and freedom to possess material objects with total freedom. It is content to suggest that technological values are tantamount to human values.
4. *Simplistic mathematical evaluation,* prevalent among enlightened technocrats and systems analysts (especially exemplified by A. Rosenstein), seeks simple mathematical functions for the maximization of values, very much in the style of Jeremy Bentham and his idea of the Felicific Calculus. This approach treats values as simple interchangeable economic commodities.

In all these approaches values are reduced to something simpler than they are: linguistic utterances, logical structures, technological commodities, mathematical functions.

To change the predominantly quantitative temper of our civilization will require a thoroughgoing change in our modes of understanding, in our institutions, and in our consciousness. The quantitative civilization that the West has developed is at once a great achievement of the human mind—for nature and her creations do not quantify—and a great aberration, for we have attempted to reduce all qualities to easily quantifiable physical quantities.

Present technology in this context will have to be viewed as obsolete, as a voyage which did not deliver us to the promised land but which showed us at least where the promised land does not lie. This voyage was not the first of its kind. Humanity has often trod a mistaken path. Our human history is still young and there is no reason to shed tears over one more wrong turning. However, there is every reason to embark on an alternative path.

Some years ago, under the auspices of AAAS and the Society for the History of Technology, a symposium devoted to the exploration of the philosophy of technology was held in San Francisco. Since then, the philosophy of technology has emerged, not so much by the laborious efforts of diligent scholars, but rather because of the urgency and magnitude of social and human problems arising from the interaction of society with technology. It is no longer a question whether there is a field called the philosophy of technology. The question is rather whether we conceive it narrowly and constrain ourselves to comfortable problems which our analytical techniques allow us to tackle, and thereby acknowledge our intellectual impotence; or whether we conceive it comprehensively and meet the challenge of our times by attempting to rethink the basic presuppositions, assumptions, and modes of thinking which, at least in part, led us to our present predicaments. The latter is not so much a choice as an imperative. In creating the philosophy of technology in the measure of our problems and on the scale of our times, we might conceivably be able to work out new tactics of living; we might be able to give a new moral impulse to students and ordinary people anesthetized by overwhelmingly quantitative ways of life; we might be able to awaken philosophers and students of philosophy from their analytical slumber and show them once more what an exciting subject philosophy is; and, last but not least, we might be able to justify our claims as thinkers and scholars who are supported by society in order to help society but who have been miserably failing society in this task. The philosophy of technology is not the panacea which will cure all ills, but it might become a most illuminating, exciting, and significant intellectual endeavor for our times.

In summary, philosophy of technology conceived as a philosophy of man insists that technology is subject to the human imperative rather than man subject to the technological imperative.[3] It insists that man respects the delicate balances of nature and permits only an instrumentation of the world that enhances these balances without undermining them. It insists that man's knowledge is not pitted against the rest of creation, and that knowledge is not power to control and manipulate, but rather to understand and to mesh into the larger scheme of things. It insists that man's concept of progress does not mean the extinction of other creatures and at the same time the deadening of man's spirituality and sensibility, but rather the enhancement of man's diversity which is mainly accomplished by the enlargement of his spirituality. It suggests that we learn a fundamental lesson from societies and civilizations which we have treated contemptuously in the past, but which have preserved sanity, unity, and coherence by consciously integrating themselves into schemes of things larger than the pursuit of material progress.

Notes

1. Oswald Spengler, *Man and Technics: A Contribution to a Philosophy of Life* (New York: Knopf, 1932), pp. 10, 11.

2. For further discussion of the views presented here, see H. Skolimowski: "Science and the Modern Predicament," *New Scientist,* Feb., 1972; "Technology versus Nature," *The Ecologist,* Feb., 1973; "The Twilight of Descriptive Systems and the Ascent of Normative Models," in *The World System: Models, Norms, Variations,* ed. Ervin Laszlo (New York: Braziller, 1973); "The Scientific World View and the Illusions of Progress," *Social Research,* Apr., 1974; "Problems of Rationality in Biology," in *Studies in the Philosophy of Biology,* ed. Francisco Ayala and Theodosius Dobzhansky (Berkeley: University of California Press, 1974); "Evolutionary Rationality," in *Proceedings of the 1974 Philosophy of Science Association,* ed. R. Cohen, C. A. Hooker, et al. (Hingham, Mass.: Reidel, 1976).

3. Ferkiss *(Technological Man),* Skinner *(Beyond Freedom and Dignity),* and Toffler *(Future Shock)* think that it should be the other way around, that man should give way to the technological imperative: that man should adjust himself to the technological system (Skinner), make himself in the image of "rational" technology (Ferkiss), consider human societies as disposable as automobiles (Toffler). One wonders whether it is the system that produces Ferkisses, Skinners, and Tofflers or whether we simply allowed Ferkisses, Skinners and Tofflers to make the system in their own image.

Part III The Future of Technology

Ion Curievici

Besieging the Fortress

Over the years, as a result of my double education, I have acquired the defects of both philosophy and technology. When contemplating technology I commit the errors of a philosopher, and when I consider philosophy my errors are those of an engineer. This remarkable symposium has helped me somewhat to reduce this double infirmity. Its participants are men of high scientific stature, its subject holds an uncommon interest at this moment in history, and its approach to the theme has been suitably complex. It seems to me that the symposium has made an admirable inventory of the problems inherent in the history and philosophy of technology and, by making use of everything that could possibly be used, it has formulated answers to an impressive set of questions—answers of great importance for both philosophers and engineers, and ultimately for everyone. Predictably, antithetical answers often emerged. Rationalism and sometimes intuitionism, historicism and sometimes antihistoricism, optimism and sometimes skepticism, anti-Marxism and sometimes Marxism—all of these voices and more have been heard. The symposium has been marked by a sincere effort to decipher technology through its history, to grasp its human significance, and to envision the ways it can be made to serve progress alone.

Although many opinions were controversial, I have sensed a general conviction that technology used in a rational manner can be a giant force toward progress, an essential component of culture, and a source of vast spiritual wealth—a source that has unfortunately been ignored for a very long time. True, many of these ideas may not be entirely new. It was, nevertheless, very encouraging to witness the common effort of so many scientists, philosophers, historians, and economists to look for solutions capable of blocking the negative potential of technology.

In addition, many ideas new to me were presented here. Among the many new approaches were references to the unifying function of technology, disclosure of the philosophic ideas implied in modern technology, and Dr. Bugliarello's suggestion of a new entity—the BIOSOMA—as a force for shaping the future.

I must admit, though, that I have often felt limited in understanding and even misled by the diversity of meanings these papers attribute to the words *technology* and *technique.* It seems to me that technology has by and large been understood to mean the set of procedures for achieving a product or an object. As Dr. Wojick said near the beginning of his paper, "Technology is usually considered to be procedures or ways of doing things," with the specification that it represents an activity at the level of intellectual operations upon an object. Dr. Bunge expands this concept bi-dimensionally. On the one hand, technology is both *technical science* and *action* and is thus subdivided into *technology* and *technical praxis.* On the other hand, technology is superimposed onto what is traditionally meant in Europe by applied science, *angewandte Wissenschaft.* Thus, medicine is also technology, and by analogy medical practice is technology in action. Finally, in some of the papers—such as Dr. Slaby's—and in some discussions (if I understood them correctly), technology also seems to subsume the machine, the physical means, the tools by which we operate in order to achieve the transformation of nature.

Strictly speaking, I would not insist on this variety of meanings if I did not have a very special motive for doing so. I would simply be content with defining technology the same way somebody as confused as I am once defined physics. "Physics," he said, "is what everybody knows it is." For me, at least, to accept that technology is only a mental operation on nature and society would be equivalent to being unable to accept that it directly changes nature, men, and society. I think, too, that no one has ever seen a mental operation building a house or an airplane. In order for this mental operation to lead to these changes, we must add to it both *action* and *tools.* Perhaps Dr. Bunge would involve the tools in the action, because without them there could be none. If we ignore action, we ignore the mass of people who in the final analysis create history. If we ignore tools we ignore the material side of the act by which man transforms nature and society. Tools have, as a matter of fact, an unsual ontological dignity about which, to my regret, nothing has been said here.

It seems to me that this intellectualistic outlook on technology largely marked the discussion of the history of technology. If such a history were only the history of technical ideas—as was repeatedly suggested here—it would be essentially incomprehensible, since technical ideas—to be more exact, the mental models of human tools—do not develop by internal logic only, but also by external conditioning which results from both the demands of society and the material transposition of prior technical ideas—in other words, the technological level existing at any one time. The history of technology, as Dr. Smith excellently expressed it, is of interest only if it is essentially the history of the relationship between technology and society. I do not deny that the history of technology may also study other interesting problems or may even satisfy simple curiosity by looking through a keyhole at our predecessors' work. These aspects, however, are of relatively minor interest.

Marx and Engels developed this kind of history of the relationship between technology and society. The history already exists, then, and it needs only to be enriched by new data. By their analysis, Marx and Engels pointed out the material side of technology. The means of production represent the more mobile elements that bring about changes in the working forces of men, while the production relations have more stability. Altered production forces reach the level of a significant discrepancy relative to existing production relations, which then, sooner or later, undergo a revolutionary transformation. In other words, social progress is not to be expected as a direct result of developing technology, but by changing the production relations. Revolutionary changes do appear, though, as a kind of regulating action to restore agreement between the production forces and the production relations. Obviously, the new production relations in turn partially influence the development of the production forces, which finally clarifies the dialectic of historical development.

Even in this extremely concentrated presentation, it is easy to see the fundamental part played in Marxist philosophy by the historical relations between technology and society.

This symposium is only a beginning; future meetings like it will undoubtedly have the participation of many qualified Marxists. It seems to me that within the set of problems under discussion here the history of technology—understood as the history of the relationships between technique and society—will represent the major point of distinction between Marxist and non-Marxist. On the other hand, I think the thesis referring to the philosophy of technology, as formulated both precisely and elegantly by Dr. Bunge, would represent a field of extremely effective collaboration. I have no doubt that the problem of relationships to the philosophy of values would lead to exciting confrontations, but I am also sure that many common conclusions could be drawn.

As we well know, technology has been used in the past for antihuman goals, but never before has its negative potential endangered the very existence of the human species. Pollution and the reduction of natural resources also began many years ago, but the values of these variables have only recently become menacing. Are engineers responsible for this state of things? I think not. The culprit is history itself—that is, society, together with the irrational forces acting in each of us. In my opinion, engineers are only executors, on a conceptual level, of orders from without. In fact, anyone would have to admit that usually the engineer does not sit in judgment on the human values implicit in his work. What he proposes is to achieve—optimally according to a point of view that is usually economic—goals set for him by others. He is not usually a generator of aims, but rather a designer of methods and ways of achieving technical aims. If the ends for which he is acting are contrary to certain human values, he will transfer responsibility to the society which fixes those aims and expresses those needs. Technology, however, comes into the conflict with certain values, reflecting the struggle among different values and interests in society.

What is to be done? Should we ask engineers to take a solemn oath, similar to that of Hippocrates, that they will not act before evaluating an assigned goal in terms of a certain scale of values? Of course some such thing might be imagined. I think, however, that such an oath would be even less effective than that of Hippocrates.

A better solution might be to introduce some humanistic components into the education of engineers—as we are doing in all socialistic countries and in various other universities around the world. Unfortunately, if a humanistic education begins only at the university level, in most cases I think it is already too late. It should begin much earlier, so that the university only has to provide, on a higher level, a conceptual structure for this education.

Let us suppose that such a humanistic engineering education were to become general. It would, of course, be a good thing both for the individual and for society. The engineer would be no longer a *bystander* but a lucid fighter for human values. Do you think the dangers threatening us would be removed? I don't believe so. Only social control of technology will be able to offer us complete security. I am convinced that, by their competence and authority, engineers could make an important contribution to the establishment of that control.

Rafael López Palanco

21

Evolution of Building Technology

As a civil engineer, I have been dealing with two types of current technology: civil engineering and the technology of teaching. This paper was prepared primarily for teaching: to try to show students how history reveals interrelations among all the different types of technology developed throughout world history. As an example, we shall trace the evolution of methods of calculation and their interaction with the evolving technology of construction. An idea I hope to demonstrate and clarify, although I do not think it will be new to anyone, is the dependence of fashionable methods of construction upon the facilities for calculation available at the time.

The latter part of my paper emphasizes relationships discussed all during this symposium, the ordered relationships between technology and society. From society, I have selected one particular, and perhaps not even very important, aspect to consider: fashion. By skimming over various periods in history, we see that in everything man touches a fashion is created. "Fashion" is to be understood in the widest sense of the word: that is, fashion grows out of the way of thinking in each period, is influenced by the technology of each age, and in turn influences and improves technology itself.

The second major thought I was trying to put into the minds of my students was that the rate—the velocity—with which things change in the course of history has been accelerating. Ortega y Gasset said that one complete historical cycle takes seventeen centuries. He based this approximate figure on the time it takes for the Sphinx at Giza to become completely covered by sand and then exposed again. Even if he is right, one cannot assume that the speed of change has been uniform during the seventeen centuries. At least in the most recent period, changes have come about at a dizzying speed. To change the first computing machine, the abacus, from its earliest form into the version now used in Japan took ten centuries, more or less. In contrast, from the very first moment I came in contact with computers until now is only twenty-five years. The velocity of change has accelerated, and the speed of change keeps increasing every day. Following the history of calculation and of structural analysis down to the present day, as I have in this paper, illustrates this phenomenal acceleration of rates of change in fashion. Finally, we shall raise the question of how mankind can adapt to the appalling increase in the velocity of change.

I. Technology of Computation

Aside from hands—for the ten fingers were certainly the first elements used—I suppose the first calculations must have been made using stones in lines. Looking at it now, when technical efficiency seems very advanced and difficult to improve on, one would think the evolution of this method might have been rapid, but in fact it must have taken a very long time. The mere operation of threading stones of the same value onto sticks, which in time was to become an abacus, took many centuries.

The earliest surviving true abacus is the Russian model, which is very similar to the Chinese one but not divided into areas of "sky" and "earth." At the end of the tenth century, Pope Sylvester II separated the beads into values. From Western Europe the abacus reached China and then Japan, but this last step took five centuries. The Chinese model, the *suanpan,* dates from the twelfth century. The *soroban,* the Japanese abacus, does not appear in its traditional form in Japan until five centuries after Pope Sylvester II gave it its present shape.

After that, in spite of the fact that mathematics made great progress during that time, it was four more centuries before the *soroban* lost its superfluous beads, one in the "sky" and another on the "earth." Even these two related simplifications did not take place at the same time; there was actually a gap of sixty years between them. The strange thing is that the Chinese *suanpan* still has the original superfluous number of beads. However, when teachers explain how to use it, they say that anyone who uses the last beads of the two bands, those of the sky and earth, is not a good computer on the *suanpan*. Indeed, the Chinese are greater traditionalists than the Japanese.

The evolution of the abacus, its modification over the years and especially the time that these successive changes took—countless centuries, five centuries, four centuries, and finally sixty years—gives us an idea of the slow pace of fashion in the mathematical world.

Adding with Roman numerals is not easy, and multiplying must have been out of the reach of ordinary mortals. Experts claim, however, that with any kind of abacus it is as easy to divide and multiply as it is to add and subtract. One must admit, however, that it is not as quick. To accumulate addition after addition, or to compare remainders to find one smaller than the divisor, takes considerable time. People who use algebraic expressions as a basic language need two or three years in a specialized school to become good *soroban* operators. So, in eastern Europe as in the Far East, we can hardly understand at first why a person who sells us ten different articles of ten different weights and at ten different prices rejects the electric adding machine at his disposal in favor of the abacus, although he uses it at quite an incredible speed.

II. Technology of Construction

The pioneers, the first technologists who were responsible for the construction of buildings, literally risked their lives and those of their children, if we are to believe the Code of Hammurabi. Even much later, building technologists used for their calculations merely abacus-like instruments, which were the only ones at their disposal, and these, as we have shown, were not in their present sophisticated form. We must admit, then, that in those early times there were no calculations, and that trial-and-error empiricism took their place. It was not the fashion to calculate, but only to build, just as it was not the fashion to theorize or measure in medicine, but only to heal. Yet people did trepanning just as they do now. The result was a great variation in the percentage of successes, both in medicine, I imagine, and in building.

The largest known brick arch without reinforcement in the world, which has a span of forty-five meters, is in the palace of Ctesiphon in Mesopotamia. It is still there after more than 1,500 years, built by the Sassanid emperors when Rome was falling, at a time and place where not even abaci existed. Apparently the builders did not know how to calculate the stresses on this arch. We do not know how many arches like this one fell down, but at least there is one still standing. Indeed, one might say it was fashionable then to build without the benefit of calculation, but some of the results were good.

Beams, which are the forms nature provides for man when he has exhausted all the possibilities of the vault, began with megalithic false cupolas. These were frequently used throughout the ancient Mediterranean world, but in the light of experience rather than mathematical theory. When stone claimed its place as being more durable than wood, it copied the methods of wood in its layout and in its decoration. Any Greek architectural design in stone reproduced what had earlier been made in wood. When these possibilities were exhausted, the architect was forced—intuitively at first and from a technical point of view later—to change the form and to invent, or to reinvent, the vault. These first steps away from intuition lost their impetus in the course of time, however, as attempts were successfully made to formulate knowledge through calculation in order to predict, before building an element, how it would withstand the load.

Recently, Spanish television has produced a series of broadcasts on the life and works of Leonardo da Vinci. There was a full account of his work: drawings, inventions, and writings. We were shown examples beyond the grasp of even a specialist in the field. Did Leonardo make calculations? The answer is that he was able to work out the geometry of his inventions but not their capacity to withstand strains and stresses. His was a compound of intuition and experience, which are almost one and the same thing. There cannot be intuition without a large dose of living experience, and *living* experience it must be, for nothing is more difficult to pass on than experience.

The first people to formulate constructive data—much of it

more destructive than constructive—were army engineers. The first printed treatise on technology, the *De re militari,* was published by Valturio in Verona in 1472. But, like Leonardo da Vinci, who was the first engineer in the modern sense of the world, these army engineers either did not use calculations or did not write them down. It was really not fashionable to calculate.

Renaissance fashion required brilliance in dress, in bearing, in painting, in one's way of life. One had to be brilliant, ingenious, witty, and bold. Clearly, calculation was not a field in which one could shine in any of these ways. As a result, architecture and engineering, which at that time were not separate sciences as they are considered to be today, reflected this fashion of brilliant ideas. Reflecting fashion, architecture in its turn influenced the nature and the character of the people. Architecture and mores at once mirrored and influenced each other in an uninterrupted series of interactions.

Let us take as an example the lives of the characters in the *Decameron,* which everyone was reading then. Many of the intrigues recounted in the novel would be inconceivable had the design of the houses in which they occurred been the same as it is today. The walls of those medieval and Renaissance houses were broken only by small windows and perhaps by the beginnings of a loggia. When the light of day faded, such houses could become the scene of all kinds of deceptions. In one story from the *Decameron,* although the confused king thinks he is in bed with his wife, the wife is in bed with another man who thinks the king is the queen and that she is the king. Such a misunderstanding couldn't happen in houses such as we have today. This is really a serious point! Our building technology allows us to flood a house with as much light as we wish to have. If Renaissance technology had permitted building such houses, then the *Decameron* could not have been written, for of course the habits of people at that time would not have been the same. In reality, the lack of calculation in that era was not the final cause either of the dark houses or of the deceptions which took place there. Precisely because calculation did not exist, however, to a certain extent all that did happen was made possible. All this gave a highly brilliant and violent character to a people who could not commit themselves to learning to calculate.

Benvenuto Cellini, that wonderful goldsmith, sculptor, and artist of the first order, was a man of tremendous sensitivity. Nevertheless, during his not very long life he personally killed fourteen people, and he writes about it in his memoirs with a certain pride. Let us see why he killed at least one of these people, and the consequences of it. He says:

> But my only wish was to watch for him continuously in the same way as if he were my beloved, the musketeer that killed my brother—only not with love but with hate. I noticed that seeing him so often was affecting my ability to sleep and eat, and I was in a very bad way. So I decided to get rid of this sorrow without thinking about any culpability in my plan. I stole up behind him skillfully with a big hunting knife. . . . I intended to take his head off with the first blow, but he turned back so quickly that my weapon only reached him in the left shoulder, and the bone was broken. He got up very quickly, staggered, dropped his sword and, stunned by the great pain, tried to run away. I ran after him more quickly and reached him in four steps. I held my knife over his head. . . .

and so on, with a full description of his victim's death. Afterwards, Cellini was called in by the Pope, not to be put in jail but only to be told, "My boy, please don't do it again." Cellini, who certainly did not expect to be punished, had already decided to soften the anticipated scolding by bringing the Pope a wonderful piece of jewelry, which is now preserved in a museum somewhere. This is the ambience in which some of the marvelous works of the Renaissance were created! A man like Cellini clearly could not dedicate himself to formulating methods of calculation, because he lacked the basis from which to begin.

In the last years of the Renaissance, at the end of the sixteenth century, a change in attitude occurred. There arose a desire to systematize which led to the habit of studying a theme in depth, abandoning the universality which had marked the previous three magic centuries of the Renaissance. Palladio was so immersed in the period that he was a disciple of Michelangelo. Between his own neat studies of architectural styles and the untidy notes of Leonardo, on subjects which range from anatomy to cannons of unusual design and form (as he describes them), there was all the difference between the Renaissance and the Baroque reaction which was to come.

In the first years of the seventeenth century, Galileo marked the path to specialization. In 1638, Galileo wrote a treatise on beams, although he used only a static concept of forces. He introduced the idea of moment, but not yet that of elasticity. This was because physics was still not really in fashion; indeed, even its name was still applied to the rudiments of medicine. The Renaissance was not long past. Educated people continued to write in Latin about anything from sheer imagination to hard practical facts. It took another two centuries for people to realize that in some ways it was better to study any branch of science in depth and in detail rather than in general terms.

Nevertheless, in the Baroque period construction technology began to make so much progress that it outstripped the scientific knowledge available. This was either because fashion was beginning to take up technology or because these first steps in technology were made very rapidly and this rapidity made it fashionable.

Outward appearances were changing, too. In Spain, the typical sobriety of Spanish taste was making itself felt. Architecture and engineering became at the same time more complicated and more simplified. In Spain the Plateresque and the Herreriano styles existed side by side, both a little later than their counterparts that were flooding the world from Italy.

The spirit of the age was also changing the conception of science. In 1678 Hooke published his *De potentia restitutiva.* In that volume appeared Hooke's Law, the famous principle of proportion, which has been reformulated but has remained the basis of all studies in structural analysis until very recently. Very soon after that, other principles appeared which extended Hooke's idea. Between 1680 and 1690 Mariotte and Leibnitz noted, among the possibilities of the new science, the study of the elasticity of a loaded beam, but it was Bernoulli in 1705 and, more especially, Euler in 1744 who succeeded in solving the problem, starting from an intuitive idea. Bear in mind what I said earlier about the debt of intuition to experience. In that same year, 1744, Euler formulated what we know as the theory of Euler-Bernoulli: moment is inversely proportional to the radius of curvature of the deformed beam.

Meanwhile, what was happening to the other sciences? They

were being outstripped by technology, at least in some aspects. Progress was not uniform. Science was not in fashion but technology was, although the sciences were trying to catch up. We can get an idea of the situation—a caricature, perhaps, but very revealing—from the speech that Montesquieu made in 1720 to the Academy of Sciences in Bordeaux, at a time when academies were beginning to be all the rage. Speaking of the transparency of solid bodies, Montesquieu said: "The majority of moderns believe that transparency is the effect of the straightness of the pores through which, in their opinion, light can be easily transmitted. . . . In my opinion, there cannot possibly exist in the world matter so condensed that it does not allow light globules to pass through." There follow such picturesque ideas as that some light always passes through every solid body, for "there exist animals who are able to see through a wall." This, then, was the level of understanding of light and transparency at a time when, in the field of calculation of structure, people were already formulating principles that are still valid today.

The development of calculating instruments had not made much progress. True, Napier had introduced his first rudimentary slide rule to facilitate the process of multiplication, which was then a considerable problem. Later he introduced logarithms, which Euler developed and applied. For ordinary people, however, slide rules appeared much later; calculation had to be done step by step by men who were almost specialists in the field. Even in 1650 an educated Englishman admitted that he was fiinding it very difficult to learn to multiply.

Specialization was beginning to bear fruit, both for good and for evil. Now one could study ideas in depth. The all-round man had fallen out of fashion. It was accepted that a man of the world might be well versed in financial problems without having heard of Bach or Voltaire. *Les Comptes rendus* and *The Philosophical Magazine* produced editions specializing in very concrete ideas. Between 1825 and 1875 all the basic theorems of Menabrea, Betti, and Maxwell were formulated and proved. Engineers were the ones working on structural analysis. Although mathematicians provided them with their framework and physicists with information about materials, methods of calculation were in the hands of specialists in building.

Let us look back to examine the pace at which changes occurred in the way a problem was tackled. Fashions have been changing more and more rapidly. The three centuries of the Renaissance, the century and a half of the Baroque, the century up to 1900 with the theorems of Euler, and the two generations up to the present day show the accelerating pace. At the end of the last century, technological evolution in our Western world began to advance like a tornado.

One can date a Renaissance dress or house to within a century. In the present age changes in style are more frequent. A specialist can fix the date of a photograph to within less than a year, merely by studying the dress of the women. Changes in the field of architecture also have taken place at an increasing rate so that in less than a century there has been a change from structural honesty to expression of the architect's emotion, and finally to a reaction against this.

Until 1875 the problem of hyperstatic frames was not even considered. Suddenly, from then on, methods were formulated which, with certain simplifications or different approaches to the basic equations, could theoretically tackle the most complicated frames. Equilibrium of the nodes of highly hyperstatic frames is the basis of all present construction methods. The almost impossibly difficult task was to solve, step by step with slide rules or mechanical calculators, the systems of the resulting equations. Approximate methods and simplifications began to appear. In 1930 Hardy Cross and Morgan began to synthesize the existing material on approximated methods and published their method of successive iterations. From then on analogous methods proliferated. Changes in fashion in this field were as swift as in any other. Philosophy and religion accepted the incredible changes in fashion very quickly. The way of life imposed these changes and in turn was a result of the rapid change. And, for the first time, fashionable methods of calculation were determined by the instruments available to execute them.

III. Relation of Calculation to Construction Today

Now we have arrived at the present day, the age of the computer. The last twenty-five years have been characterized by this machine. Although it was introduced as an aid in mathematics, it was immediately snapped up by technology, which has been decisively influenced by it. I shall primarily refer here to different generations of digital computers, leaving for another time any discussion of analog machines, which need special treatment. These marvelous instruments are a symbol of our time.

In the modern period there have been no new theories on our subject since the time of Hooke. I shall not discuss all the modifications of Cross's method because they have all been replaced by methods that can be used by a computer.

The development of more accurate measurement has shown that the behavior of concrete is not elastic, and we are beginning to use a complex definition of visco-elasto-plastic material. This development enables us to see that in steel there is a reserve capacity when the elastic limit of the material has been exceeded. All this leads to methods of calculation of efforts which are not relevant to our discussion.

Baker, Guyon, Ferry-Borges, and many other research workers are trying to systematize the use of those advantages which plastic design will eventually bring. Still, apart from some short and almost empirical rules which fundamentally affect the permissible redistributions, we have not developed any practical new methods comparable to those based on the elastic theories of 1678. Concrete poses a curious problem. We calculate the total by methods which allow for the elasticity of the material in principle, and yet we calculate sections with formulas deduced from the nonelastic behavior of the material. This illogical incongruity is tolerated for its economic advantages.

General use of the computer in the past twenty-five years has forced a return to methods of calculation which had been discarded. Equilibrium of the node was discarded as a practical method in 1890, yet that is what the computer uses now, replacing innumerable approximated methods and simplifications which are no longer necessary. In papers presented at conferences and published in specialized journals, we seem to be going backward. This is true, but it is in order to gather momentum and to forge on ahead.

For example, in designing cylindrical shells constructed in rolled steel sections, like those that roof the workshops of the

I.E.T.C.C., Florencio del Pozo used what in that period was an advanced technique, which lasted quite a number of years. Elastic characteristics were determined which allowed a discontinuous structure to be assimilated to a continuous one and to be treated as such, and thus to be calculated by the approximated methods used in the treatment of a cylindrical shell. Even as late as 1962 there was a congress in Paris on studies of this kind.

Because of the existence of the digital computer, the present-day approach has changed radically. Recent conferences present studies in which it has become fashionable to treat continuous shells as if they were made up of discontinuous elements, so that a digital computer can be used. We have been doing this in Spain for some years.

This is an indication of the modification and inversion that the techniques of calculation are undergoing. Before, the problem was to determine elastic constants which would allow the assimilation of the discontinuous to the continuous. Now we try to find elastic constants which allow us to replace the continuous elements by discontinuous elements which reproduce the behavior of our continuous elements.

Invasions of new ideas follow one another more rapidly now. In the fourth century the barbarian invasion took a century to complete. It seems that we are now on the threshold of a new fourth century. Now it is not Diocletian at the center of the Western world; there are many Wests. If there is to be another invasion seventeen centuries later (remember the Sphinx), it will certainly not take a century to complete.

The invasion of the technique based on finite elements is making very fast headway, and it seems that this technique will shortly replace the one just discussed. Of the present methods of calculation for use in the computer it is, without doubt, the most promising because of its universality.

One method of calculation cannot be superseded in its universality: model analysis. In the last twenty-five years this has been devised and perfected to produce really important practical results. Methods of measurement which are more precise and more widely applicable, new materials, and, once again, use of the computer to process information obtained from the model: all these make this new system the most compatible with the growing freedom of form of the present Baroque, which is a reaction against the immediately preceding structural expressionism.

Laboratory work plays an important part in the evolution of methods of calculation. Routine tests, basic research on proposed materials, the study of works already completed, tests on scale models, and so on—all are needed.

How much do European countries spend on structural engineering research? According to recent figures supplied by Rocha of the Civil Engineering Laboratory in Lisbon, it is, in the United Kingdom and in Portugal, 0.5 percent of what is spent on construction. Note that this is the sum spent on engineering, not on architecture, which actually spends very little on its own research. This sum is spent to guarantee financial returns, without any concessions to dilettantism.

Now I am going to look at the last few years in our field to try to forecast what the future holds. Some seven years ago at M.I.T., my friend J. M. Roësset developed the first program to connect a model of a structure with a computer, trying to get data from the model by passing it through the computer. Only two years later Hossdorf, at his laboratory in Basel, solved the problem by turning it inside out: he connected the computer to the model first, before starting to load the model. That is to say, the computer controls the process of loading and measurement, processing the information in the form of influence surface for particular forces and sections. It does not measure the tensions in the entire structure for general loading, but it measures a certain magnitude that may be a deflection at a certain point when the unit load moves over the whole structure. Memory storage of the data of the matrices of influence is a problem of capacity, and retrieval of these data to combine them lineally takes only a moment. The results of Hossdorf's method, of course, were far more useful, accurate, and indeed far more economical. And this development took only two years. This seems to be a very promising line of thought. This may be the frontier toward which we are moving, the direction in which we ought to go.

In one way or another, computers already define our way of life, and it seems certain that they will do so more and more in the future. I cannot believe but cannot forget Asimov's fifteen-year-old science fiction prophecy of "Multivac" as an instrument which governs the United States of America. By asking only three questions, in what seems to us a wild game of chance which the computer programs and carries out, the computer chooses the President of the United States. Although the story is absurd, I cannot help feeling some fear when I consider the kind of life the next two generations will live. However, man's ability to adjust is enormous, and he will perhaps be able to endure it without undue stress. Or will he? As long as we do not allow the computer to totally control us, and as long as we recognize that it is governed by that super-computer, man's brain, we shall not arrive at Orwell's fictitious year 1984.

IV. Riding the Tornado

What can we do to survive rapid change? That is the question that brought me here with extremely high hopes of learning from all of you. We must do something to help our students prepare themselves to withstand the tremendous and increasing velocity of change from which we are suffering now.

Four or five years ago we used one method of structural analysis. Now after ony five years' time we must change to other methods. This is a very narrow field of technology, of course, but I believe the problem is the same in all fields of technology. I simply do not know what to do to help students; I am absolutely sincere in this.

We are trying to rewrite our curricula in all the universities of Spain. In the technical faculties—civil engineering and architecture—we are trying to teach achitecture primarily as a technical rather than as an artistic question.

You can see that the possibility of getting our minds flexible enough to accept increasing velocity of change without trauma is something of very deep concern to me, and I really do not know how to handle it. My conclusion is just a question, What can we do?

Meetings like this one can open a little window through the fog surrounding us, clouds which really make it impossible to see our direction. Looking at the future, it appears very clear, at least to me, what we must do; unfortunately, I do not see how to do it. We must prepare ourselves for an increasing rate of change. We must become flexible enough to adapt our abilities, without excessive violence, to the new ways of life that "fashion" will impose on us. To survive, we must succeed in reaching a velocity of adaptation at least equal to the velocity of change. If we cannot adapt, then through our lack of happiness we will disappear.

Heinz Von Foerster

Where Do We Go from Here?

I am most grateful to the organizers of this symposium for injecting into the topics to be discussed concerning the philosophy of technology just enough ambiguity to stimulate rather than stifle the imagination of the discussants. One thing, however, seems to be clear. In posing the question "Where do we go from here?" the organizers ask us to polish our crystal balls and tell all we see lying ahead. Of two interpretations of our topic, either "the future philosophy of technology" or else "the philosophy of future technology," I have chosen to discuss the latter. Depending on where the accent (1, 2, 3, 4) is placed in the question "Where [1] do [2] we [3] go [2] from here [4]?" Four different contexts are created, each of which calls for different answers.

Where?

This is a request to be shown a direction. If we take it as a spatial metaphor, the three spatial coordinates give us three options. If the first, (1) up or down, is interpreted either theologically or economically the answer is clearly "up." However, we are not free to choose independently the directions in these two interpretations, for they are interrelated according to a theorem that has been with us for almost two thousand years: "It is easier for a camel to go through the eye of a needle, than for a rich man to enter into the kingdom of God." [1] Thus, if we wish to go up economically, we shall go down theologically, and vice versa. We are faced with a similar conflict if we turn to another option open to us, (2) left or right. Of course, we cannot go left, for nobody wants to be considered a communist, nor can we go right, for there we would embrace conservativism, which would not allow us to go at all. Alas, we are not a bit better off in our last option (3) forward or backward, for "forward" means "progress" and "progress" implies more industrialization, pollution, and dehumanization. We know we cannot have more of that. For those few who wish to follow Rousseau's call for a *"retour à la nature"* and want to go backward, there is no nature left to return to. These are the dilemmas of the "Where?"

Go

Sociologists who study the present state of affairs assure us that they can identify trends in today's changes of sociocultural patterns; they assure us that we do indeed "go." However, in the absence of adequate conceptual models with at least some predictive power for "pattern," "trend," "change," and so on, sociologists have gleaned from other disciplines, notably the natural sciences, some models whose predictive powers rest on principles of *constancy* and *permanence* (e.g., inertia, conservation of energy, conservation of momentum, etc.), which are, ironically, in diametrical opposition to the purpose for which they are now used, namely, to predict *change.* Consequently, with the aid of these models the future is given in terms of the past with only quantitative variations of otherwise unaltered qualities: *faster* cars, *wider* highways, *more* people, and *bigger* bombs. Nevertheless, "futurology" has become a lucrative business for entrepreneurs who sell these "future scenarios" to corporations which profit from designing for obsolescence.

Most astounding in this context is the unshakable belief of demographers [2] and others [3] in the exponential growth of the human population, that is, in the invariance of relative growth; for this hypothesis does not even work for the past! With a world population of $N = 3.78 \times 10^9$ souls and a doubling time of thirty-five years in 1972 [4] we can easily conclude, believing this doubling time to be constant, that Adam and Eve ($N = 2$) were created exactly 1080 years ago, that is, in the year 893 A.D.

It is, I hope, clear that I wish to disassociate myself from this nonsense, and I shall do this by paraphrasing the original question "Where *do* we go?" to read "Where *shall* we go?" That is, let us turn the question "Where are we pushed to?" into "Where do we wish to be?"

We

Questions containing the plural "we" of the reflexive personal pronoun "I" are suspect of being purely rhetorical. When a mother asks her child, "Shall we go to the bathroom?" she means, of course, that only the child is supposed to go to the bathroom. I wish to remove this suspicion by taking seriously the "we" in the organizers' original question. When I say "we," I do not refer to some others; I shall refer precisely to us, to you as well as to me.

Here

Even if we believe that the affairs of man are controlled by a set of partial differential equations and, moreover, even if we assume we know this set of equations, we cannot say anything about the future unless we also know the initial conditions. Consequently, if we wish to answer the question "Where shall we go from here?" we have to know the "here"; we have to perceive where we are today. Before considering this point, I have to define a few terms and concepts which may facilitate this perception.

Criteria

The roots of modern psychiatry are very young indeed. It was only slightly over one hundred years ago that Kahlbaum [5] first perceived mental diseases as functional disorders and attempted the monumental task of aiding diagnosis by classifying in the perplexing mixture of fluctuating symptoms stable criteria of functional contingencies. This brought to an end an era of superficial symptomatology which had provided no clues for therapies, and it opened the path for a diagnostics which later allowed neurological interpretations of those identified dysfunctions. Building on this school of thought, Kraepelin [6] was the first to use the notion of "sensorium" as the totality of the faculties of perception, orientation, memory, and so on as distinct from those of reasoning, volition, affectivity, etc. The concept of the sensorium being either clouded or clear, with other circumstances alike, allows for powerful diagnostic discrimination and hence provided an important conceptual tool to determine the etiology and, ultimately, the therapy of the disorder.

According to our eminent colleagues L. J. Meduna and the late W. S. McCulloch,[7] it was Bleuler,[8] however, who in 1911 first laid down the essential criteria for diagnosing schizophrenia. They are known as "Bleuler's Three Criteria" and consist of observing the absence of one disturbance in the presence of two others. To these Meduna and McCulloch added a fourth to single out schizophrenia from other disorders with other etiologies and appropriate therapies. I shall give a very brief summary of these four criteria:

1. *Breakdown of Cognitive Integration*

   Patients develop single-tracked trains of thought within a highly compartmentalized framework of topics with an increased loss of the ability to connect these topics through contextual links. Hand in hand with this narrowing of the cognitive aperture goes an impoverishment of the semantic relational structure leading to the well-known "schizophrenic speech pattern" (excessively high frequencies of normally rare words and excessively low frequencies of normally frequent words)[9] and a dramatic contraction of an otherwise rich choice structure which may lead (e.g., in the suicidal manic depressive) to states of only two choices (e.g., "I either must have a Brazilian cigar or else I will jump out of the window").

2. *Loss of Affectivity*

   The I-Thou identity is lost. In spite of their former attachment to parents, lovers, children, or friends, these patients not only lose all affection and interest in other human beings, but in progressive states of their affliction come to see others as being threats from which they unsuccessfully seek refuge in a solitude that is tormented by threatening imageries and hallucinations.

3. *Sensorium Clear*

   There is no lack of clarity of perception in these patients, nor do they lose their sense of orientation, their acuity of visual or auditory discrimination, etc. This unperturbed sensorium in conjunction with the first criterion of narrow singlemindedness prompted Warren McCulloch to observe, "If you want infallible deductions, give the problem to a schizophrenic—but watch his premises."

4. *Confusion of Symbol with Object*

While in some magic rituals the symbol of an object may be taken as the object (ju-ju), in the schizophrenic this is a consistent logical paradigm. A ten-year-old patient asked for the product of 5 x 5 answers, "It has a kitchen, a living room, two bedrooms, and is painted white." The logic is clear if it is known that he lives at 25 Main Street.

This concludes my compressed account of the four criteria for the diagnosis of a complex and profound disturbance, schizophrenia. The injustice I have done to the subtleties of some of the points can, I hope, be recovered in a more extensive study of the technical literature. Now I owe you a justification for my lengthy presentation of a topic that appears to be completely unrelated to the task at hand, namely, to discuss the philosophy of future technology. However, as I pointed out earlier, if we want to know where to go, we have to know where we are. This clinical excursion demonstrates where we are. We are, with our technology, in a state of schizophrenia.

I shall now proceed to prove this outrageous proposition by applying Bleuler's three criteria with the fourth by Meduna-McCulloch to the present state of our technological affairs.

Schizophrenia

1. *Breakdown of Cognitive Integration*

Increased compartmentalization and pressures toward higher specialization are the answers of our institutes of higher education to an acutely sensed necessity for more interdisciplinary cooperation and to a growing desire by today's students for nondepartmentally oriented curricula. The latest national trend—with some valiant exceptions—is a reduction of interdepartmental student mobility (credits accumulated in one department or school will not be counted in another department or school), and the height of the walls surrounding individual departments is constantly being increased instead of lowered. The popular excuse that mass education is incommensurate with yielding to individual whims is, of course, ridiculous in the age of giant data processors, and it is in itself a symptom of the disorder I am talking about, because the members of the departments of education know little about the potentialities of these machines and the members of the departments of computer science know little about the cognitive processes of knowledge acquisition.

Looking beyond the educational scene, the process of cognitive disintegration becomes even more dramatic and assumes global dimensions. Everybody knows that the world resources, V, of crude oil, discovered and discoverable, will under present rates of consumption, dV/dt, last for no more than forty-three years, but under the present rate of change of the rate of consumption, d^2V/dt^2, for less than twenty-three years.[10] This means that in one generation almost the entire micro- and macrostructure of our sociocultural metabolism will abruptly collapse. Without initiating today appropriate measures for a viable transition into an era of new metabolites, mankind will stand *vis à vis de rien*. What are the measures taken today? We sacrifice Alaska's ecology forever by piping oil from the North Slope to the deep sea port of Valdez to be shipped to Japan for Japanese consumption (it's good business); we resign ourselves to the whims of Arabian sheiks (it's good business); from Detroit comes the revelation that they are going to build larger and more expensive cars (it's good business).

Hand in hand with refusal or incapacity to establish contextual connections between perceptions through different social sensory modalities or "information channels," the speech pattern of spokesmen in official positions becomes progressively more schizophrenic. What follows is an abbreviated dictionary of emerging linguistic morbidities, "Officialese," with translations into English given in parentheses: [11] communist (one with whom I disagree); fascism (obstacle); food denial (poisoning men, beasts, and plants); I am not sure whether I was or not—I may very well have been (I was); incursion (invasion); inoperational statement (lie); liberation (complete destruction); pacification (complete destruction); protective reaction (aggression); zero-defect system (perfection).

Finally, the choice structure of publically perceived issues is undergoing at present a dramatic contraction so complete that it polarizes alternatives which are not mutually exclusive, but which are, nevertheless, seen as exclusive ORS: "Either economic growth or sociocultural stagnation"; "Either pollution or unemployment"; and so forth. It is not realized that we can have in these pairs of alternatives both, neither, or—beyond those—a multitude of other alternatives.

2. *Loss of Affectivity*

Science and technology share their complementarity with religion and magic. Religion is, so to say, theoretical magic, and magic, applied religion. Thus, when religious dogmas are extended beyond sustainable limits their magic begins to fail, and likewise, when science gets into trouble these troubles will be reflected in corresponding dysfunctions of the technology.

Should my analogy be rejected because of the popular belief in a fundamental distinction between the conceptual structures of religion and science, let me remind you of the progressive "esoterization" of some scientific propositions. Consider for a moment three paraphrases, each about a half century apart from the other, of perhaps the most profound law in the physical sciences, the Second Law of Thermodynamics. Here is the first formulation by Clausius (1882-88): "It is impossible that, at the end of a cycle of changes, heat has been transferred from a colder to a hotter body without at the same time converting a certain amount of work into heat." Clearly, here is an engineer talking, giving us limits of conversion of heat into work. How was the same law pronounced about fifty years later by Boltzman (1844-1906)? "For an adiabatically inclosed system the entropy can never decrease." For the uninitiated who is not completely familiar with the concept of entropy, it appears from this version that an adiabatic enclosure is most likely made of some kind of plastic stuff that is impermeable to entropy so that it cannot escape once trapped. But, peculiarly, this proposition does not specifically state that the entropy in this closed bag cannot increase as well. Hence, that plastic stuff must be semipermeable, letting entropy come in from the outside but prohibiting it from escaping when inside. However, when he is told that "a-dia-batic" means "not-through-going" in either direction, for anything, and that despite this, the entropy may still increase in that closed system, "entropy" acquires a truly magical status; such is the strength of our predilection for permanence, invariance, and constancy. Let us now look at the statement by Planck (1858-1947) thirty years later: "A perpetual motion machine (PMM) of the second kind is impossible." Let us not worry about whether a PMM of the second or the first kind is impossible. Let us worry about a PMM of any kind. Is the proposition "A PMM (of the nth kind) is impossible" a scientifically legitimate proposition? Certainly not, for we may have to wait infinitely long to disprove it.[12] It is not the *kind* but the *concept* of PMM that is impossible in the sense that it is scientifically unacceptable. It is significant that despite this deficiency this concept is fully embraced by the scientific community.

It is not these semantic hair-splittings that prompted me to suggest that science is in trouble, nor is it the waning public esteem for scientists nor the withdrawal of funds for basic research. What prompts me to say that science is in trouble is that scientists themselves have begun to sense that there is something fundamentally wrong with science.

Unfortunately, these doubts and reflections rarely reach the surface, for scientists have been conditioned to believe that they may competently talk *only* in terms of the speciality of which they are known to be competent, but *not* about what they are doing, or what it is all about. If, however, a scientist indeed talks about how his activity is supposed to fit into a more general cultural, social, and human context, he is immediately suspect of transgressing his competence. He may have an opinion. And, you know, an opinion is a most dangerous thing: it could be an original thought!

In some instances opinions are tolerated with benign condescension. When great scientists with Nobel Prizes to their names have reached their seventieth birthday and reflect upon their thoughts, motivation, and activities, people say, "Well, well, he has had his productive phase; what else could he do now?"

However, the pressure of increasing discomfort and the feeling that "there is something wrong with science" compel more and more scientists to air their doubts and reservations about the "scientific method." I have recently made it my pastime to observe the steady rise in the number of articles *about* science that appear in two important scientific journals, *Nature* and *Science*. While there was in the late forties and early fifties a slow but constant trickle of about one article every three years, this suddenly changed in the late fifties to two and in the early seventies to eight per year, with an apparent doubling time of three years at present. While the titles of these articles—for instance, "The Science Crisis," "Science and Values," "The Significance of Science"—give some clues about their main concerns, one has to read these papers in order to see in which context scientific activity is perceived. I cite one example from an eminent physicist and infinitely sensitive human being (emphasis mine): "Science is based on . . . [a] desire to improve the precarious conditions of human existence in a *hostile* world, in a *hostile* natural environment, and in *hostile* societies." [13]

Pygmies in the tropical rain forests of the Congo [14] or Bushmen of the Kalahari desert [15] would not know what this white man is talking about. To perceive the world around them as "hostile" would make as much sense to them as to think of their eyes as hostile, or their arms or legs. The world of their natural environment is beauty and plenty, food growing on trees and walking or crawling on earth on two, four, or six legs—truly a "*Tierra Generosa*." This in fact is the name of a Mexican village that I "discovered" as one discovers an orchid in the lush vegetation of the forests along the Pacific coast. If these people knew of the term *paranoia*, they would have called the perspective quoted above "the white man's paranoia."

The interpretation of our relations with others or with our surroundings in a frame of hostility rather than of affectivity and also the proscription of talking about oneself have some deeper roots. They are rooted in the fundamental belief that a prerequisite for scientific discourse is to obey this commandment, "The properties of the observer shall not enter the description of his observations." This prerequisite is generally seen as the crucial condition for a statement to be scientific, or for "objectivity."

This attitude manifests itself in a style of science writing in which the author is squeezed out: "*It* can easily be shown" or "*One* observes," rather than "*I* could easily show" or ""*You* may observe, if you care to." It is visible also in the grand scheme of natural science to squeeze all subjects out of its models of the world in order to create a "subjectless universe."

It is usually thought that objectivity arises when two observers agree, but I claim that when two observers agree, they have just discovered that they have some properties in common and, conversely, when they disagree, they have just learned that they have some properties which are not in common. I am sure you will agree with me that it is syntactically and semantically correct to say that subjective statements are made by subjects. Thus, correspondingly, we may say that objective statements are made by objects. It is too bad that objects don't make any statements.

The trouble with science, I submit, is the delusion of being able to make objective statements, that is, statements that are observer independent. This delusion forces us not only to lose sight of the other, but even to undo him should more of him become visible than our cognitive blind spot usually obscures.

How does this loss of affectivity in science project itself into today's technology? I shall give only one example, whose structure, however, will allow you to generate as many more as you wish. We Americans pride ourselves—and I think justly so—on being the people with the best know-how in the world. Why are we at the same time the people who produce the most monstrous, impractical, inefficient, and most dehumanized automobiles? This is so because the goal of our great car manufacturers is not to manufacture cars but to manufacture plants that manufacture cars. Each year plants are produced, most likely very efficiently and practically, for producing the models to be sold in the following year. That is, the attention to a commodity that is to be used by people is now shifted to the plant that is to produce some commodity, and in this process the users of this commodity are forgotten.

I may be accused of untimely naiveté for suggesting that the goal of the auto industry is making plants, while *the* goal is, of course, "making" money. However, this mental disorder is referred to as "kleptomania," and I am addressing myself to schizophrenia.

3. Sensorium Clear

I don't think I have to dwell on the last of Bleuler's three criteria, which postulates the absence of one disturbance, a clouded sensorium, in the presence of the two disturbances of above. It is, I believe, obvious to all who care to look that our technology proceeds flawlessly and with utter precision along its path of cognitive disintegration and of mounting unconcern for human values. Let me, instead, mention the last of our four criteria, the Meduna-McCulloch criterion.

4. Confusion of Symbol with Object

Symbols and tools have in common that each has a purpose other than itself. Symbols denote objects and connote concepts; [16] tools are designed to fulfill a particular function as means to an end. When attention is shifted from the car to the plant, there is a confusion of means with ends, and when the symbol (e.g., of "status") takes over the function (e.g., of safe locomotion), a student of mental disorders will become alert and will test for other criteria.

With this I have concluded my *tour de force* of a psychopathology of the state of technology now or, in terms of the initial question, of the "here." Since up or down, left or right, forwards or backwards is denied to us, where indeed shall we go?

With nowhere to go in the outside, I propose to try for a while to turn inward. Instead of using proof as paradigm for explaining my cryptic proposal, I will use a metaphor. It is more than 2400 years old, and is told by the master of Tao, Chuang Tse.[17]

THE INSTRUMENT BUILDER

Ching, the master of instrument builders, made a harp. When he had finished his work, it was of such perfection that all who saw and heard it were convinced he must have had unearthly helpers. The prince of Lu asked the master, "What is the secret of your art?"

"I am only a handicraftsman," said Ching. "What secrets could there be to that? And yet, there is something. At the time when I decided to build the harp, I started to collect, so to say, the fibers of my strength, the calmness of my mind. After three days I had forgotten about money, after three weeks about the fame that would accrue to me. My limbs, my own body slowly vanished from my thoughts, as did you, my prince, and your court who had commissioned me.

"When I felt that I and my art had become one and all, I went into the woods. Had I not found the tree with the right form, my endeavor would have been in vain. But then I saw the tree whose heavenly form corresponded to my heavenly thoughts, and I knew that my work could succeed; so I carried it out.

"One could say that the collection of my strength, the abandonment of myself and the world, these were the 'unearthly helpers' on my way inwards where I found the true vision of my goal."

Notes

1. St. Matthew, 19:24.
2. P. H. Hauser, "Implications of Population Trends for the Military," in *Science in the Sixties*, D. L. Arm, ed. (Albuquerque: University of New Mexico Press, 1965), pp. 42-57.
3. D. H. Meadows, D. L. Meadows, J. Randers, and W. W. Behrens III, *The Limits to Growth* (Washington, D.C.: Potomac Associates, 1973).
4. *1972 World Population Data Sheet* (Washington, D.C.: Population Reference Bureau, 1973).
5. K. Kahlbaum, *Die Gruppierung der Psychischen Krankheiten und die Einteilung der Seelenstörungen* (Danzig: 1863).
6. E. Kraepelin, *Psychiatrie*, 2nd ed. (Leipzig: Abel, 1887).
7. L. J. Meduna and W. S. McCulloch, "The Modern Concept of Schizophrenia," in *The Medical Clinics of North America, Chicago Number* (Philadelphia: W. B. Saunders, 1945), pp. 147-64.
8. E. Bleuler, "Dementia Praecox oder Gruppe der Schizophrenien," in *Handbuch der Psychiatrie*, Franz Deuticke, ed. (Leipzig: 1911).
9. G. K. Zipf, *Human Behavior and the Principle of Least Effort* (Cambridge, Mass.: Addison-Wesley, 1949).
10. M. K. Hubbert, "Energy Resources," in *Resources and Man*, Committee on Resources and Man, National Academy of Sciences, ed. (San Francisco: Freeman, 1969), pp. 157-241.
11. H. Von Foerster, "Perception of the Future and the Future of Perception," *Instructional Science*, vol. 1, no. 1 (1972), 31-43; S. Kanfer, "Words from Watergate," *Time* (Aug. 13, 1973), p. 20.
12. K. R. Popper, *The Logic of Scientific Discovery* (New York: Basic Books, 1959).
13. V. F. Weisskopf, "The Significance of Science," *Science*, vol. 176 (1972), 138-46.

14. C. M. Turnbull, *The Forest People* (New York: Simon and Schuster, 1961).
15. J. Marshall, film, *The Hunters* (Chicago: Films, Inc., 1970).
16. S. Langer, *Philosophy in a New Key* (New York: Mentor, 1962).
17. M. Buber, "Der Glockenspielständer," in *Reden und Gleichnisse des Tschuang Tse,* trans. M. von Foerster (Leipzig: Inselverlag, 1921), p. 69.

Iraj Zandi

Is There Anyone Else?

There are two reasons for an engineer to mingle with historians and philosophers. The first is personal: to savor "philosophy's dear delight" is intoxicating. It is good for engineers, who in the day-to-day execution of their profession fragment knowledge in search of facts, to stop and view life from another perspective in fullfillment of their desire to gain understanding and wisdom.

The second motivation is professional. For some time now the profession of engineering and its offspring, technology, have been under attack by many learned social observers of high credibility as being the root of many social ills. The most awesome accusation leveled against technology by these social critics is the assertion that technology subverts human values, culminating in a loss of freedom, and will create, finally, a monolithic world culture in which all "non-technological difference and variety is mere appearance." Man's anxiety about such a future stems less from fear of total destruction than from the fear of "total meaninglessness," to use the words of David Riesman.

The criticism is broad, deep, and historical. In one form or another it began even before technology had today's meaning. Long before the Luddites smashed the new machines in 1811, the Abbé de Saint-Pierre wrote in 1737 that "knowledge [and science] implements vice as much as it enlightens morality,"[1] anticipating the contemporary controversy on the ethical posture of technology. Is modern technology value neutral, neither good nor bad in itself, as Alfred North Whitehead,[2] David Sarnoff,[3] and a host of other technological optimists maintain? Or, on the other hand, is it either intrinsically or by manipulation the agent of debasement of the common man who is being turned into a lifeless machine? Could Jacques Ellul be right that technology is becoming a completely autonomous force threatening man himself? [4] What about Lewis Mumford's portrayal of our society as a Megatechnic Complex which, through its seemingly endless resourcefulness in concocting technocratic answers to human problems, has become blinded to the nature of these same problems, even "undermining the principles of cosmic order"? [5]

While it is easy to dispute many of the specific arguments of the technological prophets of doom, it is unwise not to heed their warnings. It is true that they have exaggerated beyond actuality and written much charming nonsense* but neither Mumford, Ellul, nor a host of other technological critics—including Oswald Spengler,[6] Thorstein Veblen,[7] Aldous Huxley,[8] and Matthew Arnold [9]—can be overlooked in their main argument. They are correct that the danger is present that technology may create a prison in the form of a technocratic society. Within this society there may be "no hope for mankind except by 'going with' its plans for accelerated technological progress, even though man's vital organs will be cannibalized in order to prolong the megamachine's meaningless existence." [10]

We engineers, as purveyors of technology, are deeply sensitive to these warnings, not because we are used as a scapegoat—as we are—and not because we think they are unfair—as they must be, when not tempered by acknowledgment of the social good that has also emanated from technology—but, primarily, because we sympathize with humanity, feel anguish over its fate, and ponder over our own roles in making the future. Our concern is not one of accepting or rejecting these criticisms; it would be easy to brush them aside by distributing culpability among others who have perhaps played an even more prominent role in creating the ills. Rather, we also search for ways to safeguard those human values most of us cherish. We deeply fear the fate of the sorcerer's apprentice who after bidding the broom to fill the water

* Note, for instance, Mumford's assertion in *The Pentagon of Power:* "Long before man himself became conscious of beauty and desirous of cultivating it, beauty existed in endless variety of forms in the flowering plants. . . ." One wonders whether Mr. Mumford is kidding! Beauty, as well as ugliness, is in the eye of the beholder. On another point, the notion of the creation of a monolithic world culture as a result of technology is very effectively challenged by Alvin Toffler in his well-documented book *Future Shock.*

barrel was unable to halt it because of ignorance of the proper spell.

Allow me to share with you a secret: we technologists are not warmongers, not lovers of mechanical hearts, not devotees to automation, not antagonists of humanism, not Dr. Strangeloves. We also, like you (perhaps to a lesser degree), are capable of thinking with our hearts, feeling the warmth of love and affection for mankind, and we are sometimes sensitive to the creations of art. And that is why we are here.

I have a feeling that our anguish has brought us here—anguish caused by a seemingly unsolvable dilemma. It appears to me that the concern bringing us together was the realization that, in the midst of triumphs too numerous to count, human societies are facing problems that are, by and large, more subtle and troublesome than we can handle. We find ourselves pondering with B. F. Skinner that "the very change which has brought some hope of a solution to issues of freedom and dignity is responsible for a growing opposition to the kind of solution proposed." [11] We almost identify ourselves with Francis Bacon who, according to Will Durant, "might have discovered in his own moral laxity the abyss created by progress of knowledge beyond the discipline of character." [12]

We are also troubled by a vague feeling that something powerful and overwhelming is brooding over human societies with the potential for either liberating man from most of his mundane chores or destroying him altogether. We came here to search for the answers to our dilemma, to seek to understand our feelings of helplessness, to strengthen ourselves with wisdom from outside our own fields, and to explore the validity of our value systems in relationship with our future in the technological society so that perhaps together we can renew civilization. Perhaps we can celebrate with Samuel Butler a civilization in which man can liberate himself from the slavery of machines, not by breaking them, but by mastering them.[13]

What then did we do?

At the outset, Dr. Kranzberg boldly brought us face to face with the difficulty of our task when he said, "It seems to me that although we are asking questions we may not really understand the problem." His challenge to the arrangement of the program, which he thought was formulated by "the logic of the engineering mind," was an exhibition of a deep frustration that social scientists and humanists feel whenever they are confronted with problems as formulated by technologists. Dr. Samuel Klausner, a renowned sociologist, some years ago brought to my attention that if we wish to make interdisciplinary endeavors succeed, we must not only ask of the various disciplines the questions that we think are relevant, but we must allow them to participate in articulating the questions that need to be asked. "To know what to ask is already to know half."

We have spent much time examining what technological historians and philosophers do and what the rest of us, the engineers, the economists, the political scientists, and the anthropologists think historians and philosophers should do. My impression is that what they see as their task and what the rest of us, very nonchalantly, see as their task is to a considerable degree different. Three points, however, have been clear: (1) historians and philosophers of technology have just scratched the surface of the problem, (2) they are handicapped by the lack of reliable data and operational definitions, and (3) more and more pressure will be exerted on them to address themselves to contemporary technology rather than examining ancient techniques.

A recurring thematic difference illustrates the last point. Whenever the majority of historians and philosophers at the conference were talking about technology, they seemed to be discussing ancient techniques practiced by carpenters, vessel makers, or smelters.* However, when technologists were talking about technology, they clearly had in mind the computer, DDT, the information satellite, television, and the hydrogen bomb. Philosophers and historians apparently felt that by studying the simple skills of the past, as a technique for deriving the principles of the relationship of man and machine, they could then extrapolate these to more complex technologies of the present day. The technologists, on the other hand, were not at all convinced that this can be done. The technologists sought to assert that global in-

* It is my unscientific observation that most of the social critics of the contemporary technology are coming from walks of life other than the academic disciplines of history or philosophy. If this is true, then perhaps the historians and philosophers need to examine the reason.

volvement, compressed response time, and the collective nature of new technological inventions and developments make this difficult, though in no manner did they intend to deny the value of such studies of the past.

The divergence of these two points of view manifested itself more profoundly when philosophy mingled with history to examine the nature of societal man himself. Bugliarello expressed the view of some of us by asserting that an important event of the twentieth century is the emergence of intimately interactive system of biological organisms, society, and the machine. Bugliarello wrote elsewhere that "the societal mechanisms complement genetics in the task of selection, and give us immortality by perpetuating our thoughts and our remembrance. Machines give ubiquity to our voice and to our image, and can store them for future generations. Also, machines can change society, and enhance most directly and, indeed, prolong our biological life."[14] This assertion has most interesting and profound implications for our civilization and our value systems. Of course, not everyone agreed with this line of reasoning. It was argued that, on the basis of strong anthropological evidence and upon examination of languages, it could be shown that the global extension of man's nervous system through machines is external to man and not internal and, therefore, the nature of man has remained unchanged, preserving his value systems intact.

How shall we proceed to define the nature of societal man? How else but by observing in what form and manner man interacts with his external world? In discussing this matter, it will be helpful to distinguish between two interacting, possibly differing, and probably interchanging entities. One is private man—an internal being known only to one's own consciousness. The other is societal man—an external being whom only others can know. I do not believe it will ever be possible to know the private man, except one knowing oneself. The societal man who is known to others by his or her interactions with elements of the environment is, in fact, the basic element of the society and, from its standpoint, all important. If technology has in any way, particularly by expanding man's knowledge and his memory system, altered this relationship, then from the societal point of view we must assume that he is changed. Throughout history, many wise and learned men with a variety of philosophical viewpoints have expounded on knowledge both as the prerequisite to thinking and as the cornerstone of the human being. "Human behavior," said Plato, "flows from three main sources: desire, emotion, and knowledge." [15] "My praise shall be dedicated to the mind itself," said Francis Bacon. "The mind is the man and knowledge mind, a man is but what he knoweth." [16] The Bergsonian notion that time holds the essence of life and in its accumulation creates duration that is the "continuous progress of the past which grows into the future and which swells as it advances" holds that memory is the vehicle of duration and "consciousness seems proportionate to the living being's power of choice." [17] Who can argue that technology has not expanded knowledge and multiplied choice? Who can, then, say assuredly that contemporary societal man is identical to the man of prewritten history?

In an operational sense, then, I submit that the tape recorder, television, and the rest have altered the range and quality of man's response to his surroundings. In this sense, I feel it is much more convincingly absurd to say that a tape recorder* or a library is not a part of man's memory system than to say it is. I feel closely akin to Victor Ferkiss who, upon examination of the character and social role of contemporary man, concludes that "one could also speak loosely of the coming into being of a man of a new type: 'technological man.' " [18]

The resolution or clarification of the above question is essential to us because it bears on our responses to what is good and what is bad. If we accept that man, machine, and society together will be the setting within which values will be developed and will be tested for conformity with societal aspirations, then we should strive to develop a system which enables us to preserve those

* At one point during the conference, I referred to my miniature tape recorder and observed that because of it my memory space is much larger than anyone before the advent of tape recorders and books could have expected, implying that, as a societal man, I can exert much vaster influence upon my relationship with the external (to me) world and therefore, I, with my machine, must be considered different than a pre–written history man from the standpoint of society. This became a point of controversy and much reference was made to "Zandi's tape recorder." How could we have known that, within a year, a president of the United States would resign because of the evidence provided by a tape recorder?

human values we cherish most. This we must do if we do not wish to give birth to Richard Lander's dream of building a machine that will have its own children,[19] because in a technological culture it is extremely important that we should be on guard against human robots. I believe that we cannot do this if we ignore the tremendous pressure that technology exerts to alter human values. In this sense, technology, and even science, are not value neutral. If we cherish our humanity, as we know it now, it is imperative for us to recognize the emergence of the new entity.

It seems to me that we cannot afford not to do our best to develop a coherent system of thought and values to make it possible for us to keep our identity in a forcefully changing environment. This, in fact, is what technologists are asking from our colleagues in philosophy, history, and sociology. You historians, philosophers, and social critics must understand our dilemma. Professional calling requires engineers to act, to design, to build, to alter nature, to attempt to expand the physical variety of human life and enrich the possibilities of human experience. In order to act, we need to identify problems, to set boundaries, to articulate goals, to set priorities, to select tools for optimization, to evaluate the various impacts of our selected tools, and, finally, to build and operate the devices and facilities. We make these decisions within the context of the time and place we happen to be. Most unhappily sometimes, we make a choice among alternatives with full appreciation of insufficient data and information. However, we also know that not acting is, in itself, an action, and its consequences may be worse than all other alternatives.

For the first time we technologists are really trying to put you on the spot. Tell us what our guiding light should be. Tell us what are the intrinsic, lasting, and noble purposes of life—in a way that is operational. Tell us what our value systems should be and tell us now—remembering that we must act now and not tomorrow. Construct for us and with us a philosophy of technology—a philosophy in the sense of Plato, who meant an active culture, wisdom, that mixes with the concrete business of life and not a closeted and impractical metaphysics. This symposium provided the forum for discussion and establishment of dialogue but did not provide us with answers.* Let's hope that this will be the beginning.

We engineers sometimes resent the fact that, almost invariably, the final judges of our actions, you philosophers and historians, use the values of a later period to judge our action of an earlier period. I can recall a hundred examples in which at their inception technological innovations were hailed as angels of mercy but damned at a later date as the enemies of mankind. DDT is a typical example. During World War II, it was used to eradicate malaria and to increase agricultural productivity and, indeed, was hailed as the most humane innovation one could hope for. Gerald Sykes is correct, "Man [meaning all of us, not only technologists] rushes first to be saved by technology and then to be saved from it."[20]

Much evidence exists to prove that technology can be either a blessing or a curse. The question is not whether it will be but which one will it be. I am reminded of Adlai Stevenson's reply to reporters' queries regarding his thoughts on the United Nations. He said, "When Adam offered marriage to Eve, Eve hesitated for a moment to accept. Very perturbed, Adam asked, 'Is there anyone else?' "

* Mario Bunge came closest to suggesting a guideline for action. In his erudite paper, he suggested a two-tiered ethical code for technology. One component is an individual ethical code emphasizing "the personal responsibility of the technologist in his professional work and his duty to decline taking part in any project aiming for antisocial goals." The other is a social code which disallows ". . . the pursuit of unworthy goals and limiting any technological processes that, while pursuing worthy goals, interfere severely with further desiderata." It would be easy to follow these pennants when the distinction between "worthy" and "unworthy" goals is clear and beyond disputation but, regrettably, most human affairs are not of that kind. It would be easy to demand that engineers decline a certain work which may aim for "antisocial goals" but it is on rare occasions that a given action can only create antisocial effects. Most often a given task can generate both desirable and undesirable end products. It would be an easy matter to establish the principle of the personal responsibility of engineers in their professional works but how is this to be implemented? Most socially significant projects are executed over a long span of time, and, more often, various components of the system are made at various locations by different companies. How is an engineer who is developing, for instance, a mechanism for triggering a laser beam to know whether his work will be used to save human life in surgery or to destroy life? These questions are not intended to diminish the significance of Dr. Bunge's suggestions, but only to point out that much clarification is required.

Notes

1. Will Durant and Ariel Durant, *The Story of Civilization,* Part IX, *The Age of Voltaire* (New York: Simon and Schuster, 1956), p. 336.

2. Alfred North Whitehead, *Science and Modern World* (New York: Mentor Books, 1923).

3. David Sarnoff, *The Fabulous Future: America in 1980* (New York: Dutton, 1965).

4. Jacques Ellul, *The Technological Society,* trans. John Wilkinson (New York: Knopf, 1964).

5. Lewis Mumford, *The Myth of the Machine—The Pentagon of Power* (New York: Harcourt Brace Jovanovich, 1970).

6. Oswald Spengler, *Man and Technic,* trans. C. F. Atkinson (New York: Knopf, 1932).

7. Thorstein Veblen, *The Instinct of Workmanship* (New York: Macmillan, 1914).

8. Aldous Huxley, *Brave New World* (New York: Doubleday 1946).

9. Matthew Arnold, *Culture and Anarchy* (London: Smith, Elder, 1875).

10. Mumford, *op cit.*

11. B. F. Skinner, *Beyond Freedom and Dignity* (New York: Knopf, 1971).

12. Will Durant and Ariel Durant, *The Story of Civilization,* Part VII, *The Age of Reason Begins* (New York: Simon and Schuster, 1961).

13. Samuel Butler, *Erewhon, or Over the Range* (London: A. C. Fifield, 1872).

14. George Bugliarello, "Rethinking Technology—Steady Earth Biosoma," paper delivered at the 138th Annual Meeting of American Association for the Advancement of Science, Philadelphia, Pa., Dec. 30, 1971.

15. Will Durant, *The Story of Philosophy* (New York: Simon and Schuster, 1961).

16. Francis Bacon, *Advancement of Learning* (New York: Dutton, Everyman's Library, 1958).

17. Henri Bergson, *Creative Evolution,* trans. Arthur Mitchell (New York: H. Holt, 1913).

18. Victor Ferkiss, *Technological Man—The Myth and the Reality* (New York: New American Library, 1969).

19. Richard R. Lander, *Man's Place in the Dybosphere* (Englewood Cliffs, N.J.: Prentice-Hall, 1966).

20. Gerald Sykes, ed., *Alienation* (New York: Braziller, 1964).

Notes on Contributors

JEAN-CLAUDE BEAUNE is a professor at the Université de Clermont-Ferrand and the author of *La technologie* and *Problèmes technologiques: Automates et technologie*. His main academic interests are in the philosophy of technology and science, sociology, and ethnology.

GEORGE BUGLIARELLO, a specialist in fluid mechanics, computer languages, and bioengineering, and former dean of the College of Engineering at the University of Illinois at Chicago Circle, is currently president of the Polytechnic Institute of New York. He is concerned with the social implications of technology and is the author of several books, including *Bioengineering: An Engineering View*, *Technology, the University and the Community*, and *The Impact of Noise Pollution: A Socio-Technological Introduction.*

MARIO BUNGE, a professor at McGill University and head of the Foundation and Philosophy of Science Unit there, is writing a seven-volume work entitled *Treatise on Basic Philosophy*. Four volumes of the work have been published; volumes five and six will deal with philosophical problems and impacts of scientific and technological knowledge.

HAROLD L. BURSTYN has been the historian of the U.S. Geological Survey since 1976. Prior to that, he taught at Harvard, Brandeis, and Carnegie-Mellon universities and at the William Paterson College of New Jersey. Burstyn has written extensively on the history of marine and earth science and technology.

DONALD S. L. CARDWELL is the author of *From Watt to Clausius: The Rise of Thermodynamics in the Early Industrial Age* (winner of the Dexter Prize of the Society for the History of Technology) and of *Turning Points in Western Technology*. He is a professor of the history of science and technology at the University of Manchester and at the university's Institute of Science and Technology.

PETER CAWS is a professor at Hunter College of the City University of New York. He is the author of *The Philosophy of Science: A Systematic Account* and of *Science and the Theory of Value.*

ION CURIEVICI is a professor at La Institutul Politehnic Iaşi in Rumania.

DEAN B. DONER is vice-president for overseas programs at Boston University and a former professor of English and dean of the College of Liberal Arts and Sciences at the University of Illinois at Chicago Circle.

ARTHUR L. DONOVAN is a member of the history department and co-founder of the Program in the History of Science and Technology at West Virginia University. He has done research on the history of the coal industry in America and is the author of *Philosophical Chemistry in the Scottish Enlightenment: The Doctrines and Discoveries of William Cullen and Joseph Black.*

DAVID EDGE has been director of the Science Studies Unit, University of Edinburgh, since 1966. He holds degress in physics and radio astronomy and is the co-author of *Astronomy Transformed: The Emergence of Radio Astronomy in Britain.*

DAVID JORAVSKY, a professor of history at Northwestern University, is the author of *Marxism and Natural Science, 1917-1932* and of *The Lysenko Affair.*

WERNER KOENNE is head of the EDP department, Osterreichische Elektrizitatswirt-schafts-Aktiengesellschaft, Vienna, Austria.

MELVIN KRANZBERG was the principal founder of the Society for the History of Technology and a major force in the development of the field of the history of technology. He is the Callaway Professor of the History of Technology at Georgia Institute of Technology and the co-editor of *Technology in Western Civilization* and of *Technology and Culture: An Anthology.*

HEATHER LECHTMAN, an associate professor of archaeology and ancient technology at Massachusetts Institute of Technology, is the director of the Center for Materials Research in Archaeology and Ethnology there. Her primary academic interest is in the archaeology of technology, with particular emphasis on the development of metallurgy in pre-Columbian South America.

CARL MITCHAM teaches philosophy and psychology at St. Catharine College in Kentucky and is an editor of *Research in Philosophy and*

Technology. He co-edited *Philosophy and Technology: Readings in the Philosophical Problems of Technology* and co-compiled *Bibliography of the Philosophy of Technology.*

RAFAEL LÓPEZ PALANCO is associate director, Escuela Tecnica Superior de Arquitectura in Seville, Spain.

NATHAN ROSENBERG is on the faculty of Stanford University, where he teaches economics. He is the editor of *The American System of Manufactures* and of *The Economics of Technological Change.*

HENRYK SKOLIMOWSKI is professor of philosophy at the University of Michigan. He is the author of *Polish Analytical Philosophy: A Survey and a Comparison with British Analytical Philosophy.*

STEVE M. SLABY, the author of books and articles on engineering geometry, statics, and strength of materials, is director of the technology and society seminars at Princeton University, where he is an associate professor. He has done post-graduate work in both labor relations and multidimensional descriptive geometry.

CYRIL STANLEY SMITH, Institute Professor Emeritus at Massachusetts Institute of Technology, is the author of numerous articles on the science of metals and the history of technology, especially in relation to art. He also wrote *A History of Metallurgy.*

ARTHUR STEINBERG is W. R. Kenan, Jr., Associate Professor of Archaeology at Massachusetts Institute of Technology. He is a specialist in the study of ancient metallurgy in the Mediterranean region and has written on aspects of the casting and welding of bronzes and on copper smelting. He is the co-editor of *Art and Technology: A Symposium on Classical Bronzes.*

E-TU ZEN SUN is a professor of Chinese history at Pennsylvania State University. His most recent book is *The Silk Industry in Ch'ing China,* an annotated translation from the Chinese. He has also written several articles on Chinese technologists for the *Dictionary of Ming Biography.*

FRANCES SVENSSON teaches political science at the University of Michigan. Her current interests are the impact of scientific and technological training on members of tribal minorities in the United States and the U.S.S.R. and the Soviet literature on the scientific and technological revolution.

HEINZ VON FOERSTER is professor emeritus of the department of biophysics and electrical engineering at University of Illinois, Urbana-

Champaign, where he established and directed the Biological Computer Laboratory. He edited the five-volume work *Cybernetics: Circular Causal and Feedback Mechanisms in Biological and Social Systems.*

David Wojick has developed the issue tree model (discussed in this book), which can be used in technology assessment, compliance planning, and complex issue analysis. Formerly on the faculty of Carnegie-Mellon University, he is now self-employed. Among the many books and articles he has written describing uses of the issue tree model is *Planning for Discourse.*

Iraj Zandi is a professor of civil and urban engineering at the University of Pennsylvania. His major research interests are environmental engineering and resource utilization. He has served as editor of several professional journals and on governmental advisory panels.